Induced Responses
to Herbivory

INTERSPECIFIC INTERACTIONS
A Series Edited by John N. Thompson

Richard Karban and
Ian T. Baldwin

Induced Responses
to Herbivory

The University of Chicago Press

Chicago and London

RICHARD KARBAN is professor of entomology at the University of
California, Davis, and is coeditor of *The Evolution of Insect Life Cycles.* IAN T.
BALDWIN is professor of biology at the State University of New York at
Buffalo and is founding director of the Max-Planck-Institut für Chemische
Ökologie in Jena, Germany.

The University of Chicago Press, Chicago 60637
The University of Chicago Press, Ltd., London

06 05 04 03 02 01 00 99 98 97   1 2 3 4 5

ISBN: 0-226-42495-2 (cloth)
ISBN: 0-226-42496-0 (paper)

Library of Congress Cataloging-in-Publication Data

Karban, Richard.
   Induced responses to herbivory / Richard Karban and Ian T. Baldwin.
      p.   cm.—(Interspecific interactions)
   Includes bibliographical references (p.   ) and index.
   ISBN 0-226-42495-2 (alk. paper)—ISBN 0-226-42496-0 (pbk. : alk.
paper)
      1. Plant defenses.   2. Animal-plant relationships.   3. Herbivores—
Ecology.   I. Baldwin, Ian T.   II. Title.   III. Series.
QK923.K37   1997
571.9′62—dc21   97-17088
                                                                    CIP

Contents

Acknowledgments

Our work on induced responses has been supported by grants from U.S. Department of Agriculture, National Research Initiative (RK) and the National Science Foundation and Mellon Foundation (IB). Parts of the manuscript were improved by Susan Abrams, Anurag Agrawal, Bob Bugg, Jenny Davidson, Sean Duffey, Greg English-Loeb, Erkki Haukioja, Mike Stout, Jennifer Thaler, John Thompson, Nora Underwood, and Emily Wheeler.

We thank our families for showing patience with us while we were preparing the manuscript. We also thank our respective institutions for granting us sabbatical leaves to work on this book. Finally, RK thanks David and Laurie Hougen-Eitzman for housing and for stimulating conversation during the early parts of this project.

1 An Introduction to the Phenomena and Phenomenology of Induction

1.1 Plants Are Defended against Many Threats

Plants cover much of our planet. Fossil evidence indicates that this condition has persisted throughout the history of multicellular life on earth. This simple observation implies that herbivores have been too few in number to consume all the food available (Hairston et al. 1960), and perhaps more important, that plants have sufficiently defended themselves against their potential herbivores and have avoided consumption (Murdoch 1966).

The diversity of potential threats to plants is impressive. Large mammalian browsers can remove branches and even entire plants in a single bite. Tall giraffes may feed preferentially on the tops of small trees while shorter browsers feed only on seedlings or on bottom branches. There are specialists on the oldest, senescent plant tissues as well as those that can feed only on newly produced leaves or buds. Herbivores sometimes remove leaf tissue, the form of damage that is most conspicuous to human observers who visit at a later time. They also parasitize the vascular transport systems of plants; cause plants to construct specialized galls in which the herbivores live and feed; cause plants to preferentially redirect nutrients in their direction; consume the pollen, seeds, and reproductive structures of plants; mine, bore, and tunnel through plant tissues above and below ground. Important herbivores include mammals, reptiles, amphibians, birds, mollusks, worms, and arthropods. Insect herbivores not only make up a very large fraction of the biomass and diversity of life on earth but are also a considerable threat to plants (Strong et al. 1984). Plants are also attacked by a diverse assemblage of parasitic microorganisms: viruses, bacteria, fungi, and others.

Most individual plants are exposed to attack by many different herbivore and disease species during their lifetimes. The list of challengers grows much larger as one considers the potential attackers encountered over the range of a plant species and over the evolutionary history of a plant lineage (see Thompson 1994). There is little doubt that these plant parasites (including herbivores as well as pathogens, sensu Price 1980) can and do kill many individuals and reduce the reproductive suc-

1

cess of others (Crawley 1983). However, it is less clear whether any given plant-parasite combination necessarily has much effect on plant fitness, because plants have evolved the ability to tolerate herbivory and disease, as well as means of protecting themselves against damaging attacks.

Plants have persisted in a world in which a wide diversity of organisms make their livings by consuming plant tissues. Plants with traits that render them (1) less preferred by herbivores or (2) relatively more successful after herbivore attack are likely to be better represented in future generations than those that succumb to their parasites. Traits that act as defenses against herbivores should be favored by selection in a world in which herbivores are a threat. Of course, these traits may have evolved, or currently function, for other reasons as well, such as protection against pathogens or generalized wound-healing responses.

Almost all forms of life possess the capability of defense against diverse attackers (Klein 1982). Higher plants defend against the possibility of fusion by rejecting genetically incompatible grafts. Plants also have a low incidence of neoplasia, in which mutant cells escape normal regulatory mechanisms. Plants heal the wounds that are inflicted upon them as they grow in harsh and unpredictable environments. Plants also have defenses against pathogens and herbivores.

The diversity of plant parasites and predators has several important consequences for plant defense. Plants are "perceived" and attacked at many different spatial and temporal scales. Therefore, plants should be defended at these many scales. A resistance mechanism such as a thorn may be effective against a large grazing mammal but is ineffective against an aphid or a microorganism that feeds intracellularly. A "defense" that is poisonous to one herbivore may be attractive and highly nutritious to an individual of a second herbivore species. Plant species that seem to have high levels of defense, such as nettles with stinging hairs or jimsonweed (sacred datura) with hallucinogenic alkaloids, are nevertheless species that commonly suffer very high levels of leaf loss to herbivores that are not deterred by those "defensive" traits. Plant parasites are ubiquitous and "not as single spies, but in battalions come." In other words, microbes are everywhere, in the saliva of every mammalian herbivore, on the mandibles of every cell-feeding homopteran, and so forth. Every herbivore has to be viewed not only as a direct threat but also as a potential Trojan horse for an army of smaller invaders. This diversity and abundance of parasites may favor plant defenses that are skewed toward universal efficacy. Similarly, elicitation by microbes of defenses against herbivores and of cross-resistance (lack of specificity) are traits that make sense when viewed against the wide variety of threats. These themes will be developed more fully later.

1.2 Definitions

1.2.1 Induced Responses Defined

Until recently, plant defenses were generally assumed to be constitutive, that is, always expressed in the plant. Constitutive defenses change over evolutionary time and even during the normal maturation of an individual plant, but they function independently of damage. Recently, we have recognized that many of the traits and processes that defend plants against herbivores change following attack.

We refer to changes in plants following damage (or stress) as induced responses (table 1.1). These are plastic traits that vary according to the plant's environment. These traits may or may not affect herbivores, and they may or may not benefit the plant when herbivores are present. Plastic phenotypes are not fixed genetically, although the plant's ability to show plastic responses may be under genetic control (Schlicting 1986; Sultan 1987; Bradshaw and Hardwick 1989).

Those induced responses that reduce herbivore survival, reproductive output, or preference for a plant are termed induced resistance (table 1.1). Induced resistance is viewed from the herbivore's point of view; it does not necessarily benefit the plant. In fact, individual plants that show induced resistance may do less well than those that do not respond. This counterintuitive situation can occur if the induced resistance is more expensive to the plant than is the risk of future damage, if induced resistance stimulates herbivores to consume more of the induced tissue, or if induced resistance makes the plant more vulnerable to other potential dangers. As with the more inclusive definition of induced responses, this definition of induced resistance makes no assumptions about the factors that led to the evolutionary origin or maintenance of the trait.

Those induced responses that currently decrease the negative fitness consequences of attacks on plants are termed induced defenses (table 1.1). Quite simply, they defend the plant. Induced defense is viewed from the plant's point of view; the plant benefits from the response even though herbivores may be unaffected. This definition includes mecha-

Table 1.1 Definitions of induced response, resistance, and defense

Term	Affect Herbivore?	Affect Plant?	Adaptation to Herbivory
Induced response	?	?	?
Induced resistance	Yes	?	?
Induced defense	?	Yes	?

nisms that reduce damage to plants as well as those that increase the tolerance of plants to a given level of damage, as long as plant fitness is increased. For example, plants that become more tolerant of herbivory following attack, measured in terms of plant fitness, are considered to have induced defenses even though the individual herbivores that induced the enhanced tolerance and the overall population of herbivores are not affected by the change. Again, this definition of induced defense makes no assumptions about selection by herbivores or the historical origin of the response, only that it currently defends the plant.

1.2.2 Comparisons with Definitions Used by Others

Although the field of induced responses has been active only in recent years, it has been marked by rancorous disputes. Unfortunately, many of these disagreements have been the result of the two sides talking past one another because they were using different definitions and making different assumptions about the connotations associated with those definitions. We have tried to outline explicitly the assumptions behind each of the terms that we use. Other people use these same terms with slightly different meanings and assumptions. One definition is not intrinsically any better than another, but it is vitally important that workers in this field are clear about what each person means so that future discussions can concentrate on scientific concepts rather than semantics.

Because other workers have had slightly different meanings for these terms, it is useful to compare our use of these terms with those found in several other influential reviews. Rhoades (1985a) defined induced resistance from the insect's point of view but assumed that such responses must have been shaped by selection. Fowler and Lawton (1985) proposed a hierarchical set of questions that correspond to our use of induced resistance and defense. However, they argued that induced changes should reduce herbivore populations before they can reduce damage suffered by the plant. This argument does not necessarily hold if the induced response serves to make the plant more tolerant to herbivory without affecting the herbivores. Schultz's review (1988) and Haukioja's more recent papers (e.g. 1990a, 1990b) used induced response and induced resistance in the same way as we have. However, Schultz mentioned that defense can refer either to any negative impact on a plant's enemies (what we term resistance) or "more strictly" to traits that reduce the plant's net losses. Similarly, Haukioja distinguished between the "colloquial" use of induced defense, meaning that the trait affects the plant's success, and the "evolutionary" use, meaning that the trait is an evolutionary response by the plant to herbivory. Some other workers (Painter 1958; Rausher 1992) described resistance from the plant's point of view, measured as the amount of damage received (a correlate of what we call defense). Defense was used by Rausher (1992) as a trait that is maintained because of selection exerted specifically by herbivores.

Plant pathologists generally use induced resistance to refer to protection from disease caused by biotic or abiotic agents that activate the host plant's physical or chemical barriers (Kloepper et al. 1992). This definition implies that the pathogen suffers reduced performance and the plant benefits, although these conditions are not specified. Plant pathologists do not imply evolutionary origins in their definitions.

Two more points are worth making with respect to definitions. There has been some discussion in the literature about whether induced responses are active processes or incidental consequences of other processes such as plant aging or nutrient stress. One reason for making this dichotomy has been to argue that only active processes should be considered as defenses resulting from selection exerted by herbivores (Haukioja and Neuvonen 1985; Bryant, Danell, et al. 1991). While an understanding of the mechanisms of induced responses is certainly of interest, this distinction implies little or nothing about whether the response should be called a defense, or whether the response resulted from selection by herbivores.

Since some other workers (perhaps a majority) use defense to mean that the trait has been shaped by natural selection by herbivores in particular, it seems appropriate to explain why we choose not to do so. First, a trait may provide defense whether or not it evolved specifically for that purpose. Second, it is extremely difficult, if not impossible, to determine the specific selective factors that shaped a trait (Endler 1986; Rausher 1992). However, phylogenetic reconstruction offers some hope that we may be able to draw reasonable inferences about the evolutionary history of induced responses (see section 4.2.5). Since we lack this information for all of the traits that currently act as effective defenses against herbivores, it seems unreasonable (though not uninteresting) to make assumptions about the evolutionary history of these traits.

1.3 A BRIEF HISTORY OF A YOUNG FIELD

Induced resistance was not widely appreciated by biologists until the last twenty years, corresponding to the rise in popularity of experimental ecology. Unlike most other natural phenomena, Darwin did not describe it. These two observations are related; to be aware of induced resistance, there must be an undamaged control for comparison, precisely the sort of thing that an experimental approach provides. Keen observation alone, the primary tool of the Victorian natural historian, does not readily or convincingly reveal induced resistance.

Several early natural historians who pondered the causes of dramatic fluctuations in numbers of herbivores suspected that changes in quality of food plants might be involved. For example, an Italian priest, Francesco Negri (1623–98), visited Norway in 1664–65 and reported that "within only one day they [lemmings] flood large areas and completely destroy grain and grasses. Therefore people are more afraid of lemmings than of the hail. But such destruction happens only rarely, and many years will pass until one will see the animals again. . . . There are no remedies against this pest, but they disappear of their own accord, because they only live until the next spring. When they begin to eat the new grass, they all die, as if they were poisoned" (Helland 1921, translated by Seldal 1994).

Darwin did not describe induced resistance, but he was certainly aware of phenotypic plasticity. "I speculated whether a species very liable to repeated and great changes of conditions might not assume a fluctuating condition ready to be adapted to either condition" (Charles Darwin to Karl Semper, 1881). (Scholars of Darwin, like biblical scholars, are able to find a reference to most any idea by looking hard enough.)

Other Victorian natural historians came much closer to articulating observations similar to our current view of induced resistance of plants to their herbivores. M. Standfuss (1896, cited in Benz 1977) described his experiences rearing caterpillars on several trees. He noted that such rearings for several successive years on the same tree would not give satisfactory results. The repeated defoliations induced changes in plant metabolism and stunted plant growth, which in turn caused malnutrition of the insects. He interpreted these observations as indications of the existence of self-protection in plants. A similar observation was made by V. G. L. van Someren in Kenya and was communicated by his friend, Sir Edward Poulton, to the Royal Entomological Society in London in 1937. He also experienced difficulty rearing larvae on trees. The high death rate and retarded larval growth were caused by "some change in the food-plants" and not by parasitism. Some of the trees used for rearing shed their leaves or withered, but van Someren speculated that "this is one method of natural control by means of a change in the chemical composition of the foliage." Both of these observers compared their failed rearing attempts with more successful "controls" that had been accomplished earlier with less damaged foliage.

Agricultural entomologists also noticed that foliage that had been damaged became less suitable as food for insects, a phenomenon that was referred to as "plant conditioning." Most descriptions of plant conditioning were anecdotal, lacking good controls, and no attempts were made to use plant conditioning as a means of protecting crops from economically important pest damage. In a particularly rigorous example Henderson and Holloway (1942) sought to explain population crashes of citrus red mites that they observed in orchards. Climatic conditions and natural enemies were not plausible explanations for the population dynamics, so they hypothesized that mite feeding rendered the citrus leaves poor hosts for spider mites. They tested their hypothesis by caging different numbers of mites on leaves to produce three levels of injury. After "several days of conditioning," these mites and their eggs were removed and the leaves were challenged with new mites. Mites that were caged on the most damaged foliage produced only 28–52% as many eggs as mites on the lightly damaged foliage. No attempt has been made to incorporate these experimental results into a management program.

Observations such as these had no theoretical basis to explain them. To the contrary, in the wake of the neo-Darwinian synthesis, biologists generally ignored the importance of phenotypic plasticity, such as induced plant responses (Sultan 1992). A legacy of this synthesis was the anti-Lamarckian assumption that the material of evolution consisted solely of morphological variants determined directly by genes and independently of environment. Phenotypic traits whose expression varied, depending on the environment, were viewed as not heritable and therefore irrelevant to evolution. The fallacious distinction between genetic and environmental causes of variation, interpreted in a strict sense, is now being replaced by a recognition of norms of reactions that depend on the environmental conditions experienced (Sultan 1992). This newer conceptual environment is much more compatible with observations about phenotypic plasticity in general and induced resistance in particular.

Plant physiologists have played a relatively small role in the development of this field. Given that the phenomena are fundamentally physiological, this observation is surprising. We predict that in ten years there will be two separate volumes of plant physiology textbooks: the volume that is already written, which covers interactions of plants with the abiotic world, and the one that is currently being written, which will cover interactions of plants in response to the biotic world.

During much of this century, plant pathologists have regarded induced resistance against viruses, bacteria, and fungi as a principle means of natural plant defense, with great potential for disease prevention (Chester 1933; Horsfall and Cowling 1980; Campbell and Macdonald

1989). Several successful vaccinations have been developed and employed on a commercial scale (see chapter 6).

Induced resistance against herbivores began to receive considerable attention in the 1970s as the result of the pioneering work of Clarence Ryan, a biochemist from Washington, and Erkki Haukioja, a population ecologist in Finland. Before Ryan and Haukioja numerous workers had developed ideas that were similar to induced resistance. However, most biologists took no notice of these observations because they had no conceptual foundation, the observations lacked adequate controls, and there was no plausible mechanism. The pioneering work of Ryan and Haukioja remedied these deficiencies. As early as 1962 Ryan and Kent Balls had found that extracts from potato tubers inhibited the activity of chymotrypsin, an enzyme that cleaves proteins (Ryan 1992). These chemicals from potato tubers that inhibit digestive enzymes were transient, mysterious in the sense that their presence or absence in plants could not be explained. By 1972 Ryan suspected that environmental factors were involved in the regulation of these chemicals and that they played a defensive role in the plant. That year, Green and Ryan (1972) reported in *Science* that wounding the leaves of young potato or tomato plants by Colorado potato beetles induced a rapid accumulation of proteinase inhibitor (figure 1.1). The response could be reproduced by damaging leaves with a paper punch or a file. This report appeared at a time when chemical interactions between plants and herbivores had captured the attention of many biologists. Green and Ryan's report was widely appreciated by ecologists, who assumed that proteinase inhibitors must act as defenses against insects, as well as mammals. This and the results that continued to come from Ryan's lab stimulated considerable effort to understand the biochemical mechanisms of induced resistance.

During this same period Erkki Haukioja, who had been trained as an ornithologist, began working on the International Biome Program, trying to understand fluctuations in productivity in northern Scandinavia. Productivity of birches cycled approximately every ten years, as did numbers of autumnal caterpillars that defoliated birches, and populations of birds that ate caterpillars. Haukioja was influenced by Swiss workers who had suggested that changes in the physiology of larches could produce periodic outbreak cycles in populations of another forest defoliator, the larch budmoth (Benz 1974). Haukioja proposed that birches responded to defoliation by producing heavily defended foliage that would not support herbivores, causing caterpillar populations to crash (figure 1.2). Gradually over several years these induced defenses would be relaxed, allowing herbivore populations to again increase to high levels (Haukioja and Hakala 1975; Haukioja 1980). This hypothesis was very appeal-

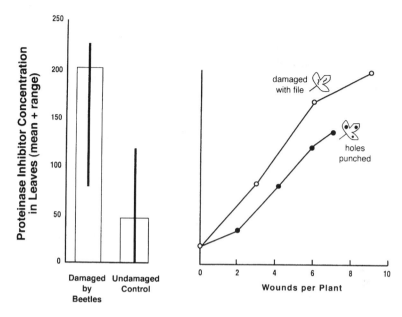

Fig. 1.1 *Left,* levels of proteinase inhibitors (PIs) (μ/ml) in leaves of damaged plants and undamaged controls. *Right,* Levels of PIs in leaves of plants with varying numbers of mechanical wounds. Data from Green and Ryan 1972.

ing because it suggested a single, density-dependent factor that could explain population cycles, a problem that had occupied the thoughts of population biologists for several decades and had defied any simple and general explanation. This scenario was supported by experiments in which caterpillars reared on foliage from trees that had been defoliated in past years were compared with caterpillars reared on foliage from trees that had received no recent defoliation (Haukioja and Niemelä 1977). Caterpillars were found to grow more slowly on foliage from trees that had been damaged in the past.

Ryan and Haukioja established two rather distinct traditions in the study of induced responses. Biochemists have worked to identify the many chemical and physiological changes that occur following plant damage. They have focused on specific chemical pathways or physiological processes. At best, they have conducted simple laboratory assays to establish whether the particular mechanism under study has biological activity. However, it has been exceedingly difficult for these laboratory-oriented scientists to link their mechanism to the biological phenomena of interest to ecologists (herbivore population cycles, etc.). At the other extreme, population biologists have documented that herbivores avoid

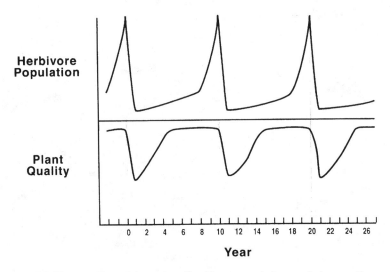

Fig. 1.2 The coordinated dynamics of herbivore populations and plant quality proposed by Haukioja and coworkers. Following peak abundance of autumnal moth caterpillars, plant quality crashes. This causes herbivore populations to crash as well. Plant quality improves slowly over the next four to five years. This allows herbivore populations to increase in response and produces a cycle with ten-year periodicity.

or suffer on damaged plant tissue. They have not convincingly demonstrated that these effects are responsible for the population cycles that made the study of induced responses so alluring two decades ago, nor do they have any good understanding of the causes for the effects that they have observed.

Critics have seized upon these shortcomings to argue that because a particular mechanism, or induced resistance in general, does not appear to provide a simple, universal explanation for population cycles then there is nothing worth talking about. Since we have taken the time to write a book about induced resistance, we obviously disagree. Induced resistance is important, in part, because it has been observed in so many plant-herbivore systems. Every plant does not show induced resistance against every parasite, just as every mammal does not become immune against every pathogen. Considering the great diversity of plants that have been found to respond to attack and the great diversity of attackers, it should not be surprising that, not one, but a great many mechanisms provide resistance and defense.

This book summarizes what we know about the mechanisms and consequences of induced resistance and induced defense, and we hope that it clearly establishes what we need to know to progress in the future.

We have attempted to integrate the two schools that Ryan and Haukioja established. We review the literature dealing with the mechanisms of induced responses (chapters 2 and 3), the consequences of induced responses on herbivores (chapter 4), the consequences and evolutionary significance of induced responses on plants (chapter 5), and possible applications of induced responses (chapter 6). Regrettably, we may have missed many opportunities to integrate other, related literature in greater detail. For example, a fuller coverage of the plant pathology literature would likely add to our understanding of plant-herbivore interactions. We have attempted to integrate the progress that has come from studying plant diseases, particularly in the sections on mechanisms of induced responses and uses in agriculture. No doubt there are other topics that could have benefited from a more thorough integration of the advances in plant pathology. However, the plant pathologists have not synthesized this vast literature, and we have had our hands full with our primary task: attempting to join the mechanistic and ecological traditions to provide some common ground for workers interested in induced responses to herbivory.

2 How a Plant Perceives Damage and Signals Other Ramets, and the Specificity of These Processes

2.1 INTRODUCTION

2.1.1 Recognizing Damage and Talking Trees

No other aspect of induced plant responses has generated as much scientific excitement and public interest as the study of plant signals. As we will see in chapter 3, plant tissues that are damaged can respond in numerous ways. Many of these responses also occur in tissues of the attacked plant at some distance from the site of actual damage. These observations raise several questions. (1) What cues are involved with plant tissue "recognizing" that it has been damaged? (2) How is this information "conveyed" to other parts of the plant? What signals does the plant use? (3) Plants may respond even when they themselves are undamaged after nearby individuals have been attacked, although the evidence for this is still a matter of controversy. How could this occur?

We feel it is necessary to put quotations around the words "recognize" and "convey" since no actual cognitive processes occur. This area of research has been widely misunderstood and misrepresented by the popular press, which refers to these phenomena by shorthand phrases such as "talking trees." At professional meetings we have heard colleagues arguing that there can be no possible advantage to plants "talking." Nobody is seriously advocating that trees talk, in the sense of choosing to dispense information to conspecifics. However, it is reasonable to entertain the notion that trees may be very sensitive to, and fully capable of responding to, signals that happen to be given off when other individuals are wounded. The shorthand phrase "listening trees" might be more appropriate and emphasizes that individuals, rather than groups, might conceivably benefit by such responses.

2.1.2 What's to Be Learned by Studying Cues and Signals

There are several reasons that this subject has generated so much scientific excitement. First, understanding the cues that initiate plant responses and the signals that convey this information will provide a wonderful tool to get at the mechanisms of induced resistance. We will then be able to experimentally begin the process and further manipulate it to follow the chains of events that ultimately result in induced resistance or defense (see section 3.3.4).

Second, knowledge of the cues and signals will allow us to experimentally induce resistance in plants that we wish to protect. This prospect is extremely attractive for crop plants where induced defenses, elicited by humans, may better protect them against their herbivores (section 6.2). One such scenario involves making defenses constitutive rather than induced, or turning them "on" and "off" at will. It may also be possible to move genes that code for inducible defenses into plants that do not normally have them (see chapter 6). Ciba-Geigy and other chemical companies have begun making commercially available elicitors that induce resistance against fungal diseases (Long 1990; Gorlach et al. 1996).

Finally, there has been interest in the cues that elicit induced responses, as a means of settling several controversies about induced resistance. One such argument concerns the evolution of induced responses (see chapter 5). A plausible explanation for induced defenses (rather than constitutive ones) is that defenses are costly to the plant. By producing the defense only when needed, the plant may save resources that it then may use to grow larger and reproduce more. By eliciting the induced responses using signal molecules, an investigator should be able to measure the costs of the response (defense) without damaging the plant. The costs of the defense can be estimated by comparing the fitness of plants that have been induced versus uninduced controls in an environment without herbivory. This approach will be particularly useful if the signal molecules are specific, activating only the induced defense.

A second controversy concerns the relationship between the specificity of induced responses and their evolutionary origin. It has been of interest to determine whether induced responses evolved specifically in response to herbivory or whether they are the incidental consequences of changes in the strengths of resource sinks or other forms of imbalances within the plant (see chapters 3 and 5). According to this line of thinking, responses that are elicited specifically by particular herbivores are likely to have evolved in response to herbivory. Responses that lack such specific elicitors are less likely to represent the evolutionary product of a history of herbivory and more likely to have resulted from incidental overflow production, essentially as a means of storage. These arguments seem only partially useful. While it may be reasonable to assume that those induced responses that are activated only by a specific damage cue evolved in response to that form of damage, the converse provides much less information. An induced response that is activated by many different cues, herbivory among them, may very well have originated or been shaped in part by herbivory. Evolutionary inferences based on knowledge of the specificity of the elicitors strike us as rather shaky.

This argument does bring attention to an important point about signals. While all mechanisms of induced resistance require cues to initiate them, not all induced responses involve signals transmitted systemically through the plant. Induced responses that are localized near the site of damage may or may not require signals, depending upon whether the actual damage is the cue that the plant recognizes. Responses that occur distant from the site of damage are more likely to involve signals, although they do not necessarily have to. Instead, it is possible that the imbalances or tissue deterioration that are associated with damage may spread from the point of attack. This distinction separates those mechanisms that rely on plant imbalances and deterioration from those that involve more actively regulated processes (sections 3.3.3.1 and 5.2).

2.2 CUES AND THE SPECIFICITY OF RESPONSES

If induced responses are to function as plant defenses, they must be activated when risk of herbivory is great. For this to happen the plant must have some way to assess this aspect of its environment. Any factor that predictably correlates with risk can function as a cue. Effective cues may include actual plant wounding, removal of plant tissue, mechanical stimuli, and chemical or physical changes associated with the presence of herbivores.

The plant must be able to respond to cues that indicate herbivores that it has never experienced before. On the other hand, plants that respond to inappropriate cues may be at a selective disadvantage. For instance, phytoalexins are, by definition, chemicals with antimicrobial properties and they are often synthesized in response to pathogen attack (see section 3.3.3.2). Production of phytoalexins after a strong wind, instead of after a pathogen attack, may be inappropriate and costly. Recognition of specific cues enables the plant to respond qualitatively and quantitatively to different abiotic and biotic challenges. Such differential response to different cues is referred to as specificity.

A response may be specific in either of two ways. It may be initiated only by specific cues generated by a restricted set of attackers. Specificity of cues is assessed by observing the responses that various stimuli induce upon one plant species. As such, characterizing cue specificity involves a detailed knowledge of the induced plant response. Specificity of cues should be distinguished from specificity of effects. Regardless of the specificity of the cue that served as the stimulus, the response may affect only those herbivores that induced it or it may have less specificity of effect. For example, the induced response may be generated by only a single species, but it may affect many different herbivores and pathogens. Specificity of effects should be measured by observing the effects that a single induced response has upon different herbivore or challenge species.

In the jargon of animal behaviorists, cues and effects would be called stimuli and responses. For many animal behaviors the stimulus and the response are directly and often tightly linked. A rattlesnake responds one way to another rattlesnake, a second way to a mouse, and a third way to a hiker. Each response is a direct consequence of the stimulus and is appropriate for each. Plant responses tend to be less often tightly linked to their stimuli (Bradshaw and Hardwick 1989). For example, seasonal leaf abscission by deciduous trees is brought about by shortened days rather than directly by freezing temperatures. The causes of indirect connections between cues and effects in plants are unclear, although indirect linkages may allow plants to anticipate environmental changes before they actually occur (Bradshaw and Hardwick 1989). For example, leaf nutrients can be resorbed before the leaf tissue freezes, at which time they would become unavailable. Several authors have argued that conditional strategies of all sorts are rare due to the unreliability of most cues. This unreliability prevents plants from determining which of several phenotypes brings higher fitness (Levins 1968; Lloyd 1984) or, in other words, when induced responses provide defense (Karban and Adler 1996).

2.2.1 Cue Specificity

Trying to determine whether plant responses are specific to particular cues is a tricky and arbitrary business. Specificity is a relative property that defies simple categorization. For example, the accumulation of phenolic compounds in plants has been found to follow herbivory, mechanical leaf damage, pathogen infection, exposure to intense light, and exposure to nutrient limitation (Coleman and Jones 1991). Having this information, one would be tempted to conclude that induced phenolic production is not cue specific. However, a more detailed look reveals that different amounts and possibly different kinds of phenolics are induced, depending on the type of damage that was inflicted to birch leaves (Hartley and Lawton 1991). Grazing by caterpillars induced greater levels of phenolics than leaf mining or artificial damage (Hartley and Lawton 1987). Mechanical cutting of leaves with clean scissors induced less accumulation of phenolics than cutting leaves with scissors dipped in insect saliva (Hartley and Lawton 1991). Having this information, one would be tempted to conclude that induced production of phenolics is very specific to particular cues. Both conclusions are probably correct.

Many examples of induced responses that are elicited by several different cues have been reported. Green and Ryan (1972) induced the synthesis of proteinase inhibitors (PIs) by allowing Colorado potato beetles (*Leptinotarsa decemlineata*) to feed on tomato leaves, although mechanical damage was an equally powerful elicitor. Other compounds (e.g. ethylene, abscisic acid) suspected of being signals that initiate systemic resistance also can be induced by mechanical wounding as effectively as by a living parasite's attack (see section 2.3). Perhaps the plant responds to herbivory as it does to many other stresses, using a centralized system of hormonal shifts that mediate a suite of physiological adjustments (Ferree and Hall 1981; Chapin 1991a, 1991b; also see section 3.3.3.1). While it may be more intuitive that diverse cues should stimulate generalized physiological responses, there are also numerous examples of diverse cues stimulating the synthesis of particular chemical end products. Feeding by insect herbivores was found to stimulate sweet potato and pea plants to accumulate phytoalexins, antimicrobial chemicals of low molecular weight, whose elicitation has often been regarded as relatively specific (Akazawa et al. 1960; Loper 1968; Uritani et al. 1975). Feeding by gall mites has been shown to trigger the accumulation of pathogenesis-related proteins, chitinase, β-1, 3-glucanase, and hypersensitive lesions, responses that are usually associated specifically with pathogen attack (Bronner et al. 1991).

On the other hand, many induced responses have been found to be specific to particular cues; damaging with different elicitors induced different responses. Perhaps the best-understood induction process involves the enzymes and reactions of phenylpropanoid metabolism (Bowles 1990). The first enzyme in this pathway, phenylalanine ammonia lyase (PAL), exists in multiple forms that are differentially induced by different cues. PAL is encoded by several different genes (at least four genes in parsley [*Petroselinum crispum*]), and each of these responds to UV light, wounding, and chemical elicitor at a different rate or by producing a different amount of end product. Similarly in beans, three classes of PAL genes respond differently, depending on whether the damage is caused by wounding of the hypocotyl or by a fungal elicitor. Insect damage has also been found to stimulate PAL activity differentially (Hartley and Lawton 1991). PAL activity does not increase in birch leaves from undamaged trees or in undamaged leaves on artificially damaged trees. Artificially damaged leaves show slight increases in PAL activity while actual insect attack causes large increases in both damaged leaves and undamaged leaves on damaged trees. Hartley and Lawton (1991) concluded that a signal passes from the site of damage to undamaged leaves and induces PAL activity following actual herbivory but not following artificial damage. Similarly, other enzymes that control reactions farther downstream in the phenylpropanoid sequence have also been found to be transcribed differently by different elicitors to produce end products appropriate for each stimulus (Bowles 1990).

Conifers, bark beetles, and the symbiotic fungi that are associated with the beetles make a second fascinating example of cue specificity. Attack by beetles and fungi induces an initial resin flow followed by formation of a necrotic lesion around the attacking beetle (Raffa and Berryman 1987). Monoterpenes, sesquiterpenes, and phenolics accumulate within this lesion. Aseptic wounding causes little monoterpene accumulation, inoculation with killed fungi causes small accumulations of monoterpenes, and natural attacks with living fungi causes massive accumulations (Raffa 1991). Different tree species respond differently to the various fungal associates of the major beetle species. Trees respond more extensively to the particular fungal symbionts of those bark beetle species that normally attack them (Raffa 1991). For most attacks the monoterpenes that are most toxic against the particular attackers are the ones that show the greatest percentage increase (Raffa and Berryman 1987). This remarkable correspondence between particular attackers and particular monoterpenes suggests cue specificity, although the biochemical chain of reactions producing this correspondence remains unknown for these systems.

Cue specificity is of interest in designing experiments to learn other things about induced responses. If artificial mechanical wounding causes the same responses as actual damage, then wounding can be used as a valuable experimental tool. Many studies have found that artificial damage causes responses in plants that affect herbivores. However, these tell us little about whether the responses caused by artificial damage are really the same as those caused by herbivores. Studies that included at least three treatments (plants damaged by herbivores, plants damaged by artificial wounding, and undamaged controls) were more informative. Several studies have found that artificial damage causes effects that are similar to those resulting from actual wounding (e.g. Green and Ryan 1972; Karban 1985; Neuvonen et al. 1987). However, these studies indicate only that the end results measured (chemicals or herbivore performance) were similar for actual and artificial damage, although the processes producing the end results may or may not have been identical (see 3.3.3.3 for such an example). Other studies have found that artificial damage induced different responses than did actual herbivory in either extent (e.g. Haukioja et al. 1985; Haukioja and Neuvonen 1985; Faeth 1986; Baldwin 1988a; Turlings et al. 1990; Stout et al. 1994) or in quality (e.g. Hartley and Lawton 1987; Neuvonen et al. 1987). In summary, artificial damage can be a useful experimental tool as long as the limitations of the technique are recognized for each particular system.

2.2.2 Specificity of Effects

A single type of damage may induce resistance against a great diversity of challenges. Plant growth regulators, hormones, and herbicides have been shown to induce resistance against many insects and mites (Kogan and Paxton 1983; Fischer, Kogan, and Greany 1990). The suspected mechanisms responsible for many of these examples involve increased accumulations of secondary chemicals in addition to changes in host-plant phenology. For example, the plant growth regulator chlormequat chloride has been found to induce resistance against many insects (Fischer, Kogan, and Greany 1990). In *Sorghum bicolor* this effect is associated with increased levels of plant pectin and methoxypectin, which are thought to interfere with feeding (Dreyer et al. 1984), thereby producing low specificity of effect.

Phytoalexins, chemicals that are regarded as primarily antifungal, have also been found to have more general effects and have been associated with resistance against insects. For example, in soybeans (*Glycine max*), phytoalexin production is associated with induced resistance against the fungal diseases *Phytophthora* rot and stem rot (Kogan and Fischer 1991). However, soybean leaf tissues that had elevated concentrations of phytoalexins were avoided by Mexican bean beetles (Hart et al. 1983). Glyceollin, one of the phytoalexins that was suspected of being involved in the induced response, was sprayed on leaf disks of soybean and broad bean in varying concentrations and then offered to three beetle species that commonly use these plants (Fischer, Kogan, and Paxton 1990). As glyceollin concentrations increased, Mexican bean beetles and southern corn rootworms became less willing to accept the leaf disks; bean leaf beetles showed no preference.

Wheeler and Slansky (1991) took a slightly different tack to the problem of specificity of effects for soybean and its herbivores. They opted for an approach that provided more realistic assays at the expense of some information about the chemicals involved. They fractionated extracts from either damaged soybean plants or from undamaged controls and then tested the fractions in artificial diets against several soybean herbivores. They found that the petroleum ether fraction is most active in reducing the growth rates of velvet-bean caterpillars, fall armyworms, and cabbage loopers. This activity was induced by either mite damage for plants grown in the greenhouse or by velvet-bean caterpillars for field-grown plants.

A third example of low specificity of effects involves young cotton plants (*Gossypium hirsutum*) that were exposed briefly to strawberry spider mites (*Tetranychus turkestani*), which made them more resistant against a variety of other challenges. Damaged plants supported much smaller populations of two-spotted spider mites (*Tetranychus urticae*) than did undamaged controls (Karban and Carey 1984). Mechanical wounding was also as effective at reducing mite populations as was actual exposure to mites, indicating low cue specificity (Karban 1985). Survival of beet armyworm caterpillars (*Spodoptera exigua*) was reduced on young plants that had previously hosted spider mites (Karban 1988). The fungal pathogen responsible for verticillium wilt was less likely to cause symptoms of this disease on plants that had been exposed to spider mites than on unexposed controls (Karban et al. 1987). However, damage by spider mites did not make plants more resistant to bacterial blight caused by *Xanthomonas campestris* (Karban and Schnathorst, personal observations). Because the mechanisms responsible for these diverse phenomena associated with mite damage are unknown, it is unclear if one or several different induced responses caused these many effects.

Other workers have found evidence of low specificity when the inducer and the challenger were different species or strains. Plant pathologists refer to this phenomenon as cross-resistance. Similar results could be presented for insects on oaks (West 1985; Faeth 1986; Hunter 1987), insects on larch (Baltensweiler 1985), and insects on lupines (Harrison and Karban 1986).

Many examples have also been found of greater specificity of effects. Induced responses may cause large effects against one species of challenger and no effects against others. Responses that induce generalized resistance against different pathogen species may cause little or no detectable effects on insects. For example, restricted inoculation of the lower leaves of cucumber with anthracnose fungus induced systemic resistance against viruses, fungi, and bacteria (Ajlan and Potter 1991). This effect was associated with phytoalexin accumulation, lignification, and increases in chitinase, peroxidase, and β-glucanase (plus other responses that were not assayed). However, induction of these effects by infection with anthracnose failed to affect population growth of spider mites, growth of fall armyworms, or progeny production of melon aphids. Similarly, cucumber plants that were infected with tobacco necrosis virus became more resistant to anthracnose fungus but were not protected against spider mites, fall armyworms, or striped cucumber beetles (Apriyanto and Potter 1990).

Several studies of the induced responses of birch foliage have found rather idiosyncratic effects on different herbivore species. Birch leaves damaged by leaf mining were avoided by four species of caterpillars, whereas leaves damaged by chewing caterpillars were avoided by one caterpillar species but no preference was detected in another two species; leaves damaged artificially were preferred by two caterpillar species and no preference was detected in another two species (Hartley and Lawton 1987). Bioassays of wound-induced effects of birch on several species of phloem feeders gave mixed results (Neuvonen and Haukioja 1991). Similarly, effects of induced responses of birch on the growth rates of eight species of sawflies ranged from slight reductions (statistically no effect) to reductions of approximately 80% (Hanhimaki 1989).

Cataloguing the diversity of effects is fine, but it would be far more satisfying to be able to generalize about the conditions that were associated with highly specific effects and those associated with generalized effects. To this end several Finnish workers (Haukioja 1990a, 1990b; Tuomi et al. 1990) have suggested that delayed induced responses (those occurring in the season[s] after damage) seem to affect many different species of herbivores on birch equally. On the other hand, rapid induced responses (those occurring during the season that the damage is inflicted) seem to affect herbivore species quite idiosyncratically. Tuomi et al. (1990) presented a theoretical basis to explain these observations, thus giving the argument more intuitive appeal. They suggested that long-term induced responses result from carbon/nutrient imbalances in the plant (see chapter 3), which are difficult for all herbivores to live with. Short-term responses may result from de novo synthesis of secondary chemicals or translocation of preexisting reserves of these chemicals from storage sites. The effects of these secondary chemicals on different herbivores may be idiosyncratic, and some herbivores may not be affected at all or may even prefer the induced state.

This simple explanation seems to work for the insects that feed on birch, which is one of the best-studied plants in terms of effects of induction on many different herbivores. However, the argument must be made considerably more complicated to account for the effects on mammalian herbivores, some of whom benefit from long-term induced responses of birch (Bryant et al. 1988; Haukioja et al. 1990).

It is interesting to ask whether this observation holds more generally for other systems. Most of the well-studied examples of rapid induced responses show high specificity in their effects, although exceptions certainly can be found. Soybean and cotton plants, mentioned above as having low specificity of effects, both have responses that are considered short-term. Long-term responses that involve dramatic redistributions of plant resources may also affect herbivores idiosyncratically. For example, leaf miners may cause early leaf abscission on some trees, although the effects of this drastic response vary considerably from species to species of miners (Kahn and Cornell 1983; Stiling and Simberloff 1989; Preszler and Price 1993).

At this point in our understanding of induced responses, it is much easier to come up with exceptions to generalizations than it is to find worthy generalizations in the first place. We wonder if a better distinction might be made between those responses that involve massive losses of plant tissue that are likely to affect almost all herbivores versus those that involve more subtle responses that are likely to affect species of herbivores idiosyncratically. In the examples cited above, long-term (delayed) and short-term (rapid) responses were correlated with massive defoliations and episodes of lesser tissue removal, respectively. The pattern noted by Haukioja, Tuomi, and coworkers could have been caused by differences associated with long-term versus short-term responses as they suggest, by the extent of tissue damage as we suggest, or by some other unrelated causal agent. Despite the exceptions that we have noted, the observations of Haukioja (1990a, 1990b) and Tuomi et al. (1990) seem to work for many systems and are worth consideration and manipulative experiments when the effects of more plant responses become known.

2.2.3 Specificity of Herbivore Counteradaptations to Induced Responses

Herbivores have evolved counteradaptations to some of the induced plant responses that they encounter. The counteradaptations that we are able to detect are often those that show exquisite specificity. Avoidance behaviors of insects correspond precisely to the architecture of the secretory canals of their host plants (figure 2.1, from Dussourd and Denno 1991). Plants with arborescent branching patterns of resin canals and lactifers are eaten by herbivores that cut the canals in the leaf veins, thus preventing the flow of resin into the leaf blade. This behavior would be ineffective against plants with netlike canal systems, which lack a main supply pipe. Herbivores that feed on plants with netlike canal systems cut a trench across the entire network, effectively severing all of the many lines of supply. The herbivores can then consume the leaf tissue without being incapacitated by plant resin. Plants of the Convolvulaceae present a special problem; they have many large veins that supply resin.

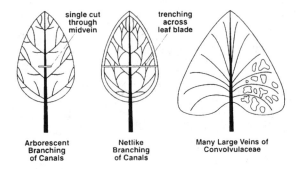

Fig. 2.1 Behaviors of herbivores are specific for the architecture of the secretory canals
of their host plants (redrawn from Dussourd and Denno 1991). Secretory canals contain
resin and latex that interfere with feeding by generalist herbivores. *Left,* specialist herbivores circumvent the defenses of plants that have arborescent branching of canals, with a
single cut through the midvein, preventing flow to the distal portion of the leaf. *Center,* a
single cut through the midvein does not stop flow for leaves with netlike branching of canals. Specialist herbivores that exploit these leaves make a trench completely across the
leaf, severing all the branches, before feeding on the distal portion. *Right,* leaves of plants
in the family Convolvulaceae have many large veins. Rather than attempting to sever all
of these, specialist herbivores feed only between the veins.

Herbivores feeding on these consume leaf tissue between the veins,
avoiding the induced resin flow entirely.

   Herbivores may use the chemicals that they induce their host plants
to produce for their own protection or to attract mates. The induced
responses that deter many herbivore species may also serve as strong
attractants for some specialized herbivores. For example, mechanical
wounding of leaves of wild cucumber plants induced the rapid accumulation of cucurbitacins and other chemicals and release of mucilaginous
sap (Carroll and Hoffman 1980; Tallamy and McCloud 1991; McCloud
et al. 1995). Cucurbitacins have been shown to be effective antifeedants
against many vertebrate and invertebrate herbivores (Metcalf and Lampman 1989). Squash beetles feed on wild cucumber leaves by first cutting
trenches, preventing the tissues that they are about to consume from
interacting with the rest of the plant (Tallamy and McCloud 1991).
Squash beetles that were prevented experimentally from trenching
suffered reduced survival and fecundity, in large part because the mucilaginous phloem sap exuded from the wound site inhibited feeding
(McCloud et al. 1995). Cucumber beetles, on the other hand, do not
trench cucumber leaves before feeding. Cucurbitacins serve as powerful
feeding stimulants for this group of beetles, which actually seek out sites

of recent damage; high concentrations of cucurbitacins also have no detrimental physiological effects on cucumber beetles. To the contrary, consumption of cucurbitacins make cucumber beetles unpalatable to invertebrate predators.

Natural selection may favor plants that reduce traits (induced responses) that specialist herbivores can use for their own advantage. Malcolm and Zalucki (1996) speculated that the timing and placement of induced responses by milkweed plants (*Asclepias* spp.) make counteradaptation by specialized monarch larvae (*Danaus plexippus*) difficult. According to the well-known milkweed-monarch story, monarch larvae are able to sequester cardenolides that protect them from their predators. In fact, later instar larvae are able to sequester cardenolides, but early instars are more sensitive to them. When *Asclepias syriaca* plants are wounded, they rapidly induce and then rapidly relax cardenolide levels. Malcolm and Zalucki argued that this induction regime allows the milkweeds to hit the larvae hard when these herbivores are small and sensitive and that relaxing their response and perhaps even negatively inducing later when larvae are large enough to sequester allows plants to reduce sequestration by older herbivores. They argue that putting the cardenolides in latex further reduces the ability of larger larvae to sequester. The latex forces larvae to trench, and trenching reduces the cardenolide content of leaves and essentially results in negative induction.

Herbivores may mount counterdefenses to those they induce in their host plants. These may be inducible and highly specific. For example, three species of swallowtails (*Papilio* spp.) that encounter xanthotoxins (furanocoumarins) in their host plants detoxify these chemicals using cytochrome P450 monooxygenases (Cohen et al. 1992). Prior ingestion of xanthotoxin induces P450 activity, which then detoxifies xanthotoxin and allows it to be consumed (Cohen et al. 1989). Three species of swallowtails that never encounter xanthotoxin did not induce P450 activity, nor did several unrelated caterpillar species that similarly do not encounter xanthotoxins (Cohen et al. 1992). The particular blend of furanocoumarins found in the host plant determines how effective the larvae are at feeding. Various furanocoumarins are found in many plant species, and each may inhibit metabolism of the other (Berenbaum and Zangerl 1993). Growth of swallowtail larvae fed parsley leaves treated with two furanocoumarins was significantly slower than growth of larvae fed equimolar concentrations of either furanocoumarin presented alone.

Results coming from the labs of Maurice Sabelis and Marcel Dicke in the Netherlands provide a fascinating example of the specificity of plant-derived signals that are used not only by herbivores but also by the predators of herbivores (see also section 3.2.3). Using a series of carefully controlled experiments in which mites were offered choices in olfactometer tubes, Dicke (1986) found that herbivorous spider mites (*Tetranychus urticae*) dispersed away from odor plumes of lima bean plants (*Phaseolus lunatus*) that were infested by high densities of conspecifics. One component of the pheromone that causes dispersal has been found to be linalool, a chemical that the plant, not the mites, produces (Dicke and Sabelis 1989). These volatiles were emitted from artificially damaged plants or spider mite–infested plants but not from the mites themselves or their feces. At low infestations, spider mites were attracted to the infested lima bean leaves, but as infestation size increased, spider mites became repelled (Dicke 1986). Predatory mites (*Phytoseilus persimilis*) that feed on spider mites were attracted to the spider mite–infested bean leaves (Dicke and Sabelis 1989). Previously infested leaves remained attractive to predators for several hours after the spider mites had been removed, but the mites themselves were not attractive to the predators if they were removed from the leaves. Plants that were damaged in a variety of ways were not attractive to predators unless feeding by spider mites had recently occurred. Predatory mites were also able to distinguish between different species of herbivorous mites based solely on the signals that were produced. Dicke and Dijkman (1992) argued that signals caused by herbivore-induced damage are useful for predators only if they contain a great deal of specific information, a condition that appears to have been satisfied in this system.

Other organisms also respond rather specifically to the cues generated by herbivory. Damaged corn seedlings release high levels of volatile terpenoids within several hours; they then become less palatable to beet armyworm caterpillars (Turlings and Tumlinson 1991) and more attractive to parasitoids that feed on caterpillars (Turlings et al. 1990). Artificially damaged seedlings did not release volatiles unless oral secretions from the caterpillars were applied to damaged sites. The damaged plant, not the caterpillar or its feces, was the main source of the cue used by parasitoids (Turlings et al. 1990). The active component that stimulated seedlings to release volatile terpenes was present in regurgitate from herbivores but not in the plant, the hemolymph of the herbivore, or the herbivore's feces (Turlings et al. 1993). Although the available evidence does not conclusively indicate that the herbivores caused the responses, rather than microbes that may have been associated with them, it is clear that other organisms use the induced responses to gain specific information about the source of the damage.

2.2.4 Specificity and Memory in Immune Responses

Specificity is considered to be one of the properties integral to any immune reaction (Klein 1982). Immune responses of vertebrates can be far more specific than those of invertebrates, and both can exceed levels of specificity observed for plants (Klein 1982; Harvell 1990a, 1990b). Some authors consider a high degree of cue specificity to be a necessary condition for an induced response to be called an immune response. This is somewhat ironic, since higher vertebrates including humans have accumulated several modes of induced defense; some of these, like phagocytosis, have very low cue specificity. Higher invertebrates have systems with less specificity than more primitive groups of invertebrates (Harvell 1990a, 1990b). In general, high specificity of recognition is associated with those phyla most vulnerable to neoplasia, that is, when a slightly different cell line multiplies within the host, potentially reducing the host's fitness. The threat of neoplasia may be much lower for plants than for animals because plants have a modular form of growth with many semiautonomous units (Watson and Casper 1984; Sprugel et al. 1991). If cue specificity is primarily a defense against neoplasia, the benefits of high cue specificity may be less for plants than for animals.

Defenses in vertebrates contain some components that possess a memory. A defense system that has a memory responds more rapidly and often more vigorously to a given stimulus the second time it is encountered (and subsequently). Memory is exquisitely specific in vertebrate cytotoxic reactions. For example, a mouse that has previously been exposed to a graft of type x rejects another graft of type x vigorously and rapidly. However, the mouse responds to a graft of type y at the normal speed of an individual that has never been exposed to any graft. Invertebrate defense systems possess memories that last for weeks as opposed to longer-lived memories known from vertebrates; memories of invertebrates show less specificity, as well, compared to those of vertebrates (Harvell 1990a, 1990b). The memory component of plant induced resistance has been virtually unexplored.

Many workers have noted that plants exposed to multiple bouts of herbivory showed stronger induced responses than plants that were exposed only once (Wallner and Walton 1979; Werner 1979; Valentine et al. 1983; Neuvonen and Haukioja 1984; Cook and Hain 1988; Baldwin 1989; Karban 1990; Clausen et al. 1991; see section 4.2.4). In these examples there was no way to separate the possible existence of a memory from the possibility that more damage induced a larger or more rapid effect. To distinguish these two possibilities, Karban and Niiho (1995) exposed cotton seedlings to a constant total amount of damage from mites but varied the timing of the damage. If induced resistance in cotton showed a memory, plants that were exposed twice were expected to respond more rapidly or more vigorously than those exposed only once to the same total amount of injury. No differences in the effects of these treatments on spider mites were found, suggesting that cotton does not possess an important memory component in its induced response.

Induced responses that affect herbivores and possess a memory may (1) generate more cue following repeated experiences with herbivores, (2) respond more rapidly, and/or (3) increase the magnitude of their response to the same level of cue. Baldwin and Schmelz (1996) tested the last two mechanisms for tobacco plants that were induced once, twice, or three times by addition of methyl jasmonate, an elicitor of nicotine synthesis (see section 3.3.3.4). The total nicotine pools induced at the end of their experiments did not differ depending on the past experience of the plant. However, they found some evidence for memory; plants with one or two prior inductions increased their total nicotine pools more rapidly than did plants with no prior inductions. Baldwin and Schmelz suggested that the speed of induction may be important in determining whether the induced response is effective, although the ecological significance of this effect has not yet been determined.

2.3 SIGNALS IN SYSTEMIC INDUCTION

Some induced responses to wounding are systemic (Green and Ryan 1972; Davies and Schuster 1981). In such cases the damaged plant tissue may produce a signal that is transmitted systemically throughout undamaged parts of the plant, causing the induction of new morphological or physiological states, the induced response. We regard induced responses as localized when they occur in the same organ (e.g. leaf or stem) and as systemic when they occur in organs other than the damaged one. Aside from being arbitrary, this definition leaves considerable variability in responses we call systemic; whenever possible, we try to specify the spatial scale involved.

Several different signals have been proposed as transporters of information within plants, including chemicals and electrical impulses (see Chessin and Zipf 1990; Ryan 1992; Enyedi et al. 1992; and Hammerschmidt 1993 for other recent reviews). In considering these, it is important to bear in mind that (1) different plants may use different signals, (2) a single plant may use different signals in response to different kinds of damage, (3) a single plant may use more than one mode of signaling for a single kind of response. Indeed, redundancy of signals makes good intuitive sense, particularly if plant parasites can deactivate particular modes of the plant's signaling capabilities. In other words, these varied signal mechanisms should not be viewed as mutually exclusive possibilities. Despite the possibility of multiple signals, the reductionist approach to finding these signals assumes that a chemical must be necessary and sufficient to induce the observed responses if it is to be considered "the signal." By necessity, the discussion that follows assumes this reductionist viewpoint.

The biochemical and physiological processes that result in induced resistance are almost certainly chains of multiple reactions involving many different compounds. Removing a particular step in one of these processes by using mutants (naturally occurring or genetically engineered) that lack the ability to synthesize a particular compound can provide powerful evidence that the compound is involved in induced resistance. When this is the case, such a mutant is incapable of showing the response. Chemical agents that block specific steps in biochemical pathways can be used in an analogous manner. If the lack of response is caused solely by the compound in question, then this condition should be corrected when the compound is experimentally added. Experiments of this sort can demonstrate that a particular compound is necessary for the chain of reactions that ultimately produces the induced response. However, it is much more difficult to establish that a particular compound is the actual signal that moves through the plant.

Any hypothetical signal must meet the following criteria to receive serious consideration: It must be (1) generated rapidly at the wound site (2) by known inducers. It must (3) travel through the plant (4) following a time course that is consistent with that of the induced response. It must (5) elicit the induced response (6) at concentrations that are consistent with those measured in damaged plants. These criteria are used below to evaluate the signals that are currently under consideration (summarized in table 2.1). For most of these candidate signals, information is available only for their ability to induce synthesis of one group of chemicals, the PIs, in a few plant species.

2.3.1 Oligosaccharide Fragments of Plant Cell Walls

Oligosaccharide fragments from the cell walls of fungi were recognized as early as the mid-1970s as being capable of acting as signals in plant tissues (Ryan 1987). When wounded, higher plants' cell walls fragment at the the wound site. However, the enzymes that initiate this process in response to wounding have not yet been found in higher plants; it is interesting to note that such enzymes are commonly found in fungi, so detection poses no great experimental problem (Ryan 1992). Feeding

Table 2.1 Signals that have been proposed to operate within plants: criteria and evidence

Signal	Generated at Wound?	Inducers	Mobility through Plant	Appropriate Time Course?	Response	Appropriate Concentration?
Oligosaccharide fragments of plant cell walls	Yes	Pathogens in soybean, possibly by aphids	Only large molecules active; only small molecules mobile	No, large molecules move too slowly.	PIs, phytoalexins, peroxidase, localized lignin induction.	Some question
Systemin, a polypeptide	Yes	Wounding	Very mobile	Yes, 30 min. throughout leaf, 2 hr to phloem.	PI; antisense plants had much lower production of PI.	Extremely low concentrations required.
Salicyclic acid	Yes	Specific; only pathogens, not wounding	In phloem	Question about speed and timing.	Pathogenesis-related proteins, peroxidases: actually induces resistance against pathogens; antisense plants showed no resistance to virus.	Some question
Ethylene	Yes	Herbivores, pathogens, wounding	Very mobile	Yes, production is greatest hours before response.	Pathogenesis-related proteins, lignification actually induces resistance against some pathogens, insects; questions since response found when ethylene was inhibited.	Some question
Abscisic acid	Yes	Wounding and other conditions not associated with PI induction	Mobile	Yes	PI but some questions; mutants that don't synthesize abscisic acid don't induce until supplied with it.	Higher concentration required than for jasmonic acid.
Jasmonic acid, methyl jasmonate	Yes	Wounding and yeast elicitor of plant cell culture	Jasmonic acid in phloem; methyl jasmonate volatile	Yes	PIs, ethylene, PAL, systemin, others.	Yes
Electrical signal	Yes	Wounding	Phloem blocked, still got response	Possibly, perhaps only for small plant.	PI; needs electrical signal for induction.	Yes

by aphids has also been found to produce some oligosaccharides of plant origin (Campbell 1986), although the consequences of this production are not currently known. Another major problem with oligosaccharides as systemic signals is the observation that large polymers of oligosaccharides (average degree of polymerization of nine to twenty-three) are more active as inducers than smaller oligomers (Ryan 1992; Hahn et al. 1993). However, labeled oligosaccharides of six or more units were not transported throughout plants when placed on wounded leaves (Baydoun and Fry 1985). The larger molecules probably move too slowly to account for the rapid induction of PIs, although this limitation has been the subject of some controversy and some workers believe that oligomers may be sufficiently mobile in the phloem to act as signals (Rigby et al. 1994). The cell walls of the plant appear to function as molecular sieves, keeping out big fragments until the cell wall is broken down by invading pathogens (Doares et al. 1995).

Oligosaccharide fragments, supplied through the plant stem, elicit synthesis of PIs. Fragments also induce synthesis of phytoalexins and peroxidases and localized production of lignin (Ryan 1987). There is some question about whether the concentrations of cell-wall fragments produced by wounding are sufficient to induce defensive responses. One oligosaccharide, chitosan, has been developed commercially and is available for control of fungal diseases (see section 6.2.2). Chitosan oligomers are found in fungal cell walls, and they are ten times more effective by mass at eliciting PIs than are fragments from plant cell walls (Reymond et al. 1995).

2.3.2 Systemin: A Polypeptide

Systemin is the only polypeptide identified thus far that is capable of acting like a plant hormone to regulate production of other proteins. Systemin is produced close to the site of wounding by mechanical damage of above-ground plant tissues (Pearce et al. 1991; McGurl et al. 1992). The kinetics of systemin induction of PIs are similar to those generated by insect feeding, suggesting that systemin can be induced by herbivores, although this has not yet been determined conclusively (Pearce et al. 1991; McGurl et al. 1992). Systemin is very mobile and can travel through the plant in the phloem; labeled material spread throughout a damaged leaf in thirty minutes and was detected in the phloem within two hours (Pearce et al. 1991). Systemin induces the synthesis of PIs. Tomato plants that were transformed with a gene for antisense prosystemin, making them unable to produce systemin, showed much reduced induction of PIs in response to wounding (McGurl et al. 1992).

Chemicals that inhibited transport of systemin from the wound site into the phloem also stopped the induction of PI (Narvaez-Vasquez et al. 1994). This suggests that systemin plays an important role in systemic induction. Minute quantities of systemin are sufficient to induce synthesis of PIs; it is approximately 100,000 times more active on a molar basis than the oligosaccharide fragments of plant cell walls (Pearce et al. 1991). The primary difficulty with the hypothesis that systemin is the signal molecule that is transported throughout the plant is the observation that wounding one leaf induces the transcription of systemin mRNA throughout the plant (McGurl et al. 1992). This raises the possibility that systemin itself may not move through the plant but rather that its accumulation in unwounded leaves may be caused by another systemic wound signal (Enyedi et al. 1992).

2.3.3 Salicylic Acid

Salicylic acid of plant origin has been recognized for its healing properties since antiquity. Its involvement with plant resistance to disease has been appreciated since 1979, and only in the past ten years has its ability as a powerful inducer of many plant pathways become known (Raskin 1992). Levels of salicylic acid increase at the site of infection by pathogens (Malamy et al. 1990). Of the signal molecules that are being considered, production of salicylic acid is probably the most specific; concentrations increase in response to pathogen infection but not as a general wound response (Enyedi et al. 1992). Salicylic acid moves readily through the phloem (Métraux et al. 1990), although there is some question about the speed of the response and whether salicylic acid leaves the infected leaf quickly enough to account for the systemic response (Rasmussen et al. 1991). Methyl salicylic acid is found in wounded plants in large quantities and is very volatile (Lee et al. 1995). Plants that are inoculated with mosaic virus apparently bleed out methyl salicylic acid, possibly in quantities that are sufficient to elicit resistance in nearby plants; this speculation is currently under investigation.

Resistance induced by salicylic acid protects plants against a great variety of pathogenic infections. Salicylic acid induces transcriptional production of a variety of chemicals thought to be involved in plant defense such as the pathogenesis-related proteins and oxidative enzymes. Tobacco leaves that are resistant to tobacco mosaic virus increase concentrations of salicylic acid following viral infection; no increases were observed in leaves of susceptible plants, and spatial patterns of accumulated salicylic acid correspond to patterns of resistance (Malamy et al. 1990). Endogenous concentrations of salicylic acid were sufficient to account for observed levels of resistance to pathogens. Transgenic plants with a gene that converts salicylic acid to catechol accumulated little salicylic acid after pathogen infection (Gaffney et al. 1993). Unlike controls, which did accumulate salicylic acid, the transformed plants failed to show induced resistance to viral attack. This result suggests that salicylic acid is an indispensable ingredient in the transduction pathway that culminates in induced resistance to pathogens, even if it is not the actual signal. Salicylic acid has not yet been shown to induce responses that affect herbivores. However, by stimulating an oxidative burst in plant tissues (Chen et al. 1995), salicylic acid may cause the hypersensitive response characteristic of induced resistance to many pathogens as well as some herbivores (see sections 2.3.8, 3.2.2)

Some questions have been raised about the concentrations of salicylic acid that are required to induce resistance (Ryals et al. 1992). Moreover, grafting experiments using combinations of rootstocks and scions that were either able or unable to accumulate salicylic acid demonstrated that salicylic acid was not responsible for endogenous signaling leading to induced resistance against tobacco mosaic virus (Lawton et al. 1993). Scions were grafted to rootstocks that were unable to transmit salicylic-acid cues; when these rootstocks were inoculated with tobacco mosaic virus, scions that were grafted to them and subsequently challenged induced resistance. This indicated that rootstocks that were unable to accumulate and transmit salicylic acid were capable of transmitting a systemically mobile signal through a graft. Scions that lacked the ability to accumulate salicylic acid were not able to induce resistance against tobacco mosaic virus even when grafted to rootstocks that could accumulate and transmit salicylic acid signal. This indicated that salicylic acid was required for induced resistance.

Exogenous applications of salicylic acid stimulate actual resistance to a variety of lesion-forming viral, bacterial, and fungal pathogens, making salicylic acid a candidate for pest-control applications (Long 1990). However, exogenously applied salicylic acid has not been found to translocate effectively throughout the plant, detracting from its potential for commercial use (Enyedi and Raskin 1993; Kessmann et al. 1994).

2.3.4 Ethylene

Ethylene is unusual as a plant hormone because it is a gas. Exposure to ethylene leads to many changes in the biochemistry and physiology of plants (Rhoades 1985b; Boller 1991). Ethylene production is generated at a wound site by damage associated with herbivores, pathogens, and mechanical injury (Enyedi et al. 1992). Ethylene production was found to be greater following insect herbivory than following mechanical damage for pine trees (Shain and Hillis 1972) and onions (Kendall and Bjostad 1990). Ethylene is very mobile, readily diffusing away from the wound site. Production of ethylene occurs several hours before the observed induced resistance, and the strength of induced responses correlates to the concentration of ethylene generated. Ethylene does not simply "escape" from wounded tissue; rather its production is regulated by induction of aminocyclopropane carboxylic acid synthase. Ethylene has been shown to induce many chemicals thought to be involved in defense, such as pathogenesis-related proteins, chitinase, and PAL enzymes, as well as the process of lignification and cell wall strengthening (Rhoades 1985b; Enyedi et al. 1992). Plants treated with ethylene become more resistant to a variety of fungal, viral, and insect attacks. However, some of these responses have been found to occur even when ethylene synthesis is inhibited, casting doubt on the role of ethylene as the signal (Mauch et al. 1984; Boller 1991; Lawton et al. 1993). In addition, questions have been raised about the possibility that at least some of the responses presumed to be induced by ethylene may be experimental artifacts. Many workers do not actually treat plants with ethylene gas but rather use ethephon, a product that breaks down to ethylene and phosphonic acid following application; activation of systemic acquired resistance in *Arabidopsis*, reportedly induced by ethylene, was actually induced by phosphonic acid (Lawton et al. 1994). In fact, ethylene is not involved in that resistance response, since mutants of *Arabidopsis* that were insensitive to ethylene developed systemic acquired resistance in response to salicylic acid. Many putatively defensive reactions can be induced by a combination of ethylene and methyl jasmonate more effectively than by either of these volatile chemicals alone (Xu et al. 1994). These results suggest that plant responses to ethylene are probably more specific than previously imagined. Questions have also been raised about whether experimentally manipulated quantities of ethylene that were necessary to induce resistance against pathogens exceeded physiologically relevant concentrations (Enyedi et al. 1992).

2.3.5 Abscisic Acid

Like ethylene, abscisic acid is a plant hormone that produces many physiological and chemical changes. As such, abscisic acid is considered not only a candidate signal molecule for induced accumulation of PIs but also a hormone that could possibly regulate numerous other responses to all kinds of plant stresses (Chapin 1991b). Local wounding of some solanaceous plants led to increased concentrations of abscisic acid at the wound site (Peña-Cortés et al. 1989). Increases in abscisic acid are not specific to processes that are associated with induction of resistance; many other conditions such as water stress also cause abscisic acid accumulation but do not produce other manifestations of induced resistance. Direct transport through plants has not been demonstrated, although endogenous levels of abscisic acid increase systemically in response to wounding, with a time course that is consistent with the hypothesis that abscisic acid is the signal molecule (Peña-Cortés et al. 1989; Hildmann et al. 1992). Plants treated with abscisic acid accumulate PIs both locally and systemically (Peña-Cortés et al. 1989). In addition, mutants that are deficient in abscisic acid do not accumulate PIs in response to wounding. This failure to accumulate PIs can be reversed by treating mutants with exogenously applied abscisic acid.

There is some concern that concentrations of abscisic acid required to induce production of PIs are unrealistically high, and certainly higher than concentrations of jasmonic acid required to produce the same effects (Hildmann et al. 1992). Even for stress responses that have classically been considered to be reactions to abscisic acid, evidence is accumulating that jasmonic acid is the primary signal; dehydration first results in changes in jasmonic acid titers and then increases in abscisic acid (Creelman and Mullet 1995). Several other problems with abscisic acid as the signal have also been expressed. The lack of correspondence between the induction of abscisic acid (i.e. by water stress) and the production of PIs was noted above. In addition, some clones of potato were found to be unresponsive to abscisic acid yet do induce synthesis of PIs upon wounding. Jasmonic acid has been found to stimulate many of the induced effects also produced by abscisic acid (Hildmann et al. 1992). Jasmonic acid also stimulated production of PIs in mutants that were unresponsive to abscisic acid, suggesting that jasmonic acid is closer to inhibitor synthesis on the chain of reactions. Hildmann et al. (1992) proposed that release of abscisic acid may be stimulated by wounding and that abscisic acid may in turn initiate a chain of reactions that releases jasmonic acid, the actual mobile signal.

2.3.6 Jasmonic Acid and Methyl Jasmonate

Jasmonic acid, a fatty acid derivative, and its volatile methyl ester, methyl jasmonate, are found in many plants and are known to elicit many different chemical and physiological responses (Staswick 1992; Reinbothe et al. 1994). Levels of jasmonic acid in leaves of *Avena sativa* and *Bryonia dioica* rose rapidly and transiently following mechanical wounding (Albrecht et al. 1993). Induction of jasmonic acid in plant cell cultures also followed elicitation by yeast (Gundlach et al. 1992). Both forms of this hormone are quite mobile; jasmonic acid moves in the phloem, and methyl jasmonate is volatile. Minute concentrations of gaseous methyl jasmonate induced the synthesis of PIs in the leaves of tomato plants (Farmer and Ryan 1990). The accumulation of these PIs depended on the dosage of methyl jasmonate applied, and accumulation exceeded levels that were recorded for wounded plants. Treatment of plants with these powerful elicitors also induced accumulation of other chemicals suspected of being involved in plant defense, including ethylene, PAL, systemin, and several alkaloids. Following attack, jasmonates set in motion a suite of plant responses that include the synthesis of diverse plant proteins thought to be involved in defense and a coincidental depression of synthesis of other plant proteins (Reinbothe et al. 1994). The synthesis of jasmonic acid and the many responses that it induces are discussed in much greater detail in section 3.3.4.

The possibility that methyl jasmonate is a volatile signaling molecule is exciting because it provides a potential mechanism for communication between plants. For example, sagebrush (*Artemisia tridentata*) releases volatile methyl jasmonate. When plants of sagebrush and tomato were placed in close proximity in enclosed chambers, concentrations of methyl jasmonate released by the unwounded sagebrush were sufficient to induce synthesis of PIs in the tomato plants (Farmer and Ryan 1992). It is unknown if processes such as these ever occur in nature, where air currents might diminish concentrations of methyl jasmonate below physiologically active levels. This question is currently under investigation (see section 3.4). Some insects and fungi are known to produce relatively large quantities of methyl jasmonate (Staswick 1992), which raises the possibility that these compounds may play other roles in plant-herbivore and plant-pathogen interactions.

2.3.7 Electrical Signals

Biologists are aware that bioelectrical signals that may be propagated through plants have amplitudes, induction thresholds, and shapes similar to those of axonic action potentials of animals (Pickard 1973). However, bioelectrical signals have been largely ignored in plants, partly because of a lack of repeatability (Pickard 1973; Roberts 1992). Mechanical wounding generated an action potential that moved through tomato seedlings at a speed of 1–4 mm/sec (Wildon et al. 1992). Chilling the petiole of tomato cotyledons was found to effectively inhibit transport through the phloem although it did not affect electrical signaling. Under these conditions, wounding a cotyledon caused synthesis of PIs in the unwounded first true leaf. This result suggests that an electrical sig-

nal, rather than a chemical signal, was involved. If the wounded cotyledon was removed before the electrical signal could have left, then no
induction of PIs occurred. If the cotyledon was removed five minutes
after wounding, the signal had already left and no reduction in the response was noted. The primary problems with the hypothesis that plant
signals are electrical involve timing and scale. Wildon et al. (1992) conducted their experiments on very small tomato seedlings. The responses
that have been observed in larger plants are probably too slow to be
attributed to the very fast moving electrical signals. For example, increased synthesis of nicotine in tobacco roots following damage to leaves
occurred at a time scale appropriate to a chemical signal but much too
slowly for an electrical signal (Baldwin, Schmelz, and Ohnmeiss 1994).
Other workers have failed to find evidence of electrical signals during
induction of systemic acquired resistance (Kessmann et al. 1994).

2.3.8 Signals Involved in Induction by Herbivores versus by Pathogens

Pathogens can influence plant fitness and can structure plant communities as much as herbivores can, if not more (Burdon 1987; Dobson
and Crawley 1994). When plants are attacked by herbivores they are frequently invaded by pathogens simultaneously as pathogens gain en-
trance to plants via herbivore saliva or through feeding wounds. It seems
plausible then that induced responses to herbivory actually might be
induced responses to the pathogens associated with herbivory. To evaluate this hypothesis, first we will consider briefly those plant responses
induced by pathogens and the signals that elicit those responses.

Induced responses to pathogens may involve any or all of the following six steps (Dixon et al. 1994; Godiard et al. 1994). The six steps often
happen in chronological order. It is not clear which (or whether all) of
the steps are correlates of induced resistance and which are necessary
(and sometimes) sufficient for resistance, although these relationships
will probably be understood in the near future.

1. Induced resistance to pathogens is correlated with transient ion
fluxes across membranes of attacked cells (Dixon et al. 1994). Generally,
elicitors cause a rapid efflux of K^+, an influx of Ca^{++}, and an alkaliniza-
tion of the membrane and, presumably, the surrounding cytoplasm. Observations suggest a correlation between ATPase-linked ion pumping
and elicitation of resistance, but since the time scales of these two events
are so different, the correlation is difficult to interpret (Dixon et al.
1994). Direct manipulations will be difficult because these processes are
central to life itself.

2. Following the ion fluxes, resistant cells exhibit an oxidative burst

and production of peroxide (Tenhaken et al. 1995). The increase in peroxide is a consequence of the inhibition of catalase by salicylic acid, which links the salicylic acid cascade to the oxidative burst and possibly to the hypersensitive response (Chen et al. 1995). Reactive oxygen species are produced that kill microbes directly and contribute to other processes that may provide defense. Reactive oxygen species cause (1) cross-linking of structural proteins in attacked cells, which may slow microbial ingress or may trap pathogens inside cells; (2) activation of a local signal eliciting the hypersensitive response (step 4, below); and (3) activation of more systemic signals, eliciting the production of antioxidants, which protect against reactive oxygen species in adjacent plant cells.

3. Next, many genes are transcriptionally activated. These are collectively called pathogenesis-related proteins; many of these proteins seem to be used in defense (PIs, lytic enzymes such as chitinases and glucanases, cell-wall reinforcements, etc.) although the functional significance of only a few have been demonstrated.

4. The damaged cells undergo a hypersensitive response in which cells around the site of infection die and effectively trap the pathogens. In most cases, systemic acquired resistance to pathogens (step 6) develops after the plant initiates a hypersensitive response, although these two processes may be regulated independently (Chessin and Zipf 1990; Jakobek and Lindgren 1993). Some plant pathologists have speculated that the hypersensitive response may be a necessary prerequisite for many forms of induced resistance (Chessin and Zipf 1990; Hammerschmidt 1993). However, in some mutants of *Arabidopsis* the hypersensitive response is not necessary for the elicitation of systemic acquired resistance.

5. Phytoalexins are produced locally and they kill microbes directly. Phytoalexins can come from any number of biosynthetic classes of compounds, including sesquiterpenes, flavonoids, furanocoumarins, polyacetylenes, and glucosinolates (Ebel 1986). There are many cases in which phytoalexin production is correlated with resistance to microbes. For soybean, de novo production of phytoalexin following infection has been shown to be induced specifically by pathogens and to provide resistance against pathogens (see section 3.3.3.3).

6. Systemic acquired resistance is induced, which is effective against many pathogens including mosaic virus, bacteria such as *Pseudomonas syringae*, and diverse fungi such as *Cercospora nicotianae, Phytophthora parasitica,* and *Peronospora tabacina*. Systemic acquired resistance is largely biologically defined at this time; the biochemical mechanisms are not well understood. Salicylic acid is clearly involved in eliciting systemic ac-

quired resistance, although its role as the systemically transported signal is unclear (see section 2.3.3).

The processes that occur earlier in the progression through the six steps listed above may elicit those later in the progression, although the steps have also been found to occur independently. Since the steps may be elicited independently, by different signals, there has been some confusion about the term "elicitor" (Dixon et al. 1994; Boller 1995); an elicitor may activate only a single step or the entire cascade. Some of the elicitors that have been identified are specific to the invading pathogen; others are more general. For example, as pathogens attack their hosts, their hydrolytic enzymes release signals (endoelicitors) that activate differently graded responses. Many of these are specific for particular pathogens and hosts; indeed, a description of all the known elicitors involved in induced responses to pathogens would be quite long.

Can we expect plant responses to herbivores to employ the same signal pathways as those involved in responses to pathogens? When we have examined this subject more thoroughly, some years from now, will we have a long and specific list of elicitors to herbivores as we do for pathogens? In order to begin to answer these questions, let us compare plant responses to salicylic acid and those to jasmonic acid. Salicylic acid activates systemic acquired resistance to pathogens, but its effects on herbivores are not yet determined; wounding alone does not activate this pathway. In contrast, wounding and jasmonic acid elicit many chemically defined responses; the biological significance of most of these is not well understood at present. Many other hormones and signals also activate the jasmonic acid cascade. As a first approximation it appears that salicylic acid may mediate a pathogen-specific suite of responses, and jasmonic acid may mediate a more general suite of responses to wounding and herbivory.

Under many conditions these two cascades do not interact, and the responses induced by salicylic acid may even inhibit the cascade induced by jasmonic acid. Recent experiments suggest a connection between these two pathways. Tobacco plants have been transformed with a GTP-encoding gene that makes them increase their levels of salicylic acid (rather than jasmonic acid) in response to wounding (Sano and Ohashi 1995). This suggests that a GTP-binding protein in the cell membrane may act as a molecular switch that can cause a signal that normally activates one cascade to activate the other. According to the model developed by Sano and Ohashi (1995), wound signals are directed to the jasmonic acid channel by this switch, and signals from pathogens are directed to the salicylic-acid channel or to both channels. In summary, two things are impressive about this finding: (1) pathogen attack and herbivore attack produce discrete signal cascades and (2) with very little modification, one stimulus is able to cross over to activate the other cascade.

Plant responses to pathogens are very specific in many cases. Recent work has validated the older notion that resistance and susceptibility in plant-pathogen interactions may be precisely controlled by specific genes (the so-called gene-for-gene model; Gabriel and Rolfe 1990; Godiard et al. 1994; Ausubel et al. 1995; Boller 1995). According to this model, plants with the R gene for resistance to particular pathogen genes respond when they react with the appropriate signal generated by the pathogen; the result is an incompatible interaction, that is, resistance. The big question has been, what do the R genes code for? In the near future R genes will be cloned and characterized, and we will be able to determine whether the mechanisms specified by these genes are the same as those already understood (the six steps outlined above).

In any case the gene-for-gene systems indicate that many induced responses to pathogens are highly specific. This level of specificity is unlikely to be found in induced responses to herbivory. Some workers have regarded induced responses to herbivory as little more than attempts to repair the physical wounds caused by herbivores. The examples described in section 2.2 have convinced us that at least some plant responses to herbivory are very specific. When pathogens attack plants, polysaccharides are released by their enzymes, making pathogen attack different from wounding. However, the same is certainly true of attack by many herbivores that secrete saliva into plants. Since oral secretions from caterpillars dramatically increase oxidative enzymes (Stout et al. 1994; Bi and Felton 1995) and trigger the jasmonate cascade (McCloud and Baldwin n.d.), plants may be responding specifically to the damage caused by caterpillar feeding with a process similar to that observed in the oxidative burst response to pathogens (see above). In summary, it is clear that plant responses to herbivores are frequently different from mechanical simulations of herbivory, so there is plenty of evidence that responses to both pathogens and herbivores can be specific.

Some of the actual mechanisms involved in responses to pathogens and to herbivores are likely to be similar. Small herbivores induce plants to produce local necrotic zones that resemble hypersensitive responses caused by pathogens (Fernandes 1990). Similarly, most described cases of postinfection resistance to nematodes also involve the hypersensitive response (Giebel 1982).

Pathogens generally injure plant tissues more slowly than do herbivores. When pathogens release diffusible toxins, they act like herbivores; for example, many homopterans inject toxins with their salivary secretions. In general, herbivores move within and between plant parts more readily than do pathogens; most pathogens must wait until the end of their life cycles to move. This difference in movement suggests that plants might evolve responses that are more likely to be systemic against herbivores and localized against pathogens. Antibiosis is more likely to be successful against pathogens because they are less mobile. However, many complications may obscure the simplicity of this prediction. For example, herbivores often vector plant diseases, making the pathogens precisely as mobile as the herbivores, and many pathogens have propagules that are so abundant as to be almost ubiquitous. In short, some induced responses to herbivores may actually be responses to pathogens. Many of the responses to herbivory and wounding are specific to particular cues, however, and appear to be distinct from plant responses to pathogens.

2.4 Communication between Individuals

The notion that plants might respond to the volatile cues released by wounding of their neighbors resulted from some observations made by David Rhoades. Rhoades was a graduate student and a postdoc at the University of Washington, conducting his thesis research on plant defenses and induced resistance. During the spring and summer of 1979 he set up several experiments in which he added caterpillars to some red alders and Sitka willows and had other control individuals to which he added no caterpillars (Rhoades 1983). Leaves from both treatments (damaged and control) were then used as food for caterpillars in the lab to assay effects of foliage quality on insect performance. However, the experiments did not go as planned. Many of the caterpillars that were placed on the trees to inflict the damage treatments ate little and died prematurely, probably due to disease. To maintain his treatments and save his experiment, Rhoades reloaded caterpillars on the trees that were scheduled to receive damage. Then a very curious thing happened. The performance of both the assay caterpillars fed foliage from the damaged trees and those fed foliage from undamaged control trees deteriorated. Rhoades was expecting a decrease in performance for caterpillars fed foliage from damaged trees if the damage induced rapid resistance, but why should caterpillars raised on foliage from the undamaged trees also do poorly all of a sudden? One possible explanation that occurred to Rhoades was that undamaged controls had responded to some signal that was released by their damaged neighbors.

Subsequent experiments conducted in 1980 and 1981 were consistent with this hypothesis (Rhoades 1983). Rhoades assigned some trees to be

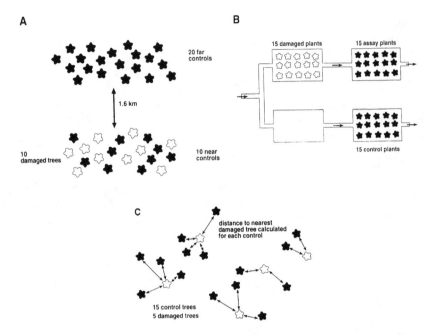

Fig. 2.2 Schematic diagrams showing the designs of experiments that test the hypothesis of communication between plants. *A*, Rhoades (1983) damaged ten trees (*open symbols*). He then compared the performance of caterpillars on ten near controls (*dark symbols*) versus performance of caterpillars on twenty far controls. The far controls were all located at a site 1.6 km away from both damaged trees and near controls. *B*, Baldwin and Schultz (1983) placed fifteen damaged plants in a growth chamber upwind from another chamber that held fifteen undamaged assay plants. (*top*). Chemicals assays were used to compare these assay plants with fifteen undamaged control plants that were not downwind from damaged plants (*bottom*). *C*, Haukioja and Neuvonen (1985) assayed the performance of caterpillars on undamaged assay trees (*open symbols*) at varying distances from experimentally damaged neighbors (*dark symbols*) in the field.

damaged and some to serve as nearby controls (figure 2.2A). He also selected other trees at some distance (1.6 or 8 km) from this test site to serve as far controls. He measured the short-term growth rate of caterpillars caged on these trees in the field. All of the test larvae grew poorly; those from the damaged and nearby control treatments lost more weight that those from the far control trees. Rhoades found no root connections between trees, and he interpreted these results as suggesting that airborne pheromonal communication might be occurring.

This suggestion generated considerable interest and a flurry of responses from colleagues (summarized in table 2.2). Other workers failed to find evidence for induced resistance in undamaged trees that were in close proximity to damaged neighbors, using the same species as Rhoades had used (Williams and Myers 1984; Myers and Williams 1984). Rhoades himself was unable to repeat his results consistently in subsequent experiments (personal communication). His results can also be explained by several mechanisms other than communication between trees. During his studies Rhoades (1983) reported that his caterpillars grew poorly and suffered from symptoms of disease. Reloading trees with caterpillars may not have caused them to release cues of damage (as Rhoades suggested) but instead may have increased the likelihood or severity of disease (Fowler and Lawton 1985). Such an increase could cause assay caterpillars to do poorly on both the trees that had just received caterpillars and on nearby neighbors, relative to the performance of assay caterpillars on more distant controls.

Rhoades's hypothesis stimulated Baldwin and Schultz (1983) to design and conduct a more controlled lab experiment to try to detect airborne communication between potted poplar cuttings and sugar maple seedlings. They constructed two airtight chambers—one held their control plants, the other was divided in half and held their damaged plants in one half and their undamaged, communication assay plants in the other half (figure 2.2B). Air from the half of the chamber holding the damaged plants was pumped into the half holding the communication assay plants. They reported that tearing leaves of the plants in the damage treatment increased levels of phenolics on damaged plants. Phenolics were also higher for the undamaged communication assay plants that had received air from the chamber holding the damaged plants. This suggested that an airborne signal may have been transmitted from the damaged plants, eliciting production of phenolics in the undamaged plants "downwind." This result is subject to other interpretations as well. Because all of the samples from each treatment were kept in a single chamber (figure 2.2B), it is not possible to determine if the results were caused by unplanned differences in the chambers or by the treatment effects (Fowler and Lawton 1985). A more definitive experiment would involve using more chambers or other independent units so that the treatments are interspersed (Hurlbert 1984). A similar criticism applies to several of the field experiments that have looked for airborne communication. A comparison between "nearby" trees, all located in site x, and "far" trees, all located in site y, does not allow the unknown effects due to differences in sites x and y to be separated from the effects caused by distance from the damaged trees (the communication effect).

A design that uses only a single chamber but places the two treatments within that chamber at different times reduces the likelihood that the chambers or sites are responsible for causing any differences that are

Table 2.2 Studies reporting communication between plants

Plant Species Producing Signal	Plant Species Responding	Inducer	Assay Species	Reference	Reported Effect	Alternative Explanation
Sitka willow	Same	Tent caterpillar	Tent caterpillar	Rhoades 1983	Biomass of caterpillars on controls decreased following addition of caterpillars on neighbors.	Disease caused effects.
Sitka willow	Same	Tent caterpillar	Tent caterpillar	Rhoades 1983	Growth of caterpillars was greater for far controls (no communication) than for near controls and damaged trees.	(1) Disease caused effects. (2) Treatments not interspersed, no replication (far controls all in one place).
Sitka willow	Same	Fall webworm	Fall webworm	Rhoades 1983	Growth of caterpillars was greater for 2 far controls (no communication) than for near controls and damaged trees.	(1) Disease caused effects. (2) Treatments not interspersed, no replication.
Poplar and sugar maple	Same	Experimental leaf tearing	Only chemicals assayed	Baldwin and Schultz 1983	Damaged plants and undamaged plants that were exposed to air from damaged plants had increased levels of phenolics.	Treatments not interspersed, no true replication.

Barley	Same	Experimental leaf removal	Powdery mildew	Fujiwara et al. 1987	Seedlings placed in a chamber with pruned seedlings became more resistant to infection by powdery mildew.	No true replication.
Cotton	Same	Spider mites	Spider mites	Bruin et al. 1992	Oviposition rates reduced on plants downwind of damaged conspecifics compared to plants downwind of undamaged controls.	No true replication.
Cotton	Same	Cotton leaves infested with *Aspergillus* fungus	Only chemicals assayed	Zeringue 1987	Undamaged leaves that were downwind of infested leaves had increased levels of phloroglucinol-reactive compounds.	Laboratory demonstration only.
Birch	Same	Autumnal moth	Autumnal moth	Haukioja et al. 1985	As distance from damaged tree to assay tree increased, pupal weight, growth rate, survival, and fecundity of moths increased.	(1) Signal may not have been airborne. (2) Neighbors near defoliated trees may experience more nutrient stress.
Sagebrush	Tomato	Chemical	Only chemicals assayed	Farmer and Ryan 1990	PIs were produced by tomato plants placed in the same chamber as sagebrush, which released methyl jasmonate.	Laboratory demonstration only.

observed. However, temporal changes, rather than the experimental treatments, could produce the differences observed in this design. A design in which the treatments were run sequentially in a single chamber was employed by Fujiwara et al. (1987), plant pathologists who were investigating interplant signals that induce resistance against powdery mildew. They placed undamaged barley seedlings in an airtight chamber with seedlings that either had been mechanically pruned or were intact controls. The seedlings placed with pruned neighbors became more resistant to infection by powdery mildew, although this resistance lasted for only about twelve hours. They had only one true replicate with damaged seedlings and one with intact controls and therefore little ability to draw statistically significant conclusions (Hurlbert 1984).

One possible design that reduces the problem caused by lack of true replication is to repeat the laboratory experiment several times, alternating the placement of the treatments. Bruin et al. (1992) placed undamaged assay cotton plants downwind of either damaged plants or undamaged controls in airtight chambers for four to eight days. The assay plants were challenged with spider mites whose rates of oviposition were measured. This experiment was run four times, and for each of these runs, they found that the rate of oviposition was greater (approximately 10% difference) for mites feeding on assay plants downwind from undamaged controls than for those downwind from damaged plants. The "replicate" plants in each run were still not independent, so that the statistics reported were subject to other interpretations (Hurlbert 1984), although alternating the placement of the treatments provided some evidence that the effect observed was not caused by differences in the chambers.

If the experiment is repeated many times with treatments randomly assigned to chambers, each experimental run can be considered a true replicate even though each plant within a chamber is not. Zeringue (1987) employed this repetitive approach and an experimental apparatus similar to the one used by Baldwin and Schultz (1983; see figure 2.2B) to test whether volatiles released by infection of *Aspergillus flavus* induced accumulations of phloroglucinol-reactive compounds in undamaged cotton leaves. He found that volatiles released from cotton leaves infested with fungus increased these compounds by 34%, although leaves downwind from liquid fungal cultures or from uninfested cut leaves showed no increases. Some of the phloroglucinol-reactive compounds were terpenoid aldehydes called heliocides that may be harmful to *Heliothis* spp. caterpillars. Wounded cotton leaves released a volatile, myrcene; when myrcene was used as the volatile source, Zeringue noted an eighteen- to twentyfold increase in heliocides relative to controls. These experiments are not subject to alternative explanations due to pseudoreplication, but laboratory demonstrations still do not answer the question of whether communication occurs in natural settings.

A possible solution to this problem of realism is to assay undamaged trees at various distances (and compass directions) from damaged neighbors (figure 2.2C). If signaling is occurring, assay trees near to damaged neighbors should respond more strongly than assay trees with only distant damaged neighbors. Such an analysis was conducted by Haukioja and Neuvonen (1985) for birch trees in northern Finland. Caterpillars were reared on assay trees at varying distances from trees that had been experimentally defoliated by autumnal moths in the previous year. As the distance from the damaged tree increased for the assay tree, the pupal weight, growth rate, survivorship, and fecundity of assay moths increased. Similar experiments have recently been performed using alders in a natural setting in Germany (R. Dolch and T. Tscharntke, personal communication). Trees that grew close to neighbors that had been mechanically defoliated (20% leaf area removed) suffered less leaf damage from alder leaf beetles during the subsequent growing season. Since mechanical damage was used, these results cannot be explained by increases in the pathogen load of herbivores. The results for these studies of birches and alders are consistent with the hypothesis of airborne communication between trees. However, alternative explanations are also plausible. Communication may have occurred via root connections or through the soil rather than by some volatile, airborne chemical (Haukioja and Neuvonen 1985). If defoliated trees are able to compensate for their losses by intensifying absorption of mineral nutrients, this effect could produce trees with lower-quality foliage for caterpillars in the immediate vicinity of defoliated neighbors (Tuomi et al. 1990).

Workers in England who looked for evidence of communication between birch trees that might affect other species of herbivores failed to find any (Fowler and Lawton 1985). This experiment also lacked true replication. Other workers who explicitly looked for airborne signals in other systems also found negative evidence (e.g. Lin et al. 1990). Reports of observed induced differences between damage and control treatments in close proximity (e.g. Karban and Carey 1984) could have occurred only if airborne signals were not swamping the responses induced by herbivory. This negative evidence is not described in much detail here because it can show only that communication does not occur in every situation. This fact is not in dispute. What is still unclear, and therefore worth examining in detail, is the question of whether communication between plants ever occurs in nature.

The exciting results described above (section 2.3.6), that methyl jasmonate released by sagebrush can induce production of PIs in tomato plants (Farmer and Ryan 1990), lends credence to the hypothesis that airborne communication is possible. This possibility is consistent with results reported by Hildebrand and coworkers (1993). They found that volatiles released by crushed tomato or tobacco leaves induced plant responses that in turn reduced the fecundity of aphids kept in small airtight containers. Similarly, methyl salicylic acid released from tobacco mosaic virus–infected plants into the air might be of sufficient quantity to elicit systemic acquired resistance in plants downwind of infected plants (Lee et al. 1995). However, it is still a long leap of faith between these effects or those reported by Farmer and Ryan (1990) conducted in airtight chambers, to interactions that might be possible under natural conditions. The evidence is increasingly convincing that volatile cues released by damaged plants are used in nature by herbivores and predators and parasites of herbivores (see sections 2.2.3 and 3.2.4). It remains to be seen whether signals released by plants that are damaged are used by unattacked, neighboring plants to induce defensive responses.

3

Mechanisms of
Induced Responses

3.1 COMPARISON OF MECHANISTIC AND BIOASSAY APPROACHES TOWARD AN UNDERSTANDING OF THE FUNCTION OF INDUCED RESPONSES

3.1.1 Different Approaches to Different Questions

The two traditions that Ryan and Haukioja established represent two separate experimental approaches to the study of induced phenomena in plants (see section 1.3). Ryan and other researchers interested in the mechanisms causing the induced plant changes have monitored the concentrations of chemicals thought to be deleterious to herbivores. This approach will be developed in this chapter. In contrast, Haukioja and other researchers interested in the population-level consequences of induced changes have used bioassays with herbivores to measure the effects of plant responses. This bioassay approach will be considered in chapter 4.

Interestingly, researchers rarely attempt to combine the two techniques in a single laboratory. One might suppose that mechanistic lines of inquiry would originate from bioassay-based observations. For example, an observation that prior damage decreased the performance of herbivores during subsequent meals might lead researchers to divide the chemicals found within a plant into various fractions. Each of these fractions could be used in a bioassay to determine which are active against herbivores. Fractionation could continue until the active compounds (or other traits) were isolated and identified. This technique follows the approach taken by Wheeler and Slansky (1991) to elucidate the mechanisms of induced resistance in soybeans. The technique is commonly applied to many other biochemical problems. Surprisingly, it is rarely used by workers studying induced plant responses.

There are at least two explanations for the lack of integration between workers following these two experimental traditions. The first explanation deals with limitations of the bioassay technique. Bioassays are clearly the most direct experimental technique to determine whether prior herbivory influences the performance of subsequent herbivores feeding on the same plant. However, bioassays with herbivores often require considerable time for the herbivores to grow, mate, and reproduce. This time requirement does not allow bioassays to keep pace with (let alone guide) chemical fractionations. In contrast, bioassays using fungi or bacteria are much quicker and more tractable than those involving herbivores. Induced phytoalexins have more often been discovered and understood by workers who have combined chemical fractionations with bioassays of microbes. Because bioassays measure the effects of plant responses on herbivore performance, they give different answers depending on the herbivore used and its own reactions to the plant responses. As such, bioassays do not offer ideal answers to research questions that demand an accurate quantification of a plant's response.

Another reason for the lack of cross-fertilization between the two research traditions may lie in the training of the researchers. Researchers from the two traditions are not only trained in different techniques (biochemical or entomological, for example) but also trained to answer questions at different levels of analysis (sensu Sherman 1988). The mechanistic tradition asks, How does it work? The functional tradition asks, What effects does it have and why is it the way it is? To pursue research efforts that cross levels of analysis requires an appreciation of the strengths and limitations of both experimental approaches. We hope that this book will help develop such an appreciation.

Collaboration between the two research traditions that Ryan and Haukioja started holds the promise of more convincing answers to many questions than either approach alone. Consider the problem of trying to understand the functional significance of induced responses. Soon researchers using molecular techniques will be able to produce transgenic plants that differ in their ability to express particular induced responses following attack. These transgenic plants will be ideal experimental systems for probing the consequences of particular components of the inducible system. The time is ripe for ecologists to develop methods to address how these studies should be conducted. Population biologists have developed methods to measure the selective costs and benefits of constitutive resistance traits of plants growing in environments with and without herbivores (Simms and Rausher 1987, 1989; Simms 1992). These techniques suffer from a lack of specific characters on which selection acts. Transgenic techniques, by providing control over specific traits, offer a marvelous tool to ask functional questions.

However, transgenic technology solves only part of the problem of determining the functional significance of particular traits. Most traits play many roles, and those that provide defense are no exception. Secondary metabolites are known to function as defenses not only against herbivores but also against pathogens and plant competitors as well as a host of abiotic stresses such as UV light and desiccation (Seigler and Price 1976). These multiple consequences of traits greatly complicate interpretation of experiments that compare the fitnesses of plants with and without the traits. Ecologists need to develop techniques that tease apart these diverse consequences. Broad-spectrum insecticides, cages, and fences are tools that have made it relatively easy for researchers to contrast correlates of fitness for plants grown in environments that are free of herbivores versus those that contain herbivores. Before the selective importance of particular traits can be fully appreciated, selective factors other than herbivory should be experimentally manipulated under field conditions. For example, when both intraspecific competition and herbivory were manipulated, competition was found to have more significant effects than induced resistance on the growth and seed production of a population of wild cotton (Karban 1993a, 1993b). In summary, as mechanistically oriented researchers develop techniques to alter plant traits precisely, field-oriented population biologists need to develop equally precise techniques to alter selective factors before the adaptive significance of these responses can be fully appreciated.

3.1.2 Mechanistic Studies and a Phytocentric View of Induced Responses

Understanding the induced responses following herbivory requires a "phytocentric" perspective, one that assesses changes in plant function by measuring plant traits and not just their effects on herbivores. Induced resistance is, after all, a plant's physiological response to having been eaten. In those cases in which plants became more susceptible to herbivory after being partly eaten (see chapter 4), explanations that consider how a plant is reorganizing itself are more useful than those that attempt to explain how the plant is "becoming more susceptible" for its herbivores. In other words, becoming more resistant or susceptible should be viewed as a consequence of the plant's reorganization. Below we develop a description of how plants reorganize following herbivory.

First, we must introduce some basic notions about how plants are thought to work. Plant ecologists describe growth in plants in much the same way economists describe growth in business firms (Bloom et al. 1985). Plants, like all organisms, acquire resources from the environment and invest these resources so as to optimize their fitness. Because fitness has been so difficult to measure, workers have assumed that optimizing plant growth is a reasonable surrogate for fitness. This is because fitness in many environments is size-dependent (and size is easy to measure). Growth optimization in plants, as in business firms, involves acquiring resources and adjusting allocations so that growth is equally limited by all resources (Bloom et al. 1985). For terrestrial plants the resources most limiting to growth are heterogeneously distributed in the environment, with light and CO_2 available above ground, and water and nutrients available below ground. These resources are acquired by the shoots and roots, respectively. As a consequence of this heterogeneous resource distribution, a plant without much storage must continually adjust its allocation so that all resources are equally limited; this allows the plant to optimize its growth rate. In economics this is called equalizing the marginal product per marginal cost for all resources. For a plant this means adjusting the allocation to shoot and root growth so that growth rates are equally limited by light, CO_2, water, and nutrients; the result is called balanced growth. This model predicts that plants should increase their root-shoot ratios in response to water and nutrient stresses, to acquire more of these limiting resources. Conversely, plants are predicted to decrease root-shoot ratios in response to light stress so as to increase their capture of this resource. These predictions have been confirmed in many plants (Bloom et al. 1985; Chapin et al. 1990).

A phytocentric perspective argues that induced responses to herbivory should be considered in the context of how a plant grows. Therefore, according to the growth-optimization model presented above, a plant that has a portion of its tissues walk off in the stomach, foregut, or crop of an herbivore is going to require a substantial amount of reconfiguration in its patterns of resource acquisition, allocation, and partitioning in order to attain balanced growth. This theory predicts that plants should have not only a suite of "resistance" responses, which would lessen the probability of the attack continuing or happening again, but also a suite of "civilian" responses, which allow plants to regrow the tissues lost to herbivory or to reconfigure remaining tissues so as to regain balanced growth (figure 3.1). Much of the research effort has been directed toward understanding the responses that provide resistance and defense, and our treatment of the subject reflects this bias. However, we would like to emphasize that "civilian" plant responses to herbivory are diverse, far-reaching, and underappreciated. Below we will briefly consider some of these "civilian" responses to one particular type of herbivory, namely, folivory.

If an herbivore removes leaf material before normal leaf senescence, a plant has at its disposal a battery of physiological responses that help it to regrow lost tissues. These physiological responses are part of a whole-plant response to damage that influences the patterns of resource allocation and partitioning. Whole-plant responses to folivory include decreases in root growth (Evans 1971; Caloin et al. 1990), movement of

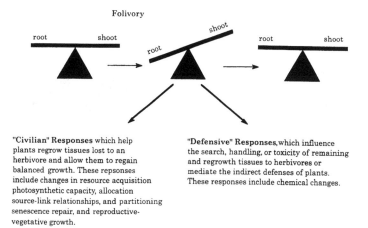

"Civilian" Responses which help plants regrow tissues lost to an herbivore and allow them to regain balanced growth. These repsonses include changes in resource acquisition photosynthetic capacity, allocation source-link relationships, and partitioning senescence repair, and reproductive-vegetative growth.

"Defensive" Responses, which influence the search, handling, or toxicity of remaining and regrowth tissues to herbivores or mediate the indirect defenses of plants. These responses include chemical changes.

Fig. 3.1 Undamaged plants adjust their root-shoot ratios so that growth is equally limited by the resources acquired by roots and shoots. Herbivory alters this balanced growth, and plants initiate a suite of "defensive" and "civilian" responses.

reserves from roots to shoots (Bokhari 1977; Caldwell et al. 1981; Ta et al. 1990), increases in resource acquisition by roots (Chapin and Slack 1979; Chapin 1980; Ruess 1988) and shoots (Caldwell et al. 1981; Welter 1989), activation of new meristems (Caldwell et al. 1981; Maschinski and Whitham 1989; Whitham et al. 1991), and alterations in patterns of leaf senescence (Nooden and Leopold 1988). One of the consequences for plant-herbivore interactions of this whole-plant reconfiguration is that the distribution and concentration of many "primary" metabolites are altered within a plant (table 3.1). In addition, the age structure and growth rates of the leaves remaining on a plant are altered. We will consider the consequences of just one of these "civilian" physiological responses in more detail.

One of the best-documented physiological responses to leaf damage is the increased rate of photosynthesis in the undamaged leaves of damaged plants as compared to that of leaves of similar age or new leaves of undamaged plants (Yoshinori et al. 1967; Wareing et al. 1968; Caldwell et al. 1981; Bassman and Dickmann 1982; Kolodny-Hirsch and Harrison 1986; Kolodny-Hirsch et al. 1986; Welter 1989). Increased rates of photosynthesis are frequently associated with increases in photosynthetic pigments and ribulose-1,5-bisphophate carboxylase-oxygenase (RuBPCase), the primary enzyme in the photosynthetic process of most plants (and the most abundant protein, representing approximately 10% of total leaf protein or 50% of the soluble protein [Evans 1989]).

Table 3.1 Primary metabolites induced by damage

Metabolites	Reference
Proteins	
Total proteins	Wagner and Evans 1985; Kolodny-Hirsch et al. 1986; Faeth 1992a
Photosynthetic proteins, pigments	Yoshinori et al. 1967
Storage proteins	Staswick 1994
Carbohydrates	Valentine et al. 1983; McNaughton 1985; Gonzalez et al. 1989; Ta et al. 1990; Dyer et al. 1991
Nutrients	
Total nitrogen	Tuomi et al. 1990
Nitric oxide	Ourry et al. 1988, 1989
Calcium	Valentine et al. 1983
Potassium	Mattson and Palmer 1988; Ruess 1988
Sodium	Mattson and Palmer 1988; Nef 1988
Phosphorous	Chapin and Slack 1979; Nef 1988
Copper	Mattson and Palmer 1988

Unfortunately for the plant and its resistance to herbivores, increased allocation to RuBPCase strongly improves the nutritional quality of the leaf for many herbivores. Not only is there a strong correspondence between leaf nitrogen content and food quality for herbivores (Scriber 1984; Slansky and Rodrigues 1987), but RuBPCase represents a large proportion of total leaf nitrogen (Evans 1989) and is probably a nutritious protein for most herbivores. Moreover, there is evidence that an herbivore's ability to deal with a plant's chemical defenses depends strongly on the amount of dietary protein that the herbivore receives. For example, both alkaloids and furanocoumarins were less detrimental to herbivores provided with relatively high levels of protein (Johnson and Bentley 1988; Berenbaum and Zangerl 1994). This interaction between dietary protein and the effects of chemical defense is to be expected if the detoxification enzymes of herbivores are substrate-induced. Since the food quality of a leaf for an herbivore is determined by the interaction of both its primary and secondary metabolites (van der Meijden et al. 1984; Duffey et al. 1986; Johnson and Bentley 1988; Berenbaum and Zangerl 1994), with nitrogen and protein being particularly important, the defensive function of induced responses may be canceled by changes in primary metabolites caused by functions related to plant growth. We wish to emphasize that whether an induced response acts as a defense depends critically on the context of the response, that is, on the conditions present in the plant, in the herbivore, and in their environment. We will return to this theme in section 5.4.1. Induced susceptibility following attack, for example, may result from selection on two or more different plant functions that have opposing consequences for future herbivores. In some habitats selection for regrowth and competitive ability may be stronger than that for induced resistance to herbivory (Herms and Mattson 1992). In these situations induced responses in the plant's ability to grow and compete with other plants for resources undoubtedly influences the resistance of tissues remaining after herbivore attack. Induced susceptibility, therefore, does not necessarily mean that plants have been selected for an increase in their susceptibility after herbivory but, rather, that selection has not been able to produce a phenotype that optimizes both regrowth and resistance against herbivores.

The recognition that plants are under multiple selection pressures raises an interesting experimental problem for the phytocentric study of induced responses, namely, what tissues should be used as controls to compare with induced tissues. Selecting control tissues requires both an understanding of the prior damage history of a plant and the relaxation times of the induced responses. Controls should be tissues of the same age or developmental stage from undamaged plants. However, some tissues produced in response to herbivory may have no counterpart on undamaged plants. For example, some determinate species may produce new leaves only when defoliated, and these leaves will have more physiological attributes in common with the first leaves produced in the growing season than with the mature leaves on undamaged plants. Hence, the documentation of induced resistance or susceptibility may in large part reflect the choice of control tissues for bioassay. Many examples of induced resistance and susceptibility are caused by the plant's producing "juvenile-type" tissues in response to losses at times when such tissues are not normally available to herbivores (see section 4.2.3).

Recognizing the complexity of the changes that occur in plant tissues after damage, many researchers have turned to artificial experimental diets, which vary only in single components from control diets. Artificial diets have little resemblance to the nutritional characteristics of an intact plant, but they do allow researchers to create a phenotype set (sensu Reeve and Sherman 1993), a set of alternative phenotypes with which to compare the induced phenotype. Artificial diets also allow researchers to combine attributes that may be outside the normal range of variation found in plants in nature. However, since the effects of many plant chemicals depend on their chemical and biological context, results from artificial diets may not reflect effects that occur in nature. Techniques that allow researchers to experimentally manipulate individual, independent variables *in planta* can provide much more realistic information. For example, transgenic technology will likely allow researchers a similar degree of control over particular traits with much more realism than artificial diets.

Since so many plant attributes change after herbivory, researchers who study the functional significance of induced responses must decide which of the many changes to examine for which functional roles. While bioassays test the defensive function of an induced response and are limited only by the choice of bioassay organisms and the choice of controls, mechanistic approaches are functionally unfocused and rely entirely on the researcher's understanding of the plant and its ecology for insights into the functional significance of induced changes in the particular traits under scrutiny. Ironically, those researchers who are least likely to have an intimate understanding of the natural history and ecology of their study organisms are the ones who need such information the most.

In summary, a coherent understanding of changes in plant function induced by herbivory must consider changes in resistance within the context of the other plant functions, namely growth, reproduction, and storage. In addition, a thorough mechanistic understanding of biochemical, physiological, and morphological responses to herbivory will lend important insights into the functional reorganization that occurs after herbivory.

3.2 OVERVIEW OF MECHANISMS

Researchers who take a mechanistic approach toward induced resistance have a problem knowing which of the many plant traits that change after herbivory should be measured. As a consequence of this complexity and the inability of bioassays to keep pace with chemical fractionations, most studies on the mechanisms of induced resistance have focused on plant traits that are already known to have broadly biocidal properties, specifically plant "secondary" metabolites. These chemicals have no known role in the primary metabolism of the plant, and are generally assumed to provide plant defense. In this section we review the progress that has been made in unraveling the mechanisms responsible for the induced chemical defenses. However, we feel that this review does not accurately represent the diversity of mechanisms potentially responsible for induced resistance in nature. While poisoning an herbivore with an increase in a toxic metabolite represents a potent and rapid means of avoiding further herbivory, other mechanisms of defense may be even more effective. For example, an induced response that moves an herbivore off a plant and onto a neighboring competitor might be a more effective defense than killing the herbivore outright. Therefore, we will develop a broad perspective on the processes that could be responsible for induced resistance in nature that includes chemical changes as well as other mechanisms, with the hope of stimulating research into this largely uncharted area. Our perspective borrows its structure from the theory that has been developed to describe predator-prey interactions.

Ecologists traditionally dissect predator-prey interactions into five steps that must be completed in sequence before the interaction has run its course (Holling 1959). The predator must detect, pursue, capture, ingest, and finally digest its prey in order to acquire the resources necessary to survive and reproduce. Predators are therefore selected for the ability to complete each step, and prey are selected to minimize a predator's ability to complete the sequence. From the prey's perspective each step (at least the first four) represents a potentially independent link that, if prevented, would provide resistance and defense.

Students of predator-prey interactions frequently put traits that decrease the ability of the predator to detect and pursue a prey into the category of "search-related" defenses, and combine defenses related to capture, ingestion, and digestion as "handling-related" defenses. The predator's rate of prey capture may be primarily limited by search-related processes at low prey densities and by handling-related processes at high prey densities. Diagrams showing the predator's rate of prey capture plotted against prey density are termed functional response curves and describe how a fixed number of predators respond to changes in prey density (figure 3.2). Functional response curves represent a useful way of categorizing the effects of plant's induced responses on herbivores, particularly those induced responses that occur before the herbivore population exhibits a numerical response (a change in birth, death, immigration, or emigration rate) to the induced plant traits. For a given herbivore, induced resistance could be viewed as a collection of traits that decreases the slope of the functional response curve by means of search-related defenses and/or increases the plant density at which handling-related defenses limit the capture rate (figure 3.2). However, not all mechanisms of induced resistance can be depicted in the construct of functional response curves. Some induced responses may act as indirect defenses. For example, an induced plant response could increase the probability that an herbivore will be discovered by a predator. We will briefly describe induced plant traits that could function as search- and handling-related defenses and as indirect defenses, as a means of broadening the discussion of mechanisms of induced resistance. Our discussion of the steps that are required for an herbivore to "capture" and "consume" its host plant successfully may at first glance seem at odds

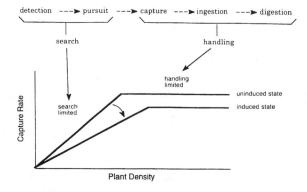

Fig. 3.2 An idealized type I functional response curve (Holling 1959) depicting a change in the capture rate of a particular herbivore as a function of plant density. The capture rate is influenced by search-related and handling-related plant responses at low and high plant densities, respectively. Induced responses can decrease the slope of the search-limited portion or increase the plant density at which the capture rate becomes handling-limited.

with our plea for a more phytocentric approach to the study of induced responses. However, it is crucial to remember that each step in this process is mediated in large part by plant traits.

3.2.1 Induced Search-Related Resistance and Susceptibility

Induced search-related resistance or susceptibility often involves changes in plant traits that influence the plant's detection by herbivores and their subsequent "pursuit." We will first consider induced responses mediated by changes in detection. The ability of an herbivore to detect a plant defines its apparency to herbivores; "apparent" plants are more likely to be detected by herbivores than are "unapparent" plants (Feeny 1976). Apparency theory proposes that a plant's apparency sculpts the types of defenses that it could use to avoid herbivory. Apparency theory was developed to explain why plants grown in monocultures are more readily discovered by insect herbivores, as well as "associative" resistance whereby plants grown in association with neighbors of some other species are less readily detected (Root 1973; Feeny 1976). Apparency theory has proven difficult to test directly, however, because it requires an understanding of what an herbivore does and does not perceive. A rudimentary understanding of the sensory capabilities of selected herbivores has only recently begun to emerge (Masson and Mustaparta 1990; Bernays and Chapman 1994; Provenza 1995).

For herbivores that rely largely on vision to detect their host plants, as humans do, it is easy to find examples of plants that mimic their backgrounds. The stone plants (*Lithops* spp.) and other unrelated species that grow in dry rocky areas and have leaves that resemble the shape and color of stones may be such an example, although drought tolerance is clearly another important selective factor influencing this morphology (Crawford 1989). A more compelling example involves the Australian mistletoes, which parasitize largely unpalatable tree species, although the mistletoes are highly palatable themselves to arboreal marsupial herbivores. The parasitic mistletoes have evolved leaf characteristics that allow them to match their particular host plants (Barlow and Wiens 1977). Unfortunately, experiments that demonstrate the effects of leaf morphologies that mimic the host on the plant's apparency to herbivores have not been conducted to our knowledge.

Induced changes in a plant's apparency following herbivory have received even less attention than apparency as a constitutive trait, although some examples suggest that the phenomenon may occur. Sensitive plants like *Schrankia microphylla*, which have thorns on their pedicles and stems, deploy their thorns more effectively after folding their leaves in response to touch (Eisner 1981). Touch-induced leaf folding could also function as a search-related defense because plants that fold their leaves in response to touch collapse their canopy and may become less apparent to folivores. Similarly, early in the season sow thistle, *Sonchus asper,* produce tall inflorescences that tower conspicuously above the foliage (Karban, personal observation). These tall inflorescences are often consumed by deer (Barbour et al. 1973). Inflorescences that are produced following this herbivory are much shorter and far less conspicuous, nestled within the tough, spiny foliage.

Plant apparency in the ecological literature has largely been defined by a plant's growth form, density, longevity, and persistence (Feeny 1976; Rhoades and Cates 1976; Chew and Courtney 1991). While it is clear that many of these traits are influenced by herbivory, it remains to be demonstrated whether prior herbivory could influence a subsequent herbivore's ability to detect a plant visually. For some herbivores, like the swallowtail butterfly *Battus philenor,* which uses leaf shape to find its host plants (Rausher 1978), prior herbivory may alter leaf shape sufficiently to alter a plant's apparency. This could occur through the loss of lamina or by processes of repair and regrowth or induced changes in heterophylly. Reflushed leaves frequently appear different than the leaves they replace, and while the physiological differences between these leaf types have received some attention, we are not aware of any studies that compare the apparency of reflushed leaves with that of mature leaves. If the idea that visually searching *Heliconius* butterflies have selected for variations in leaf shape among sympatric species of *Passiflora* is plausible (Gilbert 1975), then it appears to us to be equally reasonable to consider the direct and indirect effects of herbivory on the ability of subsequent herbivores to find plants.

Flowers, the very structures that plants have evolved to make themselves more apparent to insect pollinators, may change in response to herbivory. Many insect herbivores find their host plants and sometimes their mates by using the color, shape, and fragrances of flowers (Metcalf and Metcalf 1991). Rates of abortion of reproductive parts can increase dramatically after natural or artificial herbivory (Stephenson and Bertin 1983; Willson and Burley 1983). This wound-induced increase in rates of abortion has been interpreted traditionally as a response to the loss of resources resulting from damage. However, it may also represent a way of decreasing a plant's apparency to further attack. We are not aware of any studies that have contrasted these two alternative explanations for increases in rates of flower and fruit abortions. It would also be interesting to compare the abortion rates induced by herbivory with other manipulations that result in a loss of resources comparable in amount to that lost to herbivores. In this context it is interesting to note that the endogenously produced wound signal, jasmonic acid (discussed in section 3.3.4), has been shown to decrease the rates of flower formation and to speed flower senescence when applied exogenously to plants.

A flower's visual signals are complemented by olfactory signals that also play an important role in attracting pollinators (Dobson 1994). However, these floral volatiles function in more roles than simply as pollinator attractors. Many of these volatile compounds also function to repel potential attackers. Indeed, this role may have been the ancestral function of floral volatiles (Pellmyr and Thien 1986; Pellmyr et al. 1991). Moreover, floral volatiles are known to function by serving as orientation cues for herbivores in the location of food or oviposition sites (Metcalf and Metcalf 1991). Regardless of the sensory modality used by the herbivore to locate its host, reproductive activity may increase a plant's apparency to herbivores. For example, vertebrate herbivores do not use scarlet gilia, *Ipomopsis aggregata*, as a food plant during the time that it grows as a prostrate rosette but eat this plant only when it flowers (Paige and Whitham 1987). In addition, seed-feeding flies, *Hylemya* spp., are more likely to oviposit on fertilized *I. aggregata* flowers, suggesting that flies may be attracted to the same cues as pollinators (Brody 1992a, 1992b).

It is well-established that volatile chemicals emitted from the host play an important role in a plant's discovery by herbivorous insects (Bernays and Chapman 1994). The vast majority of herbivorous insects feed on only a few species or genera of plants (although many exceptions exist). For many of these species that have been well studied, this high degree of specificity is associated with an insect sensory system that is narrowly tuned to particular chemicals found in the host. For example, the number of flea beetles (*Phyllotreta cruciferae* and *P. striolata*) on "islands" of *Brassica nigra* and *Thlaspi arvense* plants was dramatically increased by simply increasing the emission of allyl isothiocyanate from these plants (Feeny 1976, 1977). Feeding damage quantitatively and qualitatively alters the composition of the chemicals emitted by plants (discussed in section 3.2.3). Given that the sensory systems of many herbivorous insects are highly tuned to particular volatile constituents in plants, and that plants alter their olfactory profiles when they are fed upon, it seems reasonable to suggest that prior damage may alter the olfactory apparency of a plant to its herbivores.

Another component of search-related defenses includes those defenses that interfere with the herbivore's ability to "pursue" a plant. This type of defense is generally thought to be the exclusive option of motile animals, not of rooted, immobile organisms such as plants. Alterations in growth-related processes may represent examples of plants "moving to escape" their herbivores. All plants must regrow the tissues lost to herbivory if they are going to compete successfully with other plants. How and when they initiate the regrowth has profound effects on their resistance and defense against future herbivory. Moreover, these growth responses are as likely to be evolutionary responses to herbivory as is the production of toxic metabolites.

For instance, consider grasses, which owe their tolerance for high levels of herbivory in large part to their basal meristems and to their ability to tiller. These characteristics allow them to hide or sequester sensitive tissues and stored reserves below ground, out of easy reach of mammalian grazers. An elegant physiological study of rapidly induced patterns of carbon assimilation and transport using $^{11}CO_2$ technology was conducted on the Serengeti C_4 grass *Panicum coloratum* (Dyer et al. 1991). Induced responses to grazing varied considerably between two plant populations with different grazing histories. These different responses were possible because of different modes of storing recently fixed carbon. Plants cloned from individuals collected from an area with intense grazing pressure allocated significantly more of their recently fixed carbon to below-ground storage in the roots, while plants derived from individuals collected from an area of low grazing pressure stored their recently fixed carbon above ground in stem tissue. Storage in stem tissues may be advantageous to plants growing in habitats with intense competition from other plants where vertical growth is essential for successful light capture. However, storage of reserves in the stem may represent a liability in heavily grazed habitats. This argument would be strengthened by examination of plants from other populations along the gradient from low to high grazing pressure.

Compensatory growth following damage, the allocation of reserves to storage, and other "civilian" responses to herbivory have been categorized frequently as alternative tactics to "defense" (e.g. van der Meijden et al. 1988). However, as the *Panicum coloratum* example highlights, the distinction between "civilian" and "defensive" responses is often blurred, and it may be more profitable to consider some of the "civilian" responses within a broader construct of responses that includes escape as a component of defense.

Plants can escape from a particular situation by dispersing in time or in space. While nothing is known about the relative dormancy of seeds produced by damaged and undamaged plants, or whether damage induces other seed traits that could allow seeds greater spatial dispersal characteristics, we do know that damage can alter the sexual expression of hermaphroditic plants. Many plants become functionally more male after herbivory. For example, pinyon pines (*Pinus edulis*) that were exposed to high levels of natural herbivory by boring caterpillars became functionally male plants due to a complete loss of normal female cone-bearing ability (Whitham and Mopper 1985). A shift toward more male-biased reproduction may result in increased dispersal of the plant's genetic contribution over a greater distance. This speculation is based on the assumption that pollen dispersal is more widespread in general than seed dispersal. Some plants become functionally more female following herbivory (e.g. Heslop-Harrison 1924; Hendrix and Trapp 1981), and different herbivores can produce opposing effects on the same plants (e.g. Lowenberg 1997).

3.2.2 Induced Handling-Related Resistance and Susceptibility

Handling-related resistance involves plant traits that prevent herbivores from converting a plant's tissues into their own tissues once they have found the plant. Induced handling-related resistance includes capture-, ingestion-, and digestion-related resistance. Many induced morphological and chemical traits function by increasing the time required to process plant material or by decreasing the nutritional value of the material once it has been ingested. These traits can decrease consumption and growth rates of herbivores and thereby protect plants if the herbivores are at risk from their own predators or if herbivores choose not to use induced plants that provide less nutritional benefit. Traits that affect handling-related resistance are as diverse as those affecting search-related resistance, given the many feeding modes that herbivores have evolved. Researchers studying this type of resistance need an intimate understanding of the steps necessary for successful plant capture, ingestion, and digestion. Clearly a trait that is an effective defense against a chewing folivore may not be effective against a phloem-feeding aphid, for not only are the means by which the plant material is ingested completely different, but so are the tissues consumed.

Capture and ingestion of food are influenced by how fast the herbivore can eat it. Thorns, spines, trichomes, and prickles are examples of plant morphologies that could decrease the ingestion of plant material, if the herbivore feeds on the plant at a scale where these morphologies could be influential. The role of induced changes in these structures, as they are influenced by prior herbivory, was reviewed by Myers and Bazely (1991). More recent work has developed a stronger case for the structures' inducibility and suggests a defensive function for the traits. For example, the density of nonglandular leaf trichomes of gray alder (*Alnus incana*) increased on leaves that flushed after attack by chrysomelid beetles (*Agelastica alni;* Baur et al. 1991). Similarly, the density of stinging trichomes increased on *Urtica dioica* for leaves that resprouted following damage (Pullin and Gilbert 1989). In a third example, the branches of *Acacia seyal* that regrew after browsing by giraffes had longer and more densely distributed thorns. This increase in thorn length and density protected plants from experimental browsing by goats (Milewski et al. 1991).

As these studies illustrate, some plants respond to damage by increasing the density of ornamented structures as compared to similar undamaged controls. These induced phenotypes suffer less damage when confronted with herbivores feeding at the appropriate scale, although the evidence demonstrating plant benefits is still rather weak (see section 5.1.3). Induced morphological "defenses" have not been thoroughly explored in the sense that we do not know if the induced state represents a new phenotypic state or a change to a phenotype that normally occurs during another ontogenic stage in the plant's development or under different growing conditions.

For example, trichomes originate from epidermal tissues, and the number produced by a given leaf is largely determined in the bud before leaf expansion (which is why the induced state requires new growth for its expression). It is not clear whether the induced increase in trichome density is a result of increases in the total number of trichomes and/or a decrease in leaf area; the latter might be expected if previous herbivory reduces the plant's stored reserves. For previously browsed plants in which the apical bud has been removed, regrowth leaves are frequently larger than those that they replace. If trichome number per leaf is fixed, the density and probably the protection trichomes afford will be lower in regrowth leaves. Coincidentally, leaves that regrow following removal of apical buds are often characterized as more susceptible to herbivores (Haukioja et al. 1990; Karban and Niiho 1995).

This example illustrates another important point about mechanisms. How the induced phenotype came about is irrelevant to its ecological function; however, an understanding of the mechanism by which it arose allows one to construct more informative tests of its ecological importance by including controls that represent real or imagined phenotypic states with which to compare the induced state. Since the genes controlling trichome production in *Arabidopsis* have recently been cloned (Larkin et al. 1994), the probability of constructing phenotypic states by genetic engineering is likely in the near future. Increases in leaf lamina area represent one of the most effective ways of increasing whole-plant carbon gain without increasing the number of leaves. Therefore, knowing whether a plant can alter trichome density independently of leaf area would allow one to determine whether a plant can optimize growth independently of resistance (trichome density). If the plant cannot alter these traits independently, then induced growth (production of new leaves) and induced resistance (trichome density) are likely to be negatively correlated, and this correlation may represent an important constraint.

A fascinating form of an induced morphological defense that acts by interfering with handling involves colony formation in green algae. Cladoceran grazers of phytoplankton are limited by the size of algal particles they can ingest; the size of potential food for cladocerans is linearly related to their own body length (Burns 1968). Hence for each size of cladoceran grazer, there are algal colonies that are too large to be ingested. Some algae, like *Scenedesmus* spp., vary in size from unicellular to eight-cell aggregates. Colony size is a plastic trait. A water-soluble factor, released by actively feeding *Daphnia magna* herbivores, stimulates unicellular colonies of *Scenedesmus subspicatus* to produce eight-cell colonies within forty-eight hours (Hessen and van Donk 1993). These eight-cell colonies are too large for filter-feeding *D. magna* to ingest. A similar phenomenon has been described for *S. acutus,* and the cue has been characterized in this system as a heat- and pH-stable, small, nonproteinaceous organic compound (Lampert et al. 1994). As with many induced morphological responses, induced colony formation requires actively growing colonies for its expression. It is not clear why colony size is variable instead of fixed at the relatively more defended eight-cell size; while the factor affected colony size, it did not affect colony growth rate (i.e. no cost was found in terms of growth). Larger algal colonies might have higher sinking rates than unicellular colonies, which could represent an ecological cost that might balance the fitness benefits of higher resistance to herbivores (Lambert et al. 1994).

Browsers and other folivores that use teeth or mandibles to masticate plant food are vulnerable to traits that wear down these devices prematurely. This can have important consequences; for example, until recent human activities dwarfed other natural mortality, tooth wear was an important source of mortality for elephants and set the upper limit on any individual's longevity (Truman Young, personal communication). Leaf toughness has been shown to increase mandibular wear of herbivores such as the leaf beetle, *Plagiodera versicolora* (Raupp 1985). Many of the phenolic-based biochemical responses to wounding listed in table 3.2 may make wounded leaves tougher. Silica increases leaf toughness, which reduces leaf consumption and undermines the ability of some insects to bore into plant stems. For ruminant herbivores, increased dietary silica also is thought to lower leaf digestibility, by reducing the ability of the rumen microbes to access forage carbohydrates and by causing kidney failure (Vicari and Bazely 1993). While most plants exclude silica, some grasses passively absorb or actively accumulate monosilic acid from the soil environment. These grasses have 2–3% dry mass silica, an amount ten times higher than that of most dicotyledonous plants. Moreover, the increases in silica levels after grazing among some grasses may function as induced handling-related defenses (see section 4.2.5).

Most of the effort to understand the mechanisms that might account for the changes in resistance following herbivory have focused on changes in the chemical constituents of plants. Table 3.2 lists examples of chemical constituents that are known to change. Many of these chemicals are toxins that do not directly affect the processes of resource acquisition, and they will be discussed elsewhere. However, many of the chemicals do affect ingestion and digestion and can be considered as possible "antinutrients." For example, PIs are thought to function in part by interfering with gut proteases, enzymes essential for digestion (see section 3.3.3.2). Similarly, if tannins combine with leaf proteins and render them indigestible, they would be functioning as antinutrients.

The use of "antinutrients" as defenses could be considered part of a more general tactic, a "scorched-earth" strategy whereby a plant attempts to reduce the nutritional quality of tissues it is about to lose to an herbivore. A number of processes that are rapidly activated upon wounding or infection are similar in effect to a retreating army's burning the landscape. We have already mentioned that many plants respond

Table 3.2 Secondary metabolites induced by damage

Metabolite	Reference
Phenolics	
Total phenolics	Schultz and Baldwin 1982; Edwards and Wratten 1983, 1985; Baldwin and Schultz 1984; Wagner and Evans 1985; Mattson and Palmer 1988; Nef 1988; van Alstyne 1988; Hartley and Firn 1989; Bennett and Wallsgrove 1994
Phenolic glycosides	Clausen et al. 1989, 1991
Flavonoid resins	Johnson and Brain 1985
Isoflavonoids	O'Neill et al. 1986; Kogan and Fischer 1991
Pterocarpans	Kogan and Fischer 1991
Coumestans	Kogan and Fischer 1991
Furanocoumarins	Zangerl 1990; Zangerl and Berenbaum 1995
Hydrolyzable tannins	Schultz and Baldwin 1982; Faeth 1986, 1992a; Rossiter et al. 1988
Condensed tannins	Rossiter et al. 1988; Wagner 1988; Faeth 1991
2,4-Dihydroxy-7-methoxy-1, 4-benzoxazin-3-one	Vicari and Bazely 1993
Terpenes	Gershenzon and Croteau 1991
Monoterpenes	Marpeau et al. 1989
Iridoid glycosides	Bowers and Stamp 1993
Diterpene acids	Gref and Ericsson 1985; Buratti et al. 1988; Ericsson et al. 1988; Walter et al. 1989
Triterpenes	Tallamy and Krischik 1989
Sesquiterpenes	Guedes et al. 1982; Threfall and Whitehead 1988
Alkaloids	Hartmann 1991
Tobacco	Baldwin 1991
Tropane	Khan and Harborne 1990
Quinolizidine	Wink 1983, 1984, 1987; Johnson et al. 1987; Johnson et al. 1988
Pyrrolizidine	van Dam and Vrieling 1994
Indole	Frischknecht et al. 1987; Naaranlahti et al. 1991
Glucosinolates	
Indole glucosinolates	Bodnaryk 1994
Silicates	McNaughton and Tarrants 1983; Vicari and Bazely 1993
Cyanogenic glycosides	Bennett and Wallsgrove 1994
Volatile hydrocarbons	
Linalool	Dicke 1994
Nonatrienes	Boland et al. 1992; Dicke 1994
Green-leaf aldehydes, alcohols	Hatanaka 1993
Defense-related proteins	
Hydroxyproline-rich glycoproteins	Bowles 1990
Glycine-rich proteins	Bowles 1990
Peroxidases	Felton et al. 1989; Stout et al. 1994
Cinnamyl alcohol dehydrogenase	Bowles 1990
Callosesynthase	Bowles 1990

Table 3.2 continued

Metabolite	Reference
Proteinase inhibitors	Ryan 1983, 1992; Broadway et al. 1986
Amylase inhibitors	Bowles 1990; Ryan 1992
Thaumatinlike proteins	Bowles 1990
Phenylalanine ammonia lyase	Chiang et al. 1987; Ke and Saltveit 1989; Hartley and Lawton 1991
Monoterpene cyclase	Gijzen et al. 1991
Chalcone synthase	Creelman et al. 1992
Lipoxygenase	Siedow 1991
Peroxidase	Bowles 1990; Stout et al. 1994
Polyphenol oxidase	Stout et al. 1994; Constabel et al. 1995

to pathogens by local necrosis around the site of infection, called the hypersensitive response (see section 2.3.8). Hypersensitive responses are less ubiquitous, but nonetheless common, plant reactions to attacks by small herbivores (Fernandes 1990). Wounding has been found to rapidly increase the activity of lipoxygenases (Croft et al. 1993; Siedow 1991; Felton, Bi, et al. 1994), peroxidases (Bronner et al. 1991), and phenoloxidases (Felton, Summers, et al. 1994) in wounded tissues. These enzymes produce very reactive lipid hydroperoxides from cell-membrane fatty acids and other organic constituents, and quinones from phenolic precursors. These reactive products, toxins in their own right, may also function as antinutrients and correlate with decreases in the nutritional quality of previously damaged soybean leaves to the larvae of *Helicoverpa zea* (Bi et al. 1994).

Abscission of plant parts in response to feeding damage is functionally the same process, albeit on a larger scale. In these cases the plant cuts its losses and disposes of the attacked part. Many examples of induced abscission have been reported for small plant parts, such as leaves, that are attacked by sedentary herbivores (see section 4.2).

The scorched-earth approach, exemplified by a local hypersensitive response, works for plants because they comprise cells surrounded by cell walls, which have very little motility. In other words, the fire that causes the localized scorched earth does not spread and consume the whole plant. Localized hypersensitive responses and larger-scale abscissions are possible because plants are modular organisms. In other words, no one leaf is indispensable for a tree.

Plants have an additional handling-related tactic that arises because the nutrient composition of plant tissues differs from that of their attackers. This property of herbivory differs from carnivory, in which carnivores and prey are both made of tissues with similar chemical constituents. This chemical difference means that altering the nutrient composition of plants can be quite detrimental to herbivores. However, providing herbivores with plant tissues having lower nutrient concentrations is not necessarily beneficial for plants. Herbivores can compensate for low-nutrient food by eating more of it, thereby increasing the amount of damage a "lower-nutrient" plant receives (Slansky and Feeny 1977; Moran and Hamilton 1980; Price et al. 1980). However, if feeding on a plant with decreased nutritional quality increases the probability that an herbivore will be found and consumed by its own predators (see section 5.1.5), then decreased nutritional quality could potentially represent a potent defense. The evidence for this possibility is controversial and will be discussed later.

The potential for decreased nutritional quality as a defense highlights the importance of examining variation in "primary" metabolites when considering mechanisms of induced resistance and defense. As indicated in tables 3.1 and 3.2 both primary and secondary metabolites have been found to change following damage. Overall food quality is determined by the interaction of both primary and secondary metabolites; for example, an herbivore's ability to cope with toxins also depends on its nutrition. In some circumstances, however, variations in primary metabolites can determine food quality independently of the secondary metabolites in the tissues. For example, the most abundant carbohydrate in the needles of Douglas fir is galactose, and the concentration of this unusual foliar carbohydrate varies among needles. When the western spruce budworm (*Choristoneura occidentalis*) was fed diets varying in galactose, other carbohydrates, and the terpenoid secondary metabolites found in fir, budworm mortality was found to vary most significantly with galactose content (Zou and Cates 1994). These results underscore the fallacy of associating "secondary" metabolites with defensive function and "primary" metabolites with purely growth-related functions. Variation in "primary" metabolites has been relatively understudied since Fraenkel argued in 1959 that "secondary" metabolites are the most important determinant of food quality. However, more research will likely reveal many examples of changes in induced resistance and defense attributable to changes in "primary" metabolites.

Changes in "primary" metabolites may change the efficacy of "secondary" metabolites. For example, herbivore attack increases the titer of reactive oxygen species and reduces the level of antioxidants, such as ascorbic acid (Bi and Felton 1995). These changes should combine to increase the oxidative damage directly to herbivores and to leaf nutrients that herbivores use.

3.2.3 Induced Indirect Defenses

If the world consisted only of plants and herbivores, we could consider induced resistance entirely in the context of the factors that affect these two groups of organisms. However, herbivores have their own predators, parasites, pathogens, and diseases, and a herbivore's susceptibility to these important causes of mortality can be influenced by plant traits that are induced after damage. Induced responses may act indirectly by altering a plant's indirect defenses. Here we will discuss two such indirect mechanisms that bring predators and parasites of herbivores to damaged plants.

The first traits we will consider are wound-induced changes in the chemical composition of extrafloral nectars. Many plants take a "mercenary" approach toward defense; they attract ants to their above-ground parts by secreting nectar from extrafloral nectaries (located on stems or leaves rather than on flowers). The best-described system is the ant-acacia interaction, one of the few plant defense systems included in most introductory ecology and biology texts. Leaves of *Catalpa speciosa* secrete more nectar after they have been damaged by their principal herbivore, compared to undamaged leaves (Stephenson 1982). Increased production of extrafloral nectar was associated with visitation by predators of this herbivore and reductions in rates of herbivory. Smith et al. (1990) found that the concentrations of amino acids increased in the nectar of extrafloral nectaries of *Impatiens sultani* following leaf damage. Sugar content and volume of nectar were not altered by the damage. Many ant species are particularly attracted by the amino acid contents of nectars (Lanza 1988), although it has not been documented that the induced responses caused increased visitation rates by ants in this system. It remains a matter of speculation whether these chemical alterations empower plants to attract and retain ants for longer than they otherwise would. *Cecropia obtusifolia* is another tropical tree that uses ants to favor it against competitors and herbivores. Although the mechanism is not yet known, damaged leaves of this species recruited far more mutualist *Azteca* ants than undamaged leaves (Agrawal n.d.).

The second example involves plants that after herbivory attract predators and parasites of herbivores (collectively called carnivores) by increasing volatile compounds rather than by rewards. Volatiles released by plants damaged by herbivores were found to increase the hunting efficiency of carnivores, and the carnivores learned to associate the volatiles with actively feeding herbivores (prey). Plants producing these "alarm" responses attracted more carnivores, and these in turn reduced the size of herbivore populations. These responses have been under intensive investigation by two laboratories: one at the Wageningen Agricultural University in the Netherlands, working on lima beans, herbivorous mites (*Tetranychus urticae*), and their predators; and the other at the U.S. Department of Agriculture (USDA) laboratory at Gainesville, Florida, working on corn, the armyworm (*Spodoptera exigua*), and its parasitoid wasp (*Cotesia marginiventris*). While most of the information on the phenomenon of alarm calling has come from these two systems, it has also been described in a variety of other plants, including cucumber, apple, corn, cowpea, cotton, soybean, cabbage, and *Eucalyptus globulus,* and with a variety of herbivores and their carnivores (Dicke 1994).

Arthropod carnivores, when placed in Y-tube olfactometers, showed strong preferences for volatiles emanating from herbivores, their frass, and the plants on which they were feeding. Attractants derived from herbivores themselves are likely to be the most reliable cues that a carnivore can use to guide its search for prey. Unfortunately, attractants derived from herbivores are often hard to detect and do not provide much useful information in nature, possibly because herbivores are under strong selection pressures to minimize their detectability (Vet and Dicke 1992). In contrast, volatiles derived from plants are both reliable and highly detectable, in part because of the larger mass of the plants relative to the mass of the herbivores that feed on them.

Plants release volatile compounds when they are damaged, typically mixtures of C_6 alcohols, aldehydes, and esters produced by the oxidation of membrane-derived fatty acids (see figure 3.3). These volatiles, called green-leaf odors, are produced when leaves are damaged, irrespective of the agent causing the damage, and are primarily emitted from the wounded leaf portions. Green-leaf odors are attractive to some carnivores, but the attractive effect tends to be short-lived (Steinberg et al. 1993). In contrast, the attraction of carnivores to a suite of monoterpenes, homoterpenes, and phenylpropanoids specifically emitted from plants infested with herbivores is long-lived (Turlings et al. 1991; Dicke 1994). Interestingly, two chemicals are consistently emitted from a variety of plant species after attack by herbivores, two acyclic homoterpenes, the C_{11} 4,8-dimethyl-1,3(E),7-nonatriene and the C_{16} 4,8,12-trimethyl-1,3(E),7(E),11-tridecatetraene (figure 3.4). These compounds are released either not at all or only in very small quantities in response to mechanical damage. Emission of these herbivore-induced synomones (so called because they benefit both the emitter and receiver) occurs not only from the leaf that is being eaten but also from nearby undamaged leaves on the attacked plant. The end result is that the whole plant contributes to the production of the plume (at least for small plants).

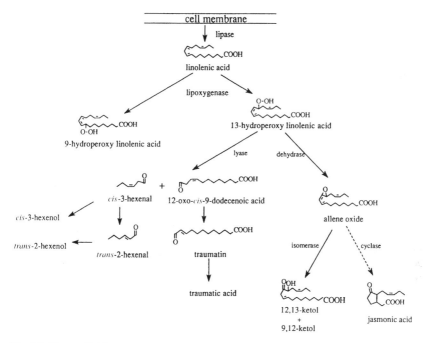

Fig. 3.3 The oxylipid pathway produces a diversity of metabolites that mediate a plant's responses to wounding.

Simply culturing unattacked leaves in the same water that previously held attacked leaves caused the unattacked leaves to emit these specific volatiles (Dicke et al. 1993). This observation suggests that a water-soluble elicitor originating from leaves infested by herbivores was responsible for the systemic production.

The C_{11} and C_{16} homoterpenes are produced *in planta* from two terpene alcohols, nerolidol and geranyllinalool, respectively, but herbivore damage is not required for this step in the biosynthetic pathway (Boland et al. 1992). This ability to produce the homoterpenes from terpene alcohols is a widespread constitutive trait reported from forty-nine species in twenty-two families (Boland et al. 1992). The process that is induced by herbivore feeding is hypothesized to be the production of the terpene alcohols from glycosidically bound or esterified precursors because the exogenous treatment of leaves with β-glucosidases results in the production of the homoterpenes. The source of the β-glucosidases is apparently the saliva of herbivores. Regurgitate of *Pieris brassicae* caterpillars was found to contain β-glucosidase (Mattiacci et al. 1995). Treatment of leaves with human saliva also caused leaves to release the homoterpenes

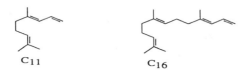

Fig. 3.4 Structures of two acyclic homoterpenes, the C$_{11}$ 4, 8–dimethyl–1,3(E),7–nonatriene and the C$_{16}$ 4,8,12–trimethyl–1,3(E),7(E),11–tridecatetraene. The emission of these volatile compounds is an indicator of herbivory in many angiosperms.

(Boland et al. 1992). Addition of exogenous β-glucosidase or regurgitate from caterpillars caused cabbage (*Brassica oleracea*) to release a blend of volatiles similar to that released by plants damaged by herbivores to which parasitic wasps (*Cotesia glomerata*) were attracted (Mattiacci et al. 1995). Hence, it appears that plants differentiate between wounding by mechanical damage and wounding by herbivores by responding to the presence of β-glucosidases or other factors in the saliva of the herbivores. Exogenous additions of jasmonic acid (JA) also caused plants to emit volatiles in a manner similar to that which followed actual herbivory, suggesting that JA may act as the signal for this response (Boland et al. 1995). Plants respond to herbivory by producing and emitting a suite of volatile "alarm" signals, components of which may be common to most angiosperms.

Many of the recent results involving this induced response are consistent with the notion that the responses represent precise tailoring of emissions for plant defense. Not only do the plant's volatile emissions depend on the presence of the herbivore, but the responses differ depending on the instar of the herbivore causing the damage. For example, female parasitic wasps (*Cotesia kariyai*) are highly attracted to the volatiles emitted by corn leaves that are being eaten by first-to-fourth-instar armyworm caterpillars (*Pseudaletia unipuncta*) but not those being fed on by fifth-to-sixth-instar caterpillars (Takabayashi et al. 1995). The differences in attractiveness were due to differences in the particular volatiles emitted by the plant. The addition of regurgitate from early-instar larvae to mechanically damaged leaves produced volatile blends that were highly attractive to wasps, but addition of regurgitate from later-instar larvae did not have this effect. The benefits that the plants can expect to receive by attracting parasites precisely match this pattern of volatile emissions. Parasitizing (or not parasitizing) the third-instar caterpillars produced the largest effects in terms of leaf area lost to the plant. For sixth-instar caterpillars the amount of leaf area lost to caterpillar feeding did not depend on whether the caterpillar was parasitized or not (Takabayashi et al. 1995). Hence from the plant's perspective, the conditional release of volatile synomones may well be adaptive (although this has not yet been demonstrated conclusively).

These plant responses can be co-opted by herbivores in some situations. For example, feeding by Japanese beetles causes volatile emissions from their crabapple host that are different from those elicited by mechanical damage (Loughrin et al. 1995). However, these emissions are used by the beetles to form large feeding aggregations.

Collectively, these studies document the potential for a great deal of specificity in the responses elicited in plants to the attack by particular herbivores. Determining what salivary factors elicit specific plant responses and whether they are transferred to plants during feeding will help establish the ecological relevance of these observations. Identification of these salivary factors will also facilitate the study of the possible benefits gained by plants in emitting volatiles. In summary, evidence is accumulating that supports the contention that plants play an active role in mediating complex trophic interactions.

3.3 MECHANISMS RESPONSIBLE FOR INDUCED INCREASES IN CHEMICALS

3.3.1 Introduction and the Problem of Autotoxicity

Plants have been especially successful in deploying toxic chemicals as a means of thwarting herbivore attacks. This success is due, in part, to the ability of plants to solve the "chemical waste dump" problem inherent in using chemicals as defenses. Simply stated, the problem is that, the more broadly effective a chemical defense is, the more likely that it is going to be toxic to its producer. Therefore organisms that use chemicals for defense must synthesize or acquire these toxins from exogenous sources, and store and deploy defenses without being poisoned in the process. Plants may be uniquely positioned to solve this problem for two reasons: (1) they have prodigious anabolic capabilities that allow them to synthesize an amazing diversity of metabolites, and (2) they differ from many of their herbivores in fundamental physiological ways and have been able to exploit these physiological differences in their "choice" of chemicals used for defense. For example, many plants infiltrate their tissues with nerve toxins that are relatively nontoxic to themselves because they lack a nervous system but are broadly toxic to herbivores.

Despite this advantage, plants are still confronted with the problem of autotoxicity. Nicotine, for example, is demonstrably toxic to tobacco plants (*Nicotiana* spp.), which normally produce this compound, despite their clear lack of an acetylcholine-based nervous system, the target of nicotine's toxic effects in herbivores (Baldwin and Callahan 1993). The solution to autotoxicity lies in the details of the mechanisms responsible for the production, storage, and release of the chemical defenses. The phenomenon of inducible production may be directly related to the cellular liabilities associated with the storage of high constitutive concentrations of toxic compounds, at least for some systems (see section 5.3). For example, tobacco plants that responded to damage by increasing production of nicotine (see section 3.3.3.4) concurrently increased their tolerance to exogenously added nicotine. This suggests that wound-induced increases in production of nicotine are associated with wound-induced increases in the plant's tolerance to this toxin (Baldwin and Callahan 1993).

Few induced chemical defenses are understood in sufficient detail to answer all the questions that ecologists might want to ask about the constraints to inducibility, but substantial progress has been made in a few model systems. In the sections that follow, we will outline the classes of mechanisms responsible for induced chemical defenses. We first consider "preformed induced" or "activated" defenses. Unlike most induced chemical defenses, preformed induced responses do not result from changes in de novo synthesis or degradation. We follow the treatment of preformed defenses with a survey of the classes of mechanisms responsible for induced chemical changes. Finally, we present in detail three examples that emphasize the importance of synthesis, degradation, and transport in determining the quantities of chemical defenses in plants.

3.3.2 Preformed Induced Chemical Responses

Preformed or activated defenses may be considered a type of induced defense, because they require wounding for the chemical or morphological defense to appear; however, they could also be considered constitutive defenses that happen to require damage for their expression. It should be noted that preformed and other forms of induced defenses represent a continuum rather than a set of discrete categories. All chemical defenses require a sequence of steps for their production, namely, regulation of the genes that code for the enzymes that catalyze chemical synthesis through transport, compartmentalization, and final deployment. Where a particular defense falls on this continuum depends on how far the plant has proceeded along this sequence of events before the herbivory occurs that activates the defense. For preformed defenses, plants rely less on transcriptional control over synthesis and more on preformed precursors; the opposite is the case for other induced defenses.

Preformed defenses, like constitutively expressed defenses, should predominate under circumstances where rapidity of response is at a premium. However, preformed defenses would be selected over constitutive defenses if the chemical defense represented a liability for the plant because it was toxic to the plant or because it attracted specialist herbivores. There is little evidence that the amount or rapidity of chemical release of most preformed defenses is altered by previous damage, nor is there evidence for systemic changes in most activated defenses as the result of prior damage. (This area has received little attention, and this last conclusion could change. For example, recent work suggests that two well-known preformed defenses, the glucosinolate releases in crucifers and cardenolides in the latex of milkweeds, change dramatically following damage [Bodnaryk 1992; Doughty et al. 1995; Malcolm and Zalucki 1996].) As such, preformed induced responses can be included in considerations of autotoxicity or counteradaptations of herbivores to plant responses, but many of the questions that have motivated research into induced defenses may not apply to preformed defenses. For example, induced defenses have been proposed as mechanisms that allow plants to save costs by timing their allocation of resources to defenses only when needed (see section 5.3). This argument does not apply to preformed defenses because it is assumed that the plant has invested resources in the preformed defense before damage occurred. Additional aspects of some preformed defenses may be induced by prior damage, but to date this remains poorly studied.

Many examples of preformed defenses involve cellular disruption resulting from tissue damage and from mixing substrates and enzymes that were localized in separate compartments prior to the damage. This mixing of previously separated components is likely operating in the following examples of preformed defenses: the production of hydrogen cyanide by the many diverse cyanogenic plants (Conn 1979); the hydrolysis of glucosinolates to form thiocyanates, isothiocyanates, or isonitriles in crucifers (Van Etten and Tookey 1979; Chew 1988); the conversion of the inactive halimedatetraacetate into the potent fish-feeding deterrent, halimedatril, in tropical seaweeds of the genus *Halimeda* (Paul and Van Alstyne 1992); and the rapid conversion of the phenolic glycosides, salicortin and tremulacin, in *Populus balsamifera* into more active feeding deterrents, salicin, tremuloiden, and 6-hydroxycyclohex-ene-1-one (Clausen et al. 1989, 1991).

Other preformed defenses involve, not the activation of precursors, but the rapid transport of compounds to the site of feeding damage. Many plants of diverse phyletic backgrounds have a network of canals or lacticifers that contain latex, gum, resin, or mucilage stored under pressure. When an herbivore breaks the canal system, it may be inundated with an immobilizing substance (Dussourd and Eisner 1987; Dussourd and Denno 1991). The canal system may contain toxins that are not activated by herbivore feeding, but it may also contain gums or latex that become viscous and sticky upon exposure to air. Many species of deciduous trees respond to cicada oviposition in their stems by producing gums and callous tissue that suffocate and entrap the cicada eggs (White 1981; Karban 1983). The "pitching out" response of conifers to the boring behavior of bark beetles as they tunnel into boles and transect resin ducts challenges the preformed-induced dichotomy. While part of this response involves the simple depressurization of a resin canal, many constituents, specifically monoterpenes and resin acids of the oleoresin, are synthesized in response to the mechanical damage and to the microbes accompanying the beetles (Raffa and Berryman 1987; Raffa 1991).

Other chemical responses involve such rapid changes in chemical titers that they are not likely to be under transcriptional control. For example, mechanical damage and herbivory to the leaves of wild parsnip (*Pastinaca sativa*) result in highly significant and highly localized increases in furanocoumarin content within three hours after damage (Zangerl and Berenbaum 1995). The speed of this response suggests that this system lies closer to the preformed side of the preformed-induced continuum of responses. Another characteristic of the induced furanocoumarin system that is consistent with expectations for a preformed defense is that the magnitude and speed of the induced increase in furanocoumarins are insensitive to the growth conditions of the plants. Plants grown under conditions of extreme nutrient and light limitation display inductions similar to those found in healthy, fast-growing plants, a result that suggests a response induced by preformed precursors. This insensitivity to the physiological condition of the plant may represent a selective advantage that some types of preformed defenses have over other induced defenses.

Induced increases in the pool sizes of chemical defenses can result from changes in synthesis and/or degradation. These induced changes in chemical pools can occur over a range of time intervals: some responses do not appear until the next growing season (delayed induced responses), while others appear within hours or days of the injury (rapidly induced responses). For obvious experimental reasons the mechanisms responsible for rapidly induced responses are much better understood on a mechanistic basis than are those responsible for the phenomenon of delayed induced resistance. In the following section we will consider some of the better-studied examples of rapidly induced chemical defenses. These are understood in varying degrees of mechanistic detail from the genetic to the whole-plant level of organization. In order to illustrate the diverse mechanisms that can account for these changes, we will first consider five mechanistic models that have been proposed to explain the diversity of induced changes in plant secondary metabolites.

3.3.3 Induced Chemical Defenses

3.3.3.1 Classes of Mechanisms

Five partially overlapping mechanistic models have been proposed to explain how secondary metabolites change after damage. The first three posit that secondary metabolites accumulate in response to imbalances between growth-related processes and metabolite production. However, the imbalance occurs differently for each of the three hypotheses.

1. The carbon/nutrient (C/N) theory attempts to explain induced changes in secondary metabolism as a result of imbalances between carbon (C) and nutrient (N) requirements for growth and the availability of these resources from the external environment (Bryant et al. 1983, 1988; Tuomi et al. 1990). This theory explicitly states that resource requirements of growth have a higher priority than those of chemical defense. Therefore, only when resources exist in excess of growth requirements are they shunted into secondary metabolism. Plants that have an excess of carbon relative to nutrients, in terms of what they can use for growth, store the excess carbon as terpenes, phenolics, and other compounds rich in carbon that coincidentally act as defenses against herbivores. A relative excess of carbon can arise as the result of herbivory or other environmental conditions (e.g. nutrient deficiency). Plants with an excess of carbon relative to nutrients are predicted to have reduced concentrations of nitrogen-based secondary metabolites such as alkaloids and PIs. Conversely, environmental factors (nutrient fertilization, shade, folivory of evergreens) that leave plants with shortages of carbon relative to nutrients are predicted to produce the opposite effects: reduced carbon-based secondary metabolites and increased nitrogen-based secondary metabolites.

According to this theory, secondary metabolites are coarsely categorized as being either carbon intensive or nutrient intensive, and the type of metabolites produced depends on whether carbon or nutrients are in excess of current requirements for growth. It is not clear if this dichotomy between carbon-intensive and nitrogen-intensive metabolites has much use. The implicit assumption that the occurrence of an element in a secondary metabolite reflects a plant's allocation or actual expenditure of that element for production of that metabolite is not well supported. For example, the amount of carbon required to synthesize a particular secondary metabolite varies substantially and is not equivalent to the amount of carbon contained in the compound (Gershenzon 1994). In other words, nitrogen-intensive metabolites may require more carbon atoms than smaller carbon-intensive metabolites and, similarly, carbon-intensive compounds may require more nitrogen for synthesis and storage than nitrogen-intensive compounds actually contain.

2. Substrate/enzyme imbalance theory argues that secondary metabolites accumulate as a result of "overflow" metabolism, and emphasizes differential enzyme compartmentalization and regulation (Haslam 1986; Waterman and Mole 1989). In terms of the plant's metabolism, secondary compounds are regarded primarily as metabolic wastes, detoxification products, or shunt metabolites produced in order to reduce "abnormal" concentrations of "normal" cellular constituents. In many ways this theory resembles the C/N theory, for it argues that changes in the production of secondary metabolites are the result of excesses of metabolites due to the plant's inability to regulate their production. While both this theory and the C/N theory present induced metabolites as being essentially waste products, neither theory precludes the "defensive" sculpting of the overflow metabolites.

3. Growth/differentiation balance theory recognizes that all secondary metabolites have an ontogenetically determined phenology and that their synthesis is emphasized during periods of plant differentiation (Loomis 1953; Lorio 1988; Herms and Mattson 1992; Lerdau et al. 1994). The process of plant growth largely occurs during different times than the process of differentiation that produces resin ducts, secretory glands, trichomes, spines, and so forth. As in the C/N theory, growth and differentiation are seen as negatively correlated. Growth dominates during favorable conditions, and differentiation is at a maximum only when conditions are suboptimal for growth. Plants that are emphasizing growth may lack the enzymatic machinery or resources necessary for the synthesis and storage of secondary metabolites. For resistance that is closely tied to a particular stage in ontogeny, such as the source-sink transition for leaves (Coleman and Jones 1991), induced resistance may be caused by herbivory that shifts the plant's ontogenetic stage.

Of these three theories the C/N theory has received the most attention and the most testing. Attempts to falsify the predictions of the C/N theory as a general explanation have been successful repeatedly, particularly with regard to the rapidly induced responses. For example, the predictions of this theory for a rapidly induced response involving the putatively nitrogen-intensive metabolite nicotine have been unambiguously falsified. Total nicotine production and the proportion of a plant's nitrogen content in nicotine increase dramatically after damage and even under nitrogen-limited growth conditions, counter to the predictions of the C/N theory (Baldwin and Ohnmeiss 1994; Ohnmeiss and Baldwin 1994). Similarly, altering the C/N ratio of plants by growing them under conditions of elevated CO_2 increases the amount of starch they contain but does not result in increased terpene or phenolic concentrations, as the theory would predict. Twenty-four other studies in which at least one prediction of the C/N theory was not supported were reviewed by Herms and Mattson (1992). The predictions of the C/N theory have fared better with regard to delayed induced responses (Tuomi et al. 1990), but these consistencies are difficult to interpret because the metabolites responsible for delayed induced responses have yet to be conclusively identified in any system.

In summary, C/N theory appears to have little value as a mechanistic model, because the explicit mechanism controlling the allocation of carbon and nitrogen to secondary metabolism has been clearly and repeatedly falsified for both carbon-intensive and nitrogen-intensive secondary metabolites. While it is true that many carbon-intensive secondary metabolites increase when plant growth is limited by nutrient availability, these increases are equally well understood as evolved responses to low-nutrient environments when tissue replacement is difficult. This does not mean that the C/N theory has no value for explaining patterns of induced secondary metabolites, but rather that it quite clearly does not explain all plant responses and is a poor mechanistic model for rapidly induced responses.

The first three theories have been termed supply-side hypotheses (Lerdau et al. 1994) because defenses are most strongly influenced by the availability or supply of secondary metabolites. According to all three hypotheses, plants do not regulate the production of secondary metabolites to any extent. The next two hypotheses differ in these two assumptions. They posit that damage results in signals that directly regulate secondary metabolism. These are demand-side hypotheses because concentrations of secondary metabolites are most strongly influenced by the plant's need or demand for defense. The two remaining theories differ from one another in the specificity of the signaling system.

4. The generalized stress-response theory posits that plants have a hormonally mediated centralized system of physiological responses for coping with many diverse stresses (Chapin 1991b). Since some stress-related plant hormones affect the production of some induced defenses,

these defenses may be part of the generalized stress response. Abscisic acid is one plant hormone that appears to play a role such as this (Pena-Cortez et al. 1989, 1991).

5. The active defense response theory is similar but posits far more specificity in the plant's signaling system (Berryman 1988; Chessin and Zipf 1990). This theory postulates that endogenously produced damage cues or cues specific to the invading organism activate specific defense responses. This theory appears to describe the induced responses found in colonial invertebrates (Harvell 1986, 1990a), many phytoalexin responses of plants to fungal pathogens (Ayers et al. 1985; Creasy 1985), as well as the majority of the rapidly induced chemical responses to herbivory that have been well studied, as we will argue below.

In recent years, research in the area of plant signal transduction, the process by which wound and pathogen responses are communicated within a plant, has been one of the fastest-growing research areas in plant physiology and biochemistry (see section 2.3). The evidence from this tremendous body of research supports variations on the latter two mechanistic theories of changes after herbivory, outlined above. Specifically, the best working model in our opinion is that rapidly induced increases in secondary metabolites result from specific signals that control the metabolic pathways that produce the chemical defenses. In some cases these same signals also cause changes in growth comparable to the changes caused by the more general stress hormones. The specificity of regulation, in some circumstances, has been shown to result from transcriptional regulation of different isomeric forms of key enzymes that control the synthesis of chemical defenses. For example, in potato both the induced synthesis of steroidal glycolalkaloids in response to wounding and the induced synthesis of sesquiterpenoid phytoalexins in response to fungal attack involve the transcriptional up regulation of the key regulatory enzyme, 3-hydroxy-3-methylglutaryl-coenzyme A reductase (HMGR; Choi et al. 1994). Moreover, different genes (*hmg1* and *hmg2*) code for different isozymes that regulate the two responses. These two genes are differentially induced by lipid-derived signals specific to the inducing agents, namely, JA for the wound-induced response and arachidonic acid for the phytoalexin response against *Phytophthora infestans*, the potato blight fungus. Molecular biological studies such as this one have given rise to a view of secondary metabolism that is precisely regulated by external stimuli. This view is difficult to reconcile with the earlier hypotheses (represented by theories 1–3 above) that many defensive products are the "flotsam and jetsam on the metabolic beach" (Haslam 1986), derived from supply-driven, metabolic overflow.

As this discussion certainly illustrates, patterns of secondary metabolism in plants have received a lot of attention and have generated diverse explanations. The multiple theories proposed to explain secondary metabolism stem in part from a reluctance of researchers working with plants to accept a defensive role for these traits; this reluctance is not shared by researchers who study chemical defenses in animals. Berenbaum (1995), who pointed out this asymmetry, attributed it partly to a question of empathy. It is easier for human researchers to accept the defensive role of metabolites that are delivered with a flurry of easy-to-observe behavioral fanfare in response to attacks on animals than to accept the defensive role of a secondary metabolite's accumulation in a leaf. Moreover, due to the nature of growth and regeneration processes, attack is more frequently fatal for nonmodular animals and less often fatal for plants and modular animals. In other words, plants are more tolerant of attacks than are animals. As a consequence, selection for dramatic, "drop-dead" defenses should be more intense in animals than in plants. Also, because plant parts have different degrees of longevity, defense clearly needs to be considered in the context of the vulnerability of various parts to attack as well as their fitness value, both of which require a sophisticated understanding of plant function to appreciate. In addition, a historical explanation may be relevant. The study of plant secondary metabolites traces its origins to the field of chemotaxonomy, which used the distribution of these metabolites to construct phylogenies and therefore had to assume that they are waste products that can be used as adaptively neutral taxonomic traits. Consequently, the notion that secondary metabolites are the products of "overflow" metabolism has deep historical roots. Most important, the concept suggests that plants have less control over their metabolic processes than animals do (Berenbaum 1995). This difference in level of control is likely to be more imagined than real. For example, excess production of metabolites in animals is stored in adipose tissue and remobilized for later use; starch plays a comparable role in plants. Yet nobody doubts that animals also possess highly regulated systems of defensive responses. Clearly, plants also have regulated metabolic options to cope with excess and to tailor effective defensive reactions.

Plants probably exhibit a greater variability in nutrient composition in response to environmental variations in resource availability than do animals. Moreover, when the mechanisms responsible for delayed induced resistance are worked out, they may be found to be due to changes in the total amount of nutrient- or carbon-based metabolites in a plant. Nevertheless, it doesn't seem logical to conclude that either the diversity or the quantity of secondary metabolites produced by a plant is the consequence of an inability to regulate its metabolism. The evidence from the mechanisms that we have found to be responsible for rapidly induced changes in secondary metabolism supports the conclusion that plants are indeed capable of considerable regulation. We will describe three case studies below to illustrate this contention.

Induced changes in a plant's chemical defense pools can result from any combination of alterations in synthesis, turnover, or transport. In the following sections we develop an example illustrating the importance of each of these three processes. These examples underscore the argument that plants are capable of regulating their metabolisms and do so in response to changes caused by herbivory.

3.3.3.2 Proteinase Inhibitors: Control by Synthesis

PIs are the best-studied induced response to date, largely due to the brilliance and perseverance of Clarence Ryan and the members of his research group at Washington State University (Ryan 1990, 1992). PIs are proteins that inactivate both endopeptidases and exopeptidases and therefore can function defensively by influencing any one of the many processes in herbivores that rely on this group of enzymes; most of the processes that PIs interfere with are involved with digestion. Inhibitors have been found in plants for each of the four classes of proteolytic enzymes (serine, cystein, aspartic, and metallo-proteases), but the best studied are the serine PIs, PI-I and PI-II, which have molecular masses of 8,000 and 12,000 kDa, respectively. PI-I is an inhibitor of chymotrypsin, while PI-II is a double-headed inhibitor with two active sites, one for chymotrypsin and the other for trypsin. These PIs are widespread, having been found in the Brassiceae, Cucurbitaceae, Fabaceae, and Salicaceae, but their inducibility is best studied in the Solanaceae.

Both serine PIs exist as transient proteins that accumulate locally and distally in response to wounding in the leaves of tomato and potato. PI-I and PI-II are coordinately expressed in leaves and can attain concentrations as high as 200 μg/g. Within three to four hours after wounding, mRNA for PI-I and PI-II increases in both wounded and unwounded leaves distal to the wound. Message levels increase for eight hours after a single wounding and have a half-life of about twelve hours. Successive wounds delivered at intervals every hour for four hours double the total mRNA and PI levels. Once translated, the PIs are stored in the central vacuole, where they have half-lives of approximately one week.

The signal-transduction pathways of this transcriptionally mediated increase in chemical defense have received much scrutiny (see section 2.3). Wounding mixes endogenous pectin-degrading enzymes with the 1, 2-galacturonic-acid oligomers that make up pectin in the cell walls of unwounded cells adjacent to the wound. Oligouronide fragments are produced and function as local wound signals. Long-distance signaling in tomato is mediated by the eighteen-amino-acid polypeptide systemin, which was the first polypeptide hormone discovered for plants. Systemin is active at femtomolar concentrations; it is proteolytically processed from a 200-amino-acid precursor protein, prosystemin. Prosystemin mRNA is found in all parts of the plants except the roots and is induced systemically in response to wounding in the leaves. The importance of systemin in regulating this induced defense was elegantly demonstrated by McGurl et al. (1992) in experiments with tomato plants that were transformed with an antisense prosystemin DNA. These plants did not produce systemin after wounding and had very little systemic PI-I and PI-II induction. Since prosystemin mRNA is wound-inducible, the damage signal pathway may provide a means of amplifying the response to continuous damage in tomato.

Both the long-distance and the short-distance signal pathways are thought to activate PI genes by means of a secondary signal pathway, the octadecanoid pathway, which results in the synthesis of JA and other jasmonates. (This pathway will be discussed more fully in section 3.3.4.) JA up regulates the transcription of both PI-I and PI-II genes, the Bowman-Birk inhibitor genes, and the accumulation of their respective proteins. In summary, the current working model for the long-distance transduction pathway that activates PI genes is that systemin is released by wounding and translocates throughout the plant. Subsequently, systemin interacts with distal cells to stimulate the synthesis of JA or a derivative of JA. These in turn interact with transcription factors activating the PI genes, culminating in the production of PI proteins. More recently, Ryan's laboratory discovered that bestatin, an inhibitor of aminopeptidases, activates PI transcription without activating the jasmonate cascade (Schaller et al. 1995). This result suggests that an aminopeptidase is involved in the signal transfer from the jasmonate cascade to the genome.

The examination of possible defensive functions of PIs has focused on the inhibitory effects of PIs on proteolytic digestive enzymes of herbivores. In other words, PIs decrease the digestibility of ingested food. Additionally, PIs may have defensive functions not directly related to digestion. For example, PIs are known to function synergistically with *Bacillus thuringiensis* (*Bt*) toxin and may regulate proteolytic processes associated with herbivore resistance to toxins produced by these bacteria.

The defensive functions of PIs are perhaps the most thoroughly established of all the secondary metabolites. High PI content correlates with decreased herbivore performance (Broadway et al. 1986; Wolfson 1991), but more direct experimental evidence exists for the defensive function of PIs from work with transgenic plants. PIs have the distinct experimental virtue of being chemical defenses that are direct gene products, unlike most secondary metabolites in which genes code for enzymes, which in turn regulate a more complicated metabolic pathway. As a consequence a number of laboratories have produced transgenic plants that express novel PIs. Transgenic manipulation allows one to test defensive functions caused by specific mechanisms independently of all the correlated changes that cloud comparisons between damaged and undamaged plants or cultivars developed by conventional breeding techniques (near-isogenic lines, etc.) For example, Hilder et al. (1987) transformed tobacco with cowpea inhibitor, which made the transgenic plants more resistant to *Heliothis virescence* larvae. Similarly, Johnson, Narvaez, et al. (1989) transformed tobacco plants with PI-I and PI-II genes under the expression of a constitutive cauliflower mosaic virus promoter, and found that PI-II proteins produced the greatest resistance to *Manduca sexta* larvae. Orozoco-Cardenas et al. (1993) showed that tomato plants transformed with an antisense construct of prosystemin had attenuated PI-I and PI-II production when damaged by *Manduca sexta* larvae. Moreover, the larvae grew better on these transgenic plants, demonstrating the effect of induced PIs by manipulating the signal-transduction pathway that activates induced resistance.

The relationships between induced PIs and herbivore performance is not quite as simple as it has been presented. Gut proteolytic activity of caterpillars feeding on diets enriched in PIs did not differ from that measured in guts of caterpillars feeding on diets without PIs (Broadway and Duffey 1986). Broadway and Duffey hypothesized that herbivores would respond to PIs in their diet by hyperproducing proteases, which would cause essential amino acids to degrade. However, the situation is even more complicated and interesting than this. In spite of the well-documented ability of PIs to inhibit larval growth and development, the effects tend to be small, extending development only a few days. The effects are not more dramatic apparently because of changes in the proteases that the insects secrete into their guts as they consume diets high in PIs (Broadway 1995, 1996; Jongsma, Bakker, Peters, Bosch, and Stiekema 1995). Larvae of beet armyworms (*Spodoptera exigua*) feeding on tobacco plants genetically engineered so that they produced 0.16% of their soluble protein as PI-II compensated for the high dietary intake of PI with a 2.5-fold increase in gut tryptic activity. Furthermore, the tryptic activity was insensitive to inhibition by PI-II. In contrast, larvae that were feeding on untransformed tobacco so that they were not exposed to high doses of PI-II did not compensate, and their gut tryptic activity was readily inhibited by experimentally added PI-II. Similar compensation was found for Colorado potato beetles (*Leptinotarsa decemlineata*) feeding on potato plants that had been sprayed with methyl jasmonate to induce high levels of papain inhibitors (Bolter and Jongsma 1995). In these cases the herbivores faced with high dietary levels of PIs increased production of proteases of a new type that were resistant to the PIs of the plants.

The process of inducing new proteases by herbivores with slight modifications in their active sites so that they are not inhibited by plant PIs has not gone unanswered by the plants. Each plant genome contains an estimated one to two hundred different PI genes, which can be grouped into different families; each family of inhibitors is hypervariable in its active domain (Hill and Hastie 1987). This is precisely the genetic signature that would be expected from a long coevolutionary arms race between plants and herbivores. Of course, other explanations for the variability of PIs are also plausible. Some of the PI genes code for PIs that function in internal processes, but apparently not very many. Furthermore, there is no reason to have a lot of variation in these endogenously functioning PIs.

3.3.3.3 Phytoalexins: Control by Synthesis and Turnover

The ability of plants to metabolize their own secondary chemicals plays an important role in determining the changes in pool size after herbivore attack, as illustrated by the example of induced phytoalexins. Moreover, the theories that address patterns of defenses in plants categorize these compounds according to how easily they can be metabolized by the plants that use them. The ability of a plant to degrade and recover the resources invested in a secondary metabolite is central to the "mobile-immobile" metabolite dichotomy of the resource availability theory (Coley et al. 1985). According to this scheme, those defensive chemicals that can be recovered are considered to be less costly than are defenses that are immobile and cannot be recycled. Differences in costs and effectiveness of these two types of defenses are posited to explain the kinds of plants and the kinds of environments where each of the defenses are found. The ability of plants to recover secondary compounds used for defense also plays a peripheral role in the "qualitative-quantitative" dichotomy of apparency theory (Feeny 1976; Rhoades and Cates 1976; Fox 1981) and optimal defense theory (Rhoades 1979; McKey 1979). In developing all of these theories, data from pulse-labeled experiments were extrapolated and were assumed to indicate that the pools of "mobile" or "qualitative" metabolites were continually degraded and resynthesized. These theories attempt to understand patterns of metabolite production in a cost-benefit framework. Accordingly, metabolic lability of these compounds as measured in pulse-labeled experiments has been assumed to affect the costs of utilizing "mobile" or "qualitative" metabolites. "Mobile" or "qualitative" metabolites have been thought to be relatively costly due to the requirements of maintaining the metabolite pool. On the other hand, recovery of resources invested in these "mobile" metabolites has been presumed to decrease their costs. Because induced defenses have been proposed to be cost-saving measures (see section 5.3), the issue of metabolite turnover is particularly germane.

Unfortunately, the pulse-labeling experiments that have been used as the basis for determining relative rates of turnover, mobility, and cost for different secondary compounds are turning out to be unreliable and misleading. While it is clear that some secondary metabolites can be rapidly metabolized, particularly those playing dual roles in defense and storage in seeds (e.g. L-canavanine [Rosenthal et al. 1988]), other metabolites are far less labile than previously thought. Both the terpenes of mint and the alkaloid nicotine have been frequently cited as the outstanding examples of "mobile" or "qualitative" plant defense metabolites; they are transported throughout the plant and have been reported to have half-lives of less than one day (Seigler and Price 1976; Fox 1981; Coley et al. 1985). Yet recent reexaminations of these two metabolites with labeled biosynthetic precursors in intact plants have not substantiated the earlier claims of high rates of turnover (Mihaliak et al. 1991; Gershenzon 1994; Baldwin, Karb, and Ohnmeiss 1994). As a consequence it is important to evaluate the techniques used to measure turnover rates and the information that these techniques have provided.

The easiest way of measuring turnover is by introducing labeled secondary metabolites and then quantifying the rate of loss of the label. However, this method may actually reflect metabolite detoxification and other processes that plants use to regulate the size of their metabolite pools, rather than turnover. Since the pool sizes of many secondary metabolites appear to be highly regulated, exogenous additions of metabolites above the plant's set points are likely to initiate detoxification processes that allow plants to regain their previous set points. Moreover, some studies of metabolite turnover have used excised plant parts to facilitate the introduction of the labeled precursors. It is now clear that the rates of metabolite turnover are not necessarily similar for excised foliage and intact foliage. In summary, the role of turnover and lability of secondary metabolites may be smaller than hypothesized previously.

Of all the different groups of induced defenses, induced phytoalexins represent the situation in which turnover most clearly plays an important role. Phytoalexins, by definition, have antimicrobial activity, and some are also active against herbivores (Kogan and Fischer 1991). Moreover, since most herbivores function as Trojan horses, carrying microbes to the wound site as they eat, phytoalexin induction is likely to be a common component of a plant's response to herbivory. Some plant-herbivore systems, such as the pine-bark beetle systems, depend entirely on the microbes that are carried by the beetles to the site of infection (Raffa 1991).

Phytoalexins are frequently synthesized de novo after a challenge and are frequently not detectable in unchallenged tissues; moreover, they frequently disappear after a challenge. As a consequence of these rapid changes in pool size, phytoalexins are a good candidate for illustrating the importance of metabolite turnover in induced defenses. We emphasize the tentative nature of this conclusion because most of the data come from highly artificial experimental systems, from cell cultures and excised plant parts. Eventually the processes need to be more firmly established in whole-plant preparations. In order to understand phytoalexin turnover, one experimental requirement must be satisfied: the response needs to be elicited by factors other than intact microorganisms because many microbes have evolved the ability to metabolize phytoalexins. As a result, the mixture of phytoalexins found in a plant induced by microbes may not reflect what the plant actually produced or degraded (Van Etten et al. 1989).

The phytoalexins of soybean are one of the better-studied systems in terms of their effects on soybean's microbial and herbivorous enemies. Of the various phytoalexins that soybean produces, the glyceollin family of phytoalexins are particularly well known (Yoshikawa et al. 1993). In hypocotyl preparations de novo glyceollin synthesis is increased approximately eightfold both by biotic factors (endogenously produced cell-wall fragments thought to be polymers of galacturonic acid and exogenous elicitors derived from fungi and thought to be β-1,3-glucans) and by aseptic wounding. However, despite the increase in de novo synthesis, only treatment with the biotic elicitors resulted in increased glyceollin pools, because glyceollin degradation processes are sufficiently stimulated by wounding such that the total pool size does not change. Abiotic elicitors (salts of heavy metals and detergents) also increase glyceollin pools; however, they do so by decreasing the glyceollin degradation activity, not by stimulating de novo synthesis (Yoshikawa et al. 1993).

Similar results highlight the importance of both synthesis and turnover in determining the induced change in the sesquiterpenoid phytoalexins, capsidiol and debneyol, in cell-suspension cultures of *Nicotiana tobacum* (Threlfall and Whitehead 1988). The elicitor, cellulase, initiates a complex and highly coordinated set of biosynthetic responses that result in the inhibition of squalene synthetase. This apparently redirects the flux of farnesyl pyrophosphate (FPP) from sterol biosynthesis and toward FPP carbocyclase and the eventual synthesis of the two sesquiterpenoid phytoalexins. However, response to the elicitor also involves the inhibition of capsidiol metabolism more than debneyol metabolism, with the net result that the former accumulates faster and attains higher concentrations than the latter (Threlfall and Whitehead 1988). Clearly, for metabolically labile compounds, induced changes in whole-plant or tissue-specific pools can result from any combination of induced changes in synthesis and degradation.

3.3.3.4 Nicotine: Control by Transport

Many different alkaloids are induced by damage (table 3.2). In this section we will discuss the mechanisms responsible for the damage-induced production of one particular alkaloid, nicotine, in two native species of tobacco, *Nicotiana sylvestris* and *N. attenuata*. Our goal is to illustrate the importance of understanding induced responses on a whole-plant basis. Although these species produce other alkaloids, nicotine represents more than 95% of the total alkaloid pool and has the greatest overall toxicity. Nicotine owes its toxicity to its ability to mimic acetylcholine and to the inability of acetylcholinesterases to free the acetylcholine receptor of nicotine. This mechanism for nicotine's effects is consistent with the symptoms that characterize nicotine toxicity, namely, excitation at low concentrations and depression, paralysis, and death at high concentrations. Moreover, this model helps explain the apparently paradoxical recreational use of this potent nerve toxin by humans. Smoking administers nicotine at low doses, and allows smokers to precisely titrate their tissues to concentrations at which nicotine has its excitatory effects.

Both *Nicotiana* species are annuals, and they accumulate nicotine at 0.1–1.0% leaf dry mass in undamaged plants. After real or simulated folivory, nicotine concentrations in remaining leaves increase dramatically, typically fourfold, and under some circumstances tenfold, to attain concentrations that are sufficient to deliver what is for many herbivores a lethal dose in a single meal. Induced increases in nicotine concentrations are sufficient in laboratory feeding trials to protect induced tissues from insect herbivores that are adapted to nicotine (Baldwin 1988a, 1988b, 1988c, 1989, 1991; Baldwin et al. 1990). The induced increases in whole-plant nicotine pools are thought to result primarily from increases in de novo synthesis as determined by ^{15}N-pulse-chase techniques. Furthermore, as mentioned in the previous section, endogenously produced nicotine is not appreciably metabolized (Baldwin, Karb, and Ohnmeiss 1994).

The results of many experiments support the model that nicotine is synthesized largely in the roots and transported to shoots in the xylem stream. For example, when the roots and shoots of nicotine-producing and non-nicotine-producing species are grafted together in different combinations, only the combinations with nicotine-producing roots accumulate this alkaloid in substantial quantities (Dawson 1941, 1942; Baldwin 1991). Leaf damage dramatically increases the amount of nicotine exported from the roots in the xylem stream of tobacco plants (Baldwin 1989; Baldwin et al. 1993; Baldwin, Schmelz, and Ohnmeiss 1994). Induced increases in nicotine accumulation are inhibited in plants whose stems have been steam-girdled or chilled below the leaf damage (Baldwin 1989; Zhang et al. n.d.). Furthermore, the distribution of nicotine's biosynthetic enzymes corroborates the model that nicotine is synthesized primarily in root tissues. Wagner et al. (1986) and Feth et al. (1986) determined that putrescine N-methyltransferase (PMT) was the key regulatory enzyme in nicotine biosynthesis, that roots had higher PMT activity than shoots, and that PMT activity was stimulated by damage to the shoot (Mizusaki et al. 1973; Saunders and Bush 1979; Wagner et al. 1986). The sequence of the PMT gene from *N. tobacum* has recently been determined: the mRNA coding for PMT is found only in root tissues, and the amount of mRNA increases after damage to the shoot (Hibi et al. 1994).

The sequestration of nicotine biosynthesis in the roots has a number of physiological and ecological consequences that highlight the role of long-distance transport in induced-defense responses. At a physiological level a large separation between the site of synthesis (roots) and the site of herbivore attack (leaves) requires a long-distance signal-transduction pathway plus a long-distance pathway to move the nicotine once it has been synthesized. After damage to leaf tissues, these pathways mediate stimulation of root nicotine biosynthesis and transport nicotine from the roots back up to the leaves. There, the nicotine probably defends the plant against folivores and other herbivores that consume above-ground parts. We review what is known about the signal-transduction pathway and the transport of nicotine from root to shoot, and consider some of the consequences of the large separation between the sites of synthesis and the sites of damage.

Some advances have been made in understanding the long-distance signal-transduction cascade in this system. The damage cue is a positive one that is correlated with the amount of damage that a plant receives (Baldwin 1989; Baldwin and Schmelz 1994). The fact that it can be blocked by steam girdling the stem of the plant below the site of damage suggests that it is phloem-borne; steam kills the phloem but leaves the xylem intact. The damage signal is relatively slow moving, taking more than one hour to leave a damaged leaf, which makes an electrical damage signal unlikely (Baldwin, Schmelz, and Onhmeiss 1994). Because the de novo rates of nicotine synthesis increase ten hours after leaf damage (Baldwin, Schmelz, and Ohnmeiss 1994; Baldwin, Karb, and Ohnmeiss 1994), the signal that causes these changes in root nicotine metabolism must travel from damaged leaves to the roots within this time period.

JA is intimately involved in the signal-transduction cascade because leaf damage rapidly (<0.5 hours) increases shoot JA pools, reaches a maximum value 90 minutes after damage, and decays to levels found in undamaged plants within 24 hours. A systemic increase in JA levels in the roots occurs more slowly, attaining maximum values approximately 180 minutes after damage to the leaves (Baldwin, Schmelz, and Ohnmeiss 1994; Baldwin et al. 1997). When the amount of leaf damage is quantitatively varied, strong positive relationships exist among the variation in leaf wounding, endogenous changes in JA concentrations, and whole-plant nicotine (Baldwin et al. 1997; Ohnmeiss et al. n.d.). In other words, the more a leaf is wounded, the larger the increase in JA in the leaf 90 minutes after wounding, and the larger the change in whole-plant nicotine, 5 days after wounding. Additions of appropriate quantities of JA to roots as methyl jasmonate increase de novo nicotine synthesis from $^{15}NO_3$ in a manner similar to that induced by leaf damage. Additions of methyl jasmonate to the shoots mimic both the timing of changes in pool sizes induced by damage and the magnitude of increased nicotine synthesis. The wound-induced increase in JA is necessary for the induced increase in nicotine; if the increase in JA is inhibited by applying either salicylic-acid methyl ester or the plant hormone indoleacetic acid directly to the wounded tissues, the nicotine response is similarly inhibited (Baldwin et al. 1997).

It is not yet known if the wound-induced increase in JA in the shoot is directly transported to the roots or if some other indirect signal pathway is involved. ^{14}C-labeled JA that is added to shoots is rapidly metabolized at the site of application, reflecting rapid detoxification of exogenously applied JA (Zhang et al. n.d.). Sufficient quantities of JA are transported to the roots to account for the change in root JA pools that are observed after wounding or after application of methyl jasmonate to the leaves. Whether endogenously produced JA is transported through the plant is not clear at this time (see section 2.3.6). In tomato, the octadecapeptide systemin stimulates an increase in JA synthesis that culminates in increased PIs (Farmer et al. 1994; see section 2.3.2). A similar transduction pathway may also be responsible for wound-induced nicotine production. Our current working model for the long-distance signal-transduction cascade is that wounding increases JA pools in shoots, which either directly through transport, or indirectly through a signal like systemin, increases JA pools in roots; these, in turn, stimulate nicotine synthesis in the roots and increase nicotine pools throughout the plant.

Regardless of the details of the transduction pathway, it is clear that quantitative variation in endogenous JA pools is significantly correlated with variation in induced increases in whole-plant nicotine (Baldwin et al. 1997; Ohnmeiss et al. n.d.). Interestingly, feeding by hornworm caterpillars (*Manduca sexta*) and careful simulation of this damage with microscissors produced dramatically different responses in JA and nicotine. Larval feeding resulted in a manyfold amplification of JA pools, and part of this was caused by oral secretions from the larvae (McCloud and Baldwin n.d.). However, this amplification of endogenous JA pools caused by larval feeding did not result in a commensurate amplification of the nicotine response. Hence, *Manduca sexta* larvae apparently feed in a "stealthy" fashion, reducing the nicotine responses below that of plants suffering a comparable amount of tissue loss from the same portions of their canopy over the same time period (Baldwin 1988c). How these larvae feed stealthily, or in other words how they interfere with the plant's signal-transduction cascade, remains to be discovered.

Once nicotine is synthesized in the roots, its transport to the shoot occurs in the xylem stream. This largely apoplastic transport route links nicotine transport to leaf transpiration rates and therefore to all the environmental factors that affect transpiration, such as humidity, temperature, and other factors that affect stomatal conductance. However, since the alkaloid is apparently stored in the central vacuole (Saunders 1979), a symplastic component of the transport must also be involved.

The long-distance transport pathways of both the signal down to the roots and nicotine back up to the leaves constrain the plant's ability to respond and therefore might be considered an ecological liability. First, the distances introduce delays in transport and in the response. It takes three to five days for a plant to attain its full induced titer of nicotine in response to a single damage episode, and many hours of this waiting period are a direct consequence of the time it takes to transport the cue to the roots and the nicotine back up to the shoot. Second, the transport pathways represent a weak link that an herbivore could short-circuit by girdling the plant. Third, locating nicotine synthesis in a tissue distal to the site of damage and transporting the defense in the xylem stream limits the plant's ability to "target" or direct the defense to tissues other than those directly connected by xylem vasculature. Last, localizing nicotine synthesis in the roots may be energetically costly. Nicotine synthesis requires a substantial quantity of reduced nitrogen because 17% of the molecule is nitrogen. Approximately 3% of an undamaged plant's total nitrogen pool is in molecules of nicotine, and this figure doubles or triples when a plant is damaged (Baldwin, Karb, and Ohnmeiss 1994). A large proportion (as high as 30%) of the nitrogen used for induced nicotine synthesis is acquired and then reduced by the plant after damage, probably in the roots. Leaf damage does not affect the amount of nitrate that is reduced in the roots. The increase in nicotine concentrations in the xylem fluid leaving roots occurs with a concurrent decrease in the export of other amino acids and amides, principally glutamate and glutamine (Baldwin et al. 1993), indicating that these plants maintain a constant amount of nitrogen reduction and assimilation in the roots in order to support the nicotine response. Since the reduction and assimilation of nitrogen is thought to be performed more efficiently in shoots than in roots (Oaks and Hirel 1985), and inhibiting the nitrogen metabolism of the roots of some plants increases plant growth rates (Oaks 1992), the advantages of sequestration of nicotine synthesis in the roots presumably also outweigh this additional biosynthetic cost.

Given all the disadvantages of sequestering nicotine production below ground, why is that the site of most production? Below-ground synthesis might be a result of constraints that inhibit nicotine synthesis in the shoots; however, given that shoots have both the enzymes and the ability to synthesize nicotine, albeit to a limited degree, why should these factors represent insurmountable constraints? The sequestration of nicotine synthesis below ground protects the plant's ability to launch a defensive response in answer to an attack that removes a large proportion of the shoot. This was demonstrated by examining the effect of removing increasing amounts of a plant's leaf area on the plant's ability to respond to damage (Baldwin and Schmelz 1994). Only when plants had lost 88% of their total leaf area was their allometrically corrected ability to increase their whole-plant nicotine pool diminished. Even when plants had lost more than 90% of their leaf area, the concentration of the last remaining leaf material continued to increase, notwithstanding the impairment of the plant's overall nicotine-synthesizing abilities. This ability to increase leaf nicotine concentrations even when the plant's allometric biosynthetic ability is impaired represents another advantage of locating nicotine synthesis below ground. If a plant's roots produce a given amount of nicotine per day and the plant's shoot suddenly gets smaller because part of it walked off in the stomach of an herbivore, the recently synthesized nicotine will be sent to a smaller shoot. Hence, the concentration of nicotine in the shoot tissues rises even if the total synthesizing ability of the root does not change or even declines. Therefore, seques-tration of nicotine synthesis in the roots buffers the induced response, making it insensitive to the plant's current photosynthetic capacity, an advantage shared with preformed induced defenses.

If the advantage of root-based nicotine synthesis is that it permits a plant to protect its ability to launch a defense response in the face of heavy herbivory, then one might expect browsers to be the most im-portant herbivores selecting this induced response. While the fitness consequences of different types of herbivory have not been formally measured in the field, mammalian browsing does account for the vast majority of the leaf area lost to herbivores in five native populations of N. *attenuata* growing in Utah over two seasons (Baldwin and Ohnmeiss 1993). Moreover, the inability of these plants to direct a chemical de-fense to particular regions of their canopy suggests that the response is more effective against herbivores that feed on plants in a "coarse-grained" fashion (selecting among plants rather than among plant parts), rather than against herbivores that feed in more "fine-grained" fashion (choosing among plant parts). By contrast, the induced accumu-lation of furanocoumarins in wild parsnip may be an example of an in-duced defense that evolved in response to herbivores feeding in a fine-grained fashion; it occurs more rapidly than the nicotine response and only in the damaged leaflets and those immediately adjacent to the dam-aged leaf (Zangerl and Berenbaum 1995).

3.3.4 Responses Mediated by Jasmonates

3.3.4.1 The Octadecanoid Pathway

By using chemicals for defense, plants can exploit the physiological and biochemical differences between themselves and their herbivores. These differences allow them to reduce the risk of autotoxicity inherent when deploying bioactive compounds. Given the way plants use these differ-ences, it is ironic that many induced chemical defenses are activated by a signal-transduction cascade that is shared to some degree by plants and their herbivores. In plants this is called the octadecanoid pathway.

Brady Vick and Don Zimmerman, working with the USDA in Fargo, North Dakota, unraveled many components of the synthesis of JA in the 1970s and 1980s. They described the octadecanoid pathway, so named because its starting material is the common 18:3 fatty acid of plants, linolenic acid. (The 18:3 refers to eighteen carbons and three double bonds in the fatty acid.) Vick and Zimmerman recognized that the octa-decanoid pathway is remarkably similar to pathways responsible for the synthesis of prostaglandins, prostacyclins, leukotrienes, and thromboxanes in animals, whose synthesis originates from the common 20:4 fatty acid in animal cells, arachidonic acid. In animal systems these potent secondary messengers play a complex suite of roles, many of which relate to the pain and inflammatory responses to wounding.

Similarly, the overall picture that is emerging from the explosion of interest in the roles played by jasmonates and other oxylipids in plants is that they mediate many of the "defensive" and "civilian" responses to wounding. We recognize that this attempt to understand the diversity of responses induced by jasmonates within a functional framework may be premature, given the diversity of systems in which they have been studied. However, we feel that it is helpful at this time to present an organizing, phytocentric framework for these diverse reactions. In this section we will cover, with broad strokes, the octadecanoid pathway, the roles that some of its metabolites play in wound-related responses, and some of the evidence for these roles. Signaling in plants mediated by fatty acids, oxylipids, and jasmonates has been reviewed recently (Sembdner and Parthier 1993; Farmer 1994; Reinbothe et al. 1994; Rhodes 1994), and frequent review will be necessary to keep pace with the rapid advances being made in this area.

Before proceeding we should note that we are not arguing that all jasmonate responses in plants relate to wounding, but, rather, that many wound-related jasmonate responses can be understood from a phytocentric perspective toward wounding. Even though jasmonates in plants may have originally evolved as a means of informing plants about wounding and its associated consequences, this putative historical function does not necessarily mean that the responses currently mediated by this pathway must be related to wounding. Jasmonates appear to be a secondary messenger, albeit a slow one compared to other secondary messengers. This role as a secondary messenger is stimulated by other signal-transduction pathways (including those involving systemin and abscisic acid) and allows jasmonate-induced responses to be uncoupled from the stimulus of wounding. For example, one of the best-studied jasmonate responses is the coiling of tendrils of *Bryonia dioica*, induced by touch. Touch on the ventral side of this species stimulates a rapid coiling reaction (Weiler et al. 1993), a clear case of a cell-specific thigmomorphogenic response that does not require wounding for its initiation.

The octadecanoid pathway (figure 3.3) starts with the release of a fatty acid, linolenic acid, from cell membranes either directly as a result of damage or from undamaged cells, mediated by lipases that in turn are induced by other signals. The pathway is essentially identical if the 18:3 fatty acid, linoleic acid, is the starting material, except that all the products would lack the appropriate unsaturation; for example, dihydrojasmonic acid, which lacks the double bond at C_9, would be produced instead of JA. In some systems, such as the induced production of PI-I and PI-II in tomato, and the tendril coiling in *Bryonia dioica*, the induced response can be elicited simply by adding the fatty acid (Falkenstein et al. 1991; Farmer and Ryan 1992). This suggests that the pathway is regulated at the lipase step for these systems. For the PI system the fact that systemin, the eighteen-amino-acid polypeptide responsible for the systemic accumulation of PIs after wounding in tomato, appears to function by increasing JA synthesis from linolenic acid emphasizes the importance of the control over the lipase step that is not a direct result of wounding (Farmer et al. 1994). Fungal elicitors that induce a suite of secondary metabolites in tissue culture without wounding apparently do so by stimulating JA synthesis (Gundlach et al. 1992). However, not all JA-induced responses can be stimulated by the addition of linolenic acid, so regulation is likely at other points in the pathway (Baldwin et al. 1996).

Once linolenic acid is released, it can be oxidized by lipoxygenases to form different reactive hydroperoxides (figure 3.3), the structures of which will depend on the type of lipoxygenase. Lipoxygenases exist in isoforms which are compartmentalized in different plant parts and which produce differing ratios of 9- and 13-hydroperoxylinolenic acid (reviewed by Siedow 1991). Nonenzymatic production of lipid hydroperoxides is likely to complement the enzymatic production in damaged tissues. These reactive hydroperoxides may function defensively in the immediate area of the wound by oxidizing nutrients or targets in the herbivore as part of the scorched-earth defense (Bi et al. 1994; Bi and Felton 1995). If the 13-hydroperoxy product is formed and this in turn is cleaved by a lyase, two types of products are formed, a C_6 group and a C_{12} group, both of which are important in the plant's response to damage. The first group, C_6, comprises the suite of metabolites generally called the green-leaf volatiles (Hatanaka 1993). Some of the green-leaf volatiles, specifically (E)-2-hexenal, are demonstrably toxic to fungi (Vaughn and Gardner 1993), bacteria (Croft et al. 1993), and aphids (Hildebrand et al. 1993). Moreover, this compound is an ingredient in the defensive sprays used by many insects, which suggests broad-based biocidal activity (Roth and Eisner 1962). These volatile C_6 metabolites may also function as indirect defenses of plants, attracting predators and parasites of herbivores or making plants more attractive to other herbivores. The second group of cleavage products, the C_{12} products, include traumatin and traumatic acid (figure 3.3). The former compound is known to stimulate cell division at the site of wounding in beans (English

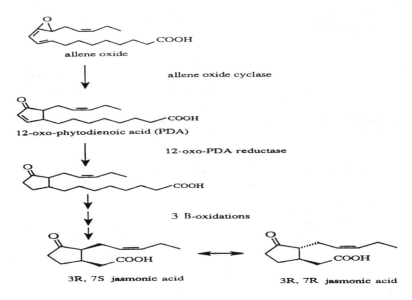

Fig. 3.5 Jasmonates are produced by a branch of the oxylipid pathway.

et al. 1939), while the latter stimulates leaf abscission in cotton (Strong and Kruitwagon 1967).

Biosynthesis of JA starts if 13-hydroperoxylinolenic acid is dehydrated to form the allene cyclase (Song et al. 1993). If allene oxide is isomerized instead of cyclized, a family of ketols is produced about which little is known other than that they are reactive metabolites. The cyclization of the allene oxide results in the formation of 12-oxo-phytodienoic acid (PDA), which is likely to be the first active jasmonate. PDA is reduced and subjected to three consecutive β-oxidations to form JA (figure 3.5). PDA and all the metabolites downstream of PDA to JA can exist in one of four epimers, due to the two chiral carbons at C3 and C7 in JA. Two of these epimers, (3R,7S)-JA and (3R,7R)-JA, occur naturally; the former epimer is synthesized after induction in tissue culture, and it rapidly epimerizes to the latter epimer (Mueller and Brodschelm 1994). The particular structures and epimers that are most active in eliciting a particular response appear to vary among systems. PDA is more active than JA in both the tendril-curling assay and transcriptional activation of the berberine-bridge enzyme assay in California poppy (*Eschscholtzia californica*) tissue culture (Weiler et al. 1994) but is less active than JA in the induction of nicotine synthesis in *N. sylvestris* (Baldwin et al. 1996). Moreover, the epimers of JA were found to be active in different ways in four different bioassay systems that examined the effects of the molecule on regulation of growth (Koda et al. 1992).

Some of the strongest evidence linking the induced changes in jasmonates *in planta* to the effects they elicit when jasmonate is added to plants, comes from the successful inhibition of JA synthesis. JA synthesis can be inhibited *in planta* by applying many of the drugs commonly used to inhibit pain and inflammatory responses in humans. For example, the immediate application of aspirin, salicylic acid, or its methyl ester directly to a wound in a lanolin paste inhibited the damage-induced increases in JA and PI proteins in tomato (Peña-Cortés et al. 1993). Similarly, the inhibition of damage-induced production of jasmonates used diethyldithiocarbamic acid (DIECA), which reduced 13-hydroperoxy-linolenic acid to hydroxylinolenic acid, concomitantly inhibited wound-induced PI production in tomato (Farmer et al. 1994). A more elegant demonstration comes from the transgenic technique when a chloroplast-specific isoform of lipoxygenase was cosuppressed in transgenic *Arabidopsis* plants (Bell et al. 1995). These plants exhibited neither wound-induced increases in JA nor wound-induced increases in vegetative storage protein message (see section 3.3.4.2).

In summary, jasmonates and their octadecanoid precursors have the characteristics of a wound signal pathway (see also section 2.3.6). They are synthesized in response to wounding, their synthesis precedes their presumed effects, and inhibitors of their synthesis block their effects. Moreover, they are rapidly metabolized and have strict structural requirements for their activity (Sembdner and Parthier 1993).

The effects of jasmonates are influenced by changes in other plant hormones. Wound-induced changes in these hormones in nature may have strong modulating effects on a plant's response to jasmonates. For example, exogenous applications of auxin (indoleacetic acid) to wounded leaves inhibited induced production of nicotine (Baldwin 1989; Baldwin et al. 1997), PI (Kernan and Thornburg 1989; Thornburg et al. 1993), and vegetative storage proteins (DeWald et al. 1994). Damage to leaves may decrease leaf auxin levels two- to threefold (Thornburg et al. 1993; Thornburg and Li 1991) and thereby amplify the response induced by increases in jasmonates. Similarly, the common wound-related hormone ethylene synergizes the effects of JA by increasing pathogenesis-related proteins 1 and 5 in tobacco (Xu et al. 1994). Not all defense-related signal pathways are likely to amplify the effects mediated by the jasmonate pathway. For example, salicylic acid is the endogenous signal that activates the systemic accumulation of pathogenesis-related proteins related to viral infection (Enyedi et al. 1992). High concentrations of salicylic acid may also be involved in the localized cell death that characterizes the hypersensitive response against pathogen viruses. However, as mentioned above, salicylic acid also is known to inhibit the lipoxygenase involved in jasmonate synthesis, and therefore it is possible that the signal-transduction pathways mediating responses induced by virus may inhibit those mediating responses induced by wounding. Last, some of the wound-induced responses mediated by jasmonates, specifically induced PI production in potatoes, are also affected by abscisic acid and gibberellic acid (Jacobson and Olszewski 1996). Abscisic acid and gibberellic acid are a pair of principal players in the centralized stress-response system, which allows plants to alter their patterns of growth in response to water and nutrient stress (Chapin 1991b).

In summary, plants have a number of signal-transduction pathways that mediate their responses to abiotic and biotic factors in their environment. Some of these systems appear to be redundant, which might be expected given their importance vis-à-vis plant fitness. Others are interactive, able to tailor the plant's responses to the situation at hand. Such flexibility is one of the advantages that conditional responses, such as induced defenses, have over nonplastic, constitutive defenses. Given that plants are rooted in the ground and generally unable to alter their environment by moving, we should expect a high degree of tailoring of induced responses.

3.3.4.2 Defensive and Civilian Responses

Jasmonates initiate many processes in plants, some of which are consistent with the expectations of a phytocentric perspective on the reorganization of plant function after herbivory. Table 3.3 lists the changes induced by jasmonates that fit into the categories of induced chemical defense, escape, storage, and growth responses. However, a few caveats are in order. First, no single species is known to exhibit all of these responses, or even a response from each of the categories, nor is any expected to. The responses of a particular plant to wounding are, like the particular suite of secondary metabolites produced by a plant, likely to depend strongly on its evolutionary history as well as on the particular ecological situation of the plant. These categories merely represent a way of organizing expectations for responses induced by wounding. Second, as is commonly the case with the discovery of a new plant-growth regulator, most of the reports of jasmonate responses involve the exogenous addition of jasmonates rather than the manipulation of endogenous pools within a plant. As a consequence, some responses may be artifacts of using unnatural isomers or of adding amounts of JA in excess of amounts found naturally. For example, jasmonates were first thought

Table 3.3 Survey of defensive and civilian response to wounding induced by jasmonate

Response	Reference
Chemical response	
Proteins	
Proteinase inhibitors	
I	Farmer and Ryan 1990
II	Farmer et al. 1992
Papain inhibitors but not wound inducible	Bolter 1993
Trypsin inhibitors	Farmer et al. 1992
Pathogenesis-related proteins	
1B and osmotin	Xu et al. 1994
Thionins	Bohlmann and Apel 1991
Jasmonate-induced protein 60	Reinbothe et al. 1994
Cell-wall proteins	Creelman et al. 1992
Lipoxygenase	Bell and Mullet 1991
Enzymes of secondary metabolism	
Phenylalanine ammonia lyase	Gundlach et al. 1992
4-Coumarate: Coenzyme ligase	Kuhn et al. 1984
Chalcone synthase	Chappell and Hahlbrock 1984
Berberine bridge enzyme	Gundlach et al. 1992
3-Hydroxy-3-methylglutaryl–coenzyme A reductase	Choi et al. 1994
1-aminocyclopropane-1-carboxylic acid	Chou and Kao 1992
Secondary metabolites	
Nicotine	Baldwin, Schmelz, and Ohnmeiss 1994
Quinine	Aerts et al. 1994
Indole monoterpenoid alkaloid of *Catharanthus*	Aerts et al. 1994
Indole glucosinolates (but not aliphatic or aromatic glucosinolates)	Bodnaryk 1994
Anthocyanins	Franceschi and Grimes 1991
Benzo(c)phenanthridine alkaloids	Gundlach et al. 1992
Isobavachalcone	Gundlach et al. 1992
Genisteine	Gundlach et al. 1992
Lettucenin A	Gundlach et al. 1992
Raucaffricine	Gundlach et al. 1992
Rubiadin	Gundlach et al. 1992
Rutacridone	Gundlach et al. 1992
Phytoalexins	Dittrich et al. 1992
Rosmarinic acid	Mizukami et al. 1993
Escape	
Bulb formation	Nojiri et al. 1992
Tuber induction	Koda et al. 1992
Decreased flower formation	Albrechtova and Ullmann 1994
Increased flower senescence	Porat et al. 1993
Leaf abscission	Ueda et al. 1991
Storage	
Vegetative storage proteins	Staswick 1994
Bark storage proteins	Davis et al. 1993

Table 3.3 continued

Response	Reference
Growth	
Growth	Koda et al. 1992
Ribulose-1, 5-bisphosphate carboxylase–oxygenase	Reinbothe et al. 1994
Light-harvesting chlorophyll a/b binding protein	Reinbothe et al. 1994
JIP 66, a ribosome-inactivating protein	Reinbothe et al. 1994

to be senescence hormones, and many plants exhibited symptoms of senescence after addition of jasmonate. However, it is now clear that the amounts required for this effect exceed the quantities found in plants (Dermastia et al. 1994). Manipulating endogenous pools is the best way to discover the roles played by hormones. With the advent of more targeted inhibitors of wound-induced jasmonate production, mutants lacking in aspects of jasmonate response or production (Staswick et al. 1992; Bell et al. 1995), as well as techniques to quantify endogenous pools, a more complete picture is likely to emerge in the near future of the role played by this ubiquitous wound signal-transduction pathway.

The list of chemicals induced by wounding or by herbivory that are also induced by jasmonates is large and continues to grow (table 3.3). Examples come from all three major classes of secondary metabolites—phenolics, alkaloids, and terpenes—as well as many of the enzymes responsible for their synthesis. In addition, a suite of defensive proteins, including the PIs, certain pathogenesis-related proteins, lipoxygenases, and the jasmonate-induced proteins (JIPs) are all induced by jasmonates. The function of the JIPs are largely unknown, with the exception of JIP 60, which may be involved in initiating cell death associated with the hypersensitive response. We do not anticipate that all chemical responses induced by jasmonates are defensive. To date, however, we know that only the papain inhibitors that are induced by jasmonates are not also induced by wounding (Bolter 1993). Similarly, not all of a plant's wound-induced responses will be found to be induced by jasmonates. All of the responses induced by jasmonates that have been studied at the molecular level involve induced changes in gene transcription or translation, and therefore it is possible that preformed defenses that do not rely on de novo synthesis are not induced by jasmonates.

Exogenously applied JA made tomato plants more resistant to naturally occurring herbivores in a recent field experiment (J. Thaler and M. J. Stout, personal communication). In this case oxidative enzyme activities and concentrations of PIs increased and correlated with field resistance. Plants treated with JA had fewer beet armyworm caterpillars (*Spodoptera exigua*), flea beetles (*Epitrix hirtipennis*), and western flower thrips (*Frankliniella occidentalis*).

Since most of the defensive responses induced by jasmonates are under transcriptional control, the resources required for these new initiatives must be made available within the responding plant. This could be accomplished through either the mobilization of stored reserves or the termination of other ongoing activities in the plant that compete for resources. In this regard it is interesting to note that jasmonates induce the down regulation of one of the quantitatively most important protein synthesis processes in green plants: the synthesis of RuBPCase (Reinbothe et al. 1994). The down regulation of this important protein and other "housekeeping" proteins occurs first at a transcriptional level and appears to be mediated by one of the JIPs, specifically JIP 66, which functions as a protein that inactivates ribosomes. Therefore, while jasmonates induce a suite of defense responses at the transcriptional level, some of these newly transcribed messages code for proteins that terminate the translation of other non-defense-related proteins. These in turn may free up the resources required for defense and, as such, may function in a budget-balancing role. The slowing of growth, which is one of the better-described effects of jasmonates (Koda et al. 1992), may be a direct result of this budget balancing or it may be a defensive response. Either way, it will be interesting to determine whether the quantities of resources liberated by growth reduction exceed requirements for the induced-defense initiatives.

Defense responses are not the only category of responses that are transcriptionally induced by jasmonates. One of the best studied civilian responses is the induced increase in vegetative storage proteins (VSP) in soybeans (Staswick 1994) and their counterparts in *Populus,* the bark storage proteins (Davis et al. 1993). While VSPs share sequence similarity with acid phosphatases, they have little phosphatase activity and function instead as temporary sinks for nitrogen storage. They are synthesized in response to sink removal such as the removal of developing pods in soybeans and are rapidly metabolized once new sinks are reestablished (Staswick 1994). Plants can be viewed as sets of physiologically integrated sources and sinks. When an herbivore walks off with either, the plant needs temporary sinks while it reconfigures its growth processes. The promoter regulating the transcription of VSP genes requires not only jasmonates for its activation but also sucrose (DeWald et al. 1994), a requisite that could provide a mechanism for determining whether a particular tissue is still a sink for photoassimilates or is capable of producing photoassimilates after an attack.

Lastly, jasmonates induce a series of other responses that we have lumped under the heading "escape" in table 3.3. At this stage it remains unknown whether they are adaptive responses to herbivory. The induction of tuber and bulb formation by jasmonate is clearly a storage response, but the shift to below-ground storage may represent a form of escape from herbivory, much the way that the grazing-adapted populations of *Panicum coloratum* of the Serengeti grasslands respond to herbivory (Dyer et al. 1991). Similarly, jasmonate-induced termination of flower formation and the rapid senescence of already-formed flowers may constitute a form of escape if it reduces a plant's apparency to herbivores. While it is easy to criticize this discussion as being overly Panglossian (Gould and Lewontin 1979), we feel that it provides a useful organizing framework for generating and testing hypotheses about the function of responses found to be induced by this ubiquitous wound signal-transduction pathway.

3.4 FUTURE DIRECTIONS: BRINGING THE PLANT BACK INTO PLANT-HERBIVORE INTERACTIONS

Mechanistic studies of plant-herbivore interactions have made substantial progress in understanding the behavioral, physiological, and biochemical responses of herbivores to the secondary metabolites found constitutively in plants. However, the same cannot be said for the plant half of the interaction. Plants have been viewed more as passive substrates on which herbivores feed than as the dynamic organisms that the study of induced responses so clearly portrays. One reason for this is that most of the researchers who study the interaction are trained as zoologists. They have been slow to develop the phytocentric perspective needed to study induced responses in the context of the functional reorganization of a plant that occurs after herbivory. Interestingly, plant physiologists have not played a central role in this research effort until recently, following the discovery of wound-induced signal-transduction pathways.

The recognition that plants are a dynamic component of the interaction suggests that components of the feeding behavior of herbivores, such as feeding duration, rate, choice of tissue, and interval between feeding bouts, may be sculpted in part by plant responses. For example, the feeding rhythms of many herbivorous insects may reflect the herbivore's digestive physiology and the probability of being discovered by predators while feeding, but these rhythms may also be influenced by the plant's signal-transduction pathway and by the responses to wounding activated by that pathway. The interval between feeding episodes could be influenced by the decay time of the wound signals, functionally the equivalent of a memory of attack within the plant. Herbivores that distribute feeding bouts over time may decrease the defensive responses of their host plants to a given amount of wounding. A study of a plant's wound-induced signal-transduction pathway would therefore be the functional equivalent to a study of a plant's sensory perception of its world. Whether or not this level of detail is informative for the study of a plant's interactions with free-living herbivores has yet to be determined. However, it is clear that those herbivores with less mobility, such as leaf-mining caterpillars or gall-forming insects, owe their success as herbivores to their ability to manipulate their host plant's physiology, particularly the mechanisms by which the plant establishes source-sink relationships.

Understanding plant-herbivore interactions clearly requires a detailed mechanistic knowledge of a plant's responses; acquiring this understanding has been thwarted by the sheer complexity of responses that occur. This overwhelming complexity makes it difficult to establish that a particular plant trait is relevant for the interaction. In the past artificial diets were used but there are now two more elegant solutions to this experimental problem. These techniques should allow us to link specific plant responses to biological effects in the near future.

The first solution involves using transgenic techniques to genetically manipulate particular traits, by either overexpressing or inhibiting the expression of the genes coding for those traits. These techniques hold great promise for giving experimenters control over the inducible traits. The most tractable use of this technology, for which the greatest number of successes have been reported, involves the expression of traits that are direct gene products, such as defensive proteins. Manipulating secondary metabolism with transgenic techniques is far more difficult, since the products of these pathways result from complicated biosynthetic reactions requiring many enzymatically mediated steps. However, successes have been reported in changing the concentrations of particular secondary metabolic pathways with the transgenic manipulation of single regulatory enzymes in the biosynthetic pathway of a secondary metabolite (DeLuca 1993; Yao et al. 1995). For example, one of the most elegant demonstrations of the defensive functions of phytoalexins against pathogens involves the transgenic expression of a stilbene, a class of phenylpropanoid-derived phytoalexins. One stilbene, resveratrol, is a major phytoalexin produced by the enzyme stilbene synthase, which catalyses the condensation of p-coumaroyl coenzyme A and malonyl co-enzyme A. The stilbene synthase gene from grape, *Vitis vinifera,* when cloned and expressed in tobacco (which lacks this condensation enzyme) under control of a constitutive promoter, produced plants that made large quantities of resveratrol (a new metabolite for tobacco). These plants were protected when challenged with the fungal pathogen *Botrytis cinerea* (Hain et al. 1993). This allowed researchers to link a specific chemical change to resistance to a particular organism.

Advances in the control over secondary metabolism as well as the identification of wound-inducible and jasmonate-inducible elements in promoters controlling important plant traits will allow researchers to ask even more interesting questions using this technology. Transgenic technology offers the opportunity to explore the functional significance of particular plant traits while maintaining the rigor of a double-blind experiment. That is, plants manipulated to express particular traits could be grown in various environments and subjected to a suite of different selection regimes; prior expectations for the functional role of the manipulated traits would not be confounding factors. Transgenic techniques are not without their limitations. Novel genes that are expressed in transgenic plants are not in the environments that they evolved in and therefore may not be expressed in the same way. In addition, expression of novel genes may not have the same costs as in the original organism. However, transgenic techniques offer large advances in our ability to experimentally manipulate individual plant traits, compared to the techniques that preceded them.

A second solution to the problem of the complexity of using whole plants is to manipulate plant traits by means of the plant's signal-transduction pathway. In this and the preceding chapters, we have described two pathways that appear to be widespread among higher plants and are readily activated by simple manipulations. The systemic release of the acyclic homoterpenes occurs following the addition of β-glucosidases to mechanical wounds. Similarly, jasmonate-induced responses are activated by the exogenous addition of jasmonates. Each plant's particular response to these treatments needs to be carefully characterized. However, we expect that it may well be worth the effort, for once these responses are well understood, a researcher can manipulate a plant's response to herbivory independently of the actual act of herbivory.

These techniques will allow workers to simplify the complexity of the chemical responses, to manipulate one component at a time in a realistic manner. However, it is also important to recognize that induced responses are complex and to consider the diversity of chemicals concurrently induced by a single form of damage. Studies should expand from their past focus on individual phytochemicals to include the breadth of the induced response and to consider how it is influenced by the physical and chemical milieu of the plant and herbivore. As a starting place, the basic "natural history" of the timing (how quickly induction occurs and relaxes) and the spatial extent of induction should be determined.

In summary, the benefits of incorporating a thorough mechanistic understanding of a plant's responses to herbivore attack will repay the effort. Moreover, the incorporation of mechanisms into considerations of induced responses mirrors a similar trend in the development of plant-herbivore theory over the past two decades. As plant-herbivore theory has been improved from its origins in apparency theory (Feeny 1976) and optimal defense theory (Rhoades and Cates 1976), more and more mechanistic and physiological information has been included in subsequent theoretical developments (Haukioja et al. 1994). For example, the effects of variable resources (resource allocation theory of Coley et al. 1985) and of modular growth of plants (Tuomi et al. 1990; Haukioja et al. 1990) on patterns of secondary-metabolite allocation are recent modifications to plant defense theory. These recent efforts represent refinements on the original cost-benefit model that include information on how plants work. Further refinements that provide more realistic and informed views of plant function would surely be beneficial. Similarly, a more thorough working knowledge of the mechanisms of induced responses will facilitate our understanding of the effects that these responses have on herbivores, the kinds of selective pressures and evolutionary constraints that molded these responses, and the useful applications of plant responses in agriculture.

4 Induced Resistance against Herbivores

4.1 EFFECTS ON PERFORMANCE OF BIOASSAY HERBIVORES AS EVIDENCE OF INDUCED RESISTANCE AND SUSCEPTIBILITY

Changes that occur in a plant after it has been attacked by herbivores may affect those attackers or other herbivores that attempt to use the plant at a later time. The easiest and most direct way to determine whether responses provide induced resistance—that is, whether they affect herbivores—is to challenge damaged plants with living herbivores. The behavior, performance (survival, growth rate, fecundity), or population size of herbivores feeding on plants that have been attacked can be compared with those parameters for herbivores on unattacked control plants. Such comparisons are called bioassays because some measure of the success of a living challenger is used to characterize the induced response.

Often investigators use indirect measures of behavior, performance, or population size because these indirect measures are much easier to obtain. For example, the amount of damage may be measured as an indirect correlate of behavioral preference, rate of growth during some small portion of the life cycle may be measured as a correlate of performance, or weight at pupation may be measured as a correlate of fecundity. The use of these indirect correlates is based on convenience and experimental tractability, and the measures vary in the degree to which they present an accurate estimate of herbivore success.

Bioassays may be conducted to compare plants that the investigator found damaged versus those found undamaged. Such comparisons (or "natural experiments") are quite realistic in terms of the type, timing, and extent of the damage. However, herbivores may nonrandomly select plants. Any differences that the investigator finds in the comparison described above may be due to herbivore damage (as the investigator generally assumes) or may be due to other, preexisting differences between plants that herbivores chose or didn't choose. This ambiguity can be eliminated if the experimenter randomly assigns plants to receive different levels of damage. Designs involving random assignment of treatments are generally preferred, although they may sacrifice realism; every effort should be taken to apply the experimental damage treatments as realistically as possible, using actual herbivores.

Bioassays measure induced resistance, effects on herbivores, rather than induced plant responses. Herbivores sometimes mount counter-defenses to a plant's induced responses (see section 3.2.2). Variability in induced resistance may be caused by variability in the induced response and/or by variability in the effects of those responses on herbivores. The "resistance" that we measure in bioassays is actually the sum of the plant's responses plus the herbivore's responses to the plant's responses. For example, earlier we described the resin flows that result from herbi-vores' tearing leaf tissue of some plants and the specialized adaptations that some herbivores have to deal with such flows (section 2.2.3). Resin flows have been found to be very effective against some generalist herbi-vores, although other herbivores are able to circumvent this defense, though not completely (Dussourd and Denno 1994; Becerra 1994).

Another potential problem with bioassays is that the assay herbivores that are used to challenge damaged and undamaged plants may them-selves induce some level of resistance. In other words, it may be difficult to get a genuinely undamaged control. This problem is much more se-vere for those systems in which the challengers or bioassay herbivores cause a considerable level of damage. Similarly, this problem can be min-imized, though not eliminated, by using herbivores to assay induced re-sponses that cause a minimum amount of damage themselves. For some systems in which genuinely undamaged controls cannot be achieved or maintained, an alternative method is to feed herbivores artificial diets containing previously damaged or undamaged plants that were har-vested after damage treatments; obviously this technique sacrifices some realism.

Induced responses may affect the individual herbivores that cause the damage or herbivores that feed subsequently on the damaged plants. Some workers use this distinction, based on characteristics of the herbi-vores, to provide a frame of reference to differentiate rapid induced re-sistance from delayed induced resistance (e.g. Haukioja 1990b). The induced response may affect herbivores during the season when the damage occurs or it may affect herbivores in subsequent seasons. This distinction, which is based on the plant's life cycle to provide a frame of reference, has been used by others (e.g. Bryant et al. 1988; Clausen et al. 1991) to differentiate between short-term and long-term induction. The latter distinction provides less useful information about the conse-quences of the responses on herbivore population dynamics, but it is not conditional on the life histories of the particular herbivores in-volved. These distinctions are made because many workers regard rapid or short-term responses to be different from delayed or long-term responses in both their mechanisms and in the effects they cause on herbivores, particularly herbivore population dynamics.

Recent interest in using bioassays to evaluate whether damage has induced resistance in plant-herbivore systems has been enormous. Evidence for induced resistance has been found in a very wide variety of plant species (table 4.1). These examples include very disparate results involving spatial scales ranging from single leaves to entire trees and temporal scales ranging from seconds to years. The list in table 4.1 is intended to present an overall impression of where induced resistance has been observed; it is not a complete or systematic review, although it includes many more examples than previously published lists. Many of the plant species have been the subject of several studies, but only one study is included in the table in most cases. Since only a minuscule fraction of species have been tested thus far, the list will no doubt grow considerably.

All of the studies listed in table 4.1 included effects on the performance or behavior of herbivores. When unusual mechanisms were indicated (resin production, hypersensitive reaction, tissue abscission, callus formation), these are listed. In most cases there was no convincing evidence about the probable mechanism. Several of the listings involved situations that were arguably constitutive, rather than induced, resistance. In the cases of activated precursors, as in cyanogenesis and latex flow, the chemicals responsible for resistance were preformed. These may be considered induced responses in a broad sense because the plant holds the precursors in isolated vacuoles until it is damaged (see section 3.3.2 for a discussion of preformed induced mechanisms). Only then are the precursors released to react and provide resistance; damage is required for expression of the resistance. On the other hand, in these examples prior feeding did not necessarily induce a change in the chemistry or amount of the precursors; they were already formed and in place. This ambiguity applies to only a few of the examples listed in table 4.1 (these examples are indicated in the column "Mechanism/Type of Effect"), and inclusion or exclusion of these examples does not influence the general patterns.

Although the list of plant species for which induced resistance has been found is sizable, the impression that induced resistance is everywhere is certainly incorrect. Many studies have failed to find evidence of induced resistance, although most of these negative findings are probably not reported in the published literature. The plants that have been most intensively studied, such as birch (*Betula* spp.) and cotton (*Gossypium* spp.), present stories that are inconsistent. When the same controlled experiments have been conducted repeatedly, the strength of resistance has been found to vary greatly; some experiments produce statistically significant indications of induced resistance and others do not (Karban 1987; Ruohomaki et al. 1992).

Compared to undamaged controls, some plants may become more preferred by herbivores or better hosts in terms of increased herbivore performance or population size after they have been damaged. This effect, which we call induced susceptibility, also has been reported for many plant species (table 4.2). (Induced susceptibility is analogous to what Haukioja [1990a, 1990b, 1990c] termed induced amelioration.)

Equipped with these lists, we might now be in a position to ask, What kinds of plants show induced resistance? What herbivores are affected by it? What factors influence where it is found? Unfortunately, the diversity of examples found in tables 4.1 and 4.2 makes comparisons difficult. Whether an investigator found induced resistance or not often depended on the methods, the spatial and temporal scale, the bioassay, the sample size, environmental conditions other than damage, the extent of genetic variability, and the ontogenetic states of the plants and herbivores (Coleman and Jones 1991). This makes comparisons between results of different studies uninterpretable because so many factors are not controlled and each of the factors could strongly influence the results. Whenever possible we looked for patterns in the lists presented in tables 4.1 and 4. 2. However, we also looked for studies that considered several species simultaneously, in an attempt to reduce some of the uncontrolled variance associated with comparisons from several different studies.

4.2 Where Is Induced Resistance Found?
4.2.1 Plant Life History

Induced resistance has been reported from many diverse families and habitats, including plants from marine intertidal zones, freshwater aquatic habitats, deserts, grasslands, and tropical, temperate, and boreal forests (table 4.1). The plant families that are well represented by temperate and boreal tree species are also best represented in the list of examples of induced resistance. The bias toward temperate and boreal environments probably reflects the historical development of the field more than actual biological patterns.

The bias toward long-lived plants is more likely to reflect a true pattern. We put forth this conjecture because induced resistance makes more sense for longer-lived species. Individual long-lived plants may be

Table 4.1 Plants for which induced resistance has been reported

Family	Species	Growth Form	Relative Rate of Growth	Tissue Responding
Fucaceae	*Fucus distichus*	Intertidal alga	Fast	Leaves
Pinaceae	*Abies balsamea*	Woody perennial	Intermediate-fast	Needles
	Abies grandis	Woody perennial	Slow	Stem
	Larix decidua	Woody perennial	Fast	Needles
	Picea excelsa	Woody perennial	Slow	Stem
	Picea glauca	Woody evergreen	Intermediate	Stem
	Picea lutzii	Woody evergreen	Fast	Stem
	Picea pungens	Woody perennial	Slow	Stem
	Picea sitchensis	Woody perennial	Very fast	Stem
	Pinus contorta	Woody evergreen	Intermediate-fast	Stem
	Pinus echinata	Woody evergreen	Intermediate	Phloem
	Pinus nigra	Woody evergreen	Very fast	Shoot
	Pinus ponderosa	Woody evergreen	Fast	Needles
	Pinus radiata	Woody evergreen	Very fast	Needles
	Pinus sylvestris	Woody evergreen	Fast	Shoot
	Pinus taeda	Woody evergreen	Fast	Phloem
Cupressaceae	*Juniperus monosperma*	Woody evergreen	Slow	Fruit
	Juniperus virginiana	Woody evergreen	Slow	Stems
Lauraceae	*Persea indica*	Evergreen tree	Crop	Leaves
Nymphaeaceae	*Nuphar lutea*	Aquatic perennial	Fast	Leaves
Tiliaceae	*Grewia flavescens*	Woody perennial	Fast	Leaves
	Tilia spp.	Woody perennial	Intermediate	Leaves
Malvaceae	*Gossypium australe*	Woody perennial	Fast	Leaves
	Gossypium hirsutum	Woody perennial	Crop	Leaves
	Gossypium thurberi	Woody perennial	Fast	Leaves
Zygophyllaceae	*Larrea tridentata*	Evergreen shrub	Slow	Stems
Euphorbiaceae	*Omphalea diandra*	Perennial liana		Leaves
Brassicaceae	*Brassica nigra*	Annual	Fast	Leaves
	Raphanus raphanistrum	Annual	Fast	Leaves
	Sinapis alba	Annual	Crop	Leaves
Polygonaceae	*Rumex obtusifolius*	Herbaceous perennial	Fast	Leaves
Chenopodiaceae	*Beta vulgaris*	Biennial	Crop	Leaves
Oleaceae	*Fraxinus americana*	Woody perennial	Slow-intermediate	Stems
Asclepiadaceae	*Asclepias syriaca*	Herbaceous perennial	Fast	Leaves

108

Same/ Subsequent Season	Mechanism/ Type of Effect	Herbivore Affected	Reference
Same	Behavior	Mollusks	Van Alstyne 1988
Same	Resistance	Caterpillars	Mattson, Lawrence, et al. 1988
Same	Resin	Beetles	Lewinsohn et al. 1991
Subsequent	Resistance	Caterpillars	Baltensweiler et al. 1977
Same	Hypersensitive	Gall aphids	Rohfritsch 1981
Same	Hypersensitive	Beetles	Werner and Illman 1994
Same	Hypersensitive	Beetles	Werner and Illman 1994
Same	Hypersensitive	Beetles	Lewinsohn et al. 1991
Subsequent	Numbers and behavior	Weevils	Overhulser et al. 1972
Subsequent	Resistance	Caterpillar	Leather et al. 1987,
Same	Resin	Beetles	Raffa and Berryman 1987
Same	Resin, hypersensitive	Beetles	Cook and Hain 1988
Same	Resin	Shoot moth	Harris 1960
Same	Resistance	Sawflies	Wagner 1988
Subsequent	Resistance	Mites	Karban 1990
Same	Resin	Shoot moth	Harris 1960
Same	Resin, hypersensitive	Beetles	Cook and Hain 1988
Same	Abscission	Insects	Fernandes and Whitham 1989
Same	Resin	Cicada eggs	White 1981
Same	Resistance	Mites	McMurtry 1970
Same	Morphology	Beetles	Kouki 1991a, 1991b
Subsequent	Behavior, % damage	Insects	Bryant, Heitkonig, et al. 1991
Same	Resistance	Aphids	Dixon and Barlow 1979
Same	Resistance	Leaf miners	Karban 1991
Same	Resistance	Mites	Karban and Carey 1984
Same	Resistance	Leaf miners	Karban 1993b
Subsequent	Behavior, damage	Jackrabbits	Ernest 1994
Subsequent	Resistance	Caterpillars	Smith 1983
Same	Hypersensitive	Lepidoptera eggs	Shapiro and DeVay 1987
Same	Resistance	Caterpillars	Agrawal et al. n.d.
Same	Behavior	Beetles	Palaniswamy and Lamb 1993
Same	Resistance	Beetles	Benz 1977; Hatcher et al. 1994
Same	Resistance	Flies	Rottger and Klinghauf 1976
Same	Callus	Cicada eggs	White 1981
Same	Latex/behavior	Caterpillars, chewers	Dussourd and Eisner 1987

Table 4.1 continued

Family	Species	Growth Form	Relative Rate of Growth	Tissue Responding
Convolvulaceae	*Ipomoea purpurea*	Annual	Fast	Leaves
Boraginaceae	*Cynoglossum officinale*	Biennial	Intermediate	Leaves
Solanaceae	*Lycopersicon esculentum*	Annual	Crop	Leaves
	Nicotiana sylvestris	Annual	Fast	Leaves
	Physalis alkehengi	Woody perennial	Fast, ornamental	Leaves
	Solanum dulcamara	Herbaceous perennial vine	Fast	Leaves
	Solanum tuberosum	Herbaceous perennial	Crop	Leaves
Lamiaceae	*Dicerandra fructescens*		Slow	Leaves
Rosaceae	*Crataegus monogyna*	Woody perennial	Slow	Leaves
	Fragaria × ananassa	Herbaceous perennial	Crop	Leaves
	Fragaria grandiflora	Herbaceous perennial	Crop	Leaves
	Malus pumila	Woody perennial	Intermediate, crop	Leaves
	Prunus domestica	Woody perennial	Crop	Leaves
	Prunus padus	Woody perennial	Very fast	Leaves
	Prunus serotina	Woody perennial	Very fast	Stems
	Sorbus aucuparia	Woody perennial	Slow	Leaves
Leguminosae	*Acacia drepanolobium*	Woody perennial	Slow	Stems
	Acacia seyal	Woody perennial	Slow	Stems
	Acacia tortilis	Woody perennial	Fast	Leaves
	Acacia xanthophloea	Woody perennial	Slow	Stems
	Dichrostachys cinerea	Woody perennial	Fast	Leaves
	Glycine max	Annual	Crop	Leaves
	Lupinus arboreus	Woody perennial	Fast	Leaves
	Phaseolus lunatus	Perennial	Crop	Leaves
	Trifolium repens	Herbaceous perennial	Fast	Leaves
Betulaceae	*Alnus crispa*	Woody perennial	Fast	Shoots
	Alnus glutinosa	Woody perennial	Fast	Leaves
	Alnus incana	Woody perennial	Fast	Leaves

Same/Subsequent Season	Mechanism/Type of Effect	Herbivore Affected	Reference
Same	Resistance	Caterpillars	Rausher et al. 1993
Same	Behavior	Caterpillars	Prins et al. 1987
Same	Resistance	Caterpillars	Broadway et al. 1986
Same	Resistance	Caterpillars	Baldwin 1988c
Same	Behavior	Caterpillars	Beuz and Abivardi 1991
Same	Hypersensitive	Mites	Westphal et al. 1990
Same	Resistance	Beetles, leafhoppers	Tomlin and Sears 1992
Same	Preformed precursors, behavior	Ants, roaches	Eisner et al. 1990
Same	Behavior	Caterpillars	Edwards et al. 1991
Same	Resistance	Mites	Shanks and Doss 1989
Same	Resistance	Mites	Kielkiewicz 1988
Same	Resistance	Mites	Croft and Hoying 1977
Same	Resistance	Mites	Kuenen 1948
Subsequent	Resistance	Caterpillars, aphids	Neuvonen et al. 1987; Leather 1993
Same	Resin	Cicada eggs	White 1981
Same and subsequent	Behavior	Caterpillars	Edwards et al. 1986; Neuvonen et al. 1987
Same and subsequent	Thorns, behavior	Mammals	Milewski et al. 1991
Same and subsequent	Thorns, behavior	Mammals	Milewski et al. 1991
Subsequent	Behavior, % damage	Insects	Bryant, Heitkonig, et al. 1991
Same and subsequent	Thorns, behavior	Mammals	Milewski et al. 1991
Subsequent	Behavior, % damage	Insects	Bryant, Heitkonig, et al. 1991
Same	Behavior, resistance	Many	Hart et al. 1983; Hildebrand, Rodriguez, Brown, and Volden 1986
Same	Resistance	Caterpillars	Harrison and Karban 1986
Same	Behavior	Mites	Dicke 1986
Same	Activation of precursors	Mollusks	Dirzo and Harper 1982a, 1982b
Same and subsequent	Resistance	Caterpillars	Clausen et al. 1991
Same	Behavior	Caterpillars	Edwards et al. 1986
Same and subsequent	Resistance, behavior	Beetles	Jeker 1983; Baur et al. 1991

Table 4.1 continued

Family	Species	Growth Form	Relative Rate of Growth	Tissue Responding
	Alnus rubra	Woody perennial	Fast	Leaves
	Betula papyrifera	Woody perennial	Very fast	Shoots
	Betula pendula	Woody perennial	Fast	Leaves
	Betula populifolia	Woody perennial	Fast	Leaves
	Betula pubescens	Woody perennial	Fast	Leaves
	Carpinus betulus	Woody perennial	Slow	Leaves
	Corylus americana	Woody perennial	Intermediate-fast	Stems
	Corylus avellana	Woody perennial	Intermediate	Leaves
Fagaceae	*Castanea dentata*	Woody perennial	Intermediate-fast	Stems
	Fagus sylvatica	Woody perennial	Slow	Leaves
	Quercus calliprinos	Woody evergreen	Slow	Leaves
	Quercus emoryi	Woody evergreen	Slow	Leaves
	Quercus garryana	Woody perennial	Slow	Leaves
	Quercus hemisphaerica	Woody evergreen	—	Leaves
	Quercus nigra	Woody semievergreen	Fast	Leaves
	Quercus robur	Woody perennial	Intermediate	Leaves
	Quercus rubra	Woody perennial	Fast	Leaves
	Quercus velutina	Woody perennial	Intermediate	Leaves, stems
Salicaceae	*Populus angustifolia*	Woody perennial	Fast	Leaves
	Populus balsamifera	Woody perennial	Very fast	Shoots
	Populus fremontii	Woody perennial	Fast	Leaves
	Populus tremuloides	Woody perennial	Very fast	Shoots
	Salix alba	Woody perennial	Ornamental	Leaves
	Salix alaxensis	Woody perennial	Fast	Shoots
	Salix babylonica	Woody perennial	Ornamental	Leaves
	Salix cinerea	Woody perennial	Fast	Leaves
	Salix lasiolepis	Woody perennial	Fast	Leaves
	Salix sitchensis	Woody perennial	Fast	Leaves
Ulmaceae	*Ulmus americana*	Woody perennial	Intermediate-fast	Stems
	Ulmus rubra	Woody perennial	Intermediate	Stems
Vitaceae	*Vitis vinifera*	Perennial vine	Fast, crop	Leaves
Rutaceae	*Citrus sinensis*	Woody perennial	Crop	Leaves

Same/Subsequent Season	Mechanism/Type of Effect	Herbivore Affected	Reference
Same and subsequent	Resistance	Caterpillars	Rhoades 1983; Myers and Williams 1987
Subsequent	Resistance, behavior	Caterpillars, hares	Werner 1979; Bryant 1981
Same	Behavior	Caterpillars	Hartley and Lawton 1987
Same and subsequent	Resistance	Caterpillars	Wallner and Walton 1979
Same and subsequent	Resistance	Many	Haukioja and Niemelä 1979
Same	Behavior	Caterpillars	Edwards et al. 1986
Same	Resin, callus	Cicada eggs	White 1981
Same	Behavior	Caterpillars	Edwards et al. 1986
Same	Callus	Cicada eggs	White 1981
Same	Behavior	Caterpillars	Edwards et al. 1986
Same	Abscission	Leaf miners	Auerbach and Simberloff 1989
Same	Abscission, numbers	Leaf miners	Faeth n.d.
Same	Resistance	Caterpillars	Roland and Myers 1987
Same	Abscission	Leaf miners	Faeth et al. 1981
Same	Abscission	Leaf miners	Faeth et al. 1981
Same and subsequent	Resistance	Caterpillars	West 1985; Hunter 1987
Same	Resistance	Caterpillars	Rossitor et al. 1988
Same	Resistance, resin, callus	Caterpillars, cicada eggs	Wallner and Walton 1979, White 1981
Same	Abscission	Gall aphids	Williams and Whitham 1986
Same and subsequent	Resistance	Caterpillars	Clausen et al. 1991
Same	Abscission	Gall aphids	Williams and Whitham 1986
Same and subsequent	Resistance	Caterpillars	Clausen et al. 1991
Same	Resistance	Leaf beetles	Raupp and Denno 1984
Same	Behavior	Mammals	Bryant 1987
Same	Resistance	Leaf beetles	Raupp and Denno 1984
Same	Behavior	Caterpillars	Edwards et al. 1986
Same	Abscission	Caterpillars	Preszler and Price 1993
Same	Resistance	Caterpillars	Rhoades 1983
Same	Callus	Cicada eggs	White 1981
Same	Callus	Cicada eggs	White 1981
Same	Resistance	Mites	English-Loeb and Karban 1988
Same	Resistance	Mites	Henderson and Holloway 1942

Table 4.1 continued

Family	Species	Growth Form	Relative Rate of Growth	Tissue Responding
Burseraceae	*Bursera schlechtendalii*	Succulent, deciduous shrub	Slow	Leaves
Anacardiaceae	*Rhus glabra*	Perennial shrub	Fast	Shoots
Umbelliferae	*Pastinaca sativa*	Biennial	Fast	Leaves
Caprifoliaceae	*Sambucus nigra*	Woody perennial	Fast	Leaves
Cucurbitaceae	*Cucumis sativus*	Perennial	Crop	Leaves
	Cucurbita moschata	Annual or perennial	Crop	Leaves
	Cucurbita pepo	Annual	Crop	Leaves
Asteraceae	*Chromolaena colorata*	Perennial	Fast	Leaves
	Chrysanthemum carinatum	Herbaceous perennial	Crop	Leaves
	Chrysothamnus nauseosus	Woody perennial	Slow	Shoots
	Cirsium kagamontanum	Herbaceous perennial	Fast	Leaves
	Erigeron glaucus	Herbaceous perennial	Slow	Shoots
	Helianthus annuus	Annual	Fast	Leaves
	Solidago altissima	Herbaceous perennial	Fast	Stems
Cyperaceae	*Carex bigelowii*	Perennial sedge	—	Leaves
	Eriophorum angustifolium	Perennial sedge	—	Stems
Poaceae	*Andropogon greenwayi*	Graminoid	—	Leaves
	Eustachys paspaloides	Graminoid		Leaves
	Panicum coloratum	Graminoid		Leaves
	Phragmites australis	Graminoid reed	Fast	Shoots
	Zea mays	Graminoid	Crop	Leaves

expected to experience more radical environmental changes during their lifetimes compared to the relatively less temporally variable conditions experienced during a lifetime by short-lived individuals (Levins 1968; Lloyd 1984). If the environment changes often relative to the generation time of the plant, adaptation to the best character state may not be possible. Selection may instead favor the ability to switch between states during development in response to current conditions. In other

Same/ Subsequent Season	Mechanism/ Type of Effect	Herbivore Affected	Reference
Same	Resin	Beetles	Becerra 1994
Subsequent	Damage	Deer	Strauss 1991
Same	Resistance	Caterpillars	Zangerl 1990
Same	Behavior	Caterpillars	Edwards et al. 1986
Same	Resistance	Mites	A. A. Agrawal, personal communication
Same	Behavior	Beetles	Carroll and Hoffman 1980
Same	Behavior	Beetles	Tallamy 1985
Same	Resistance	Caterpillars	Marutani and Muniappan 1991
Same	Behavior	Flies, mites	van de Vrie et al. 1988
Same	Hypersensitive	Galling fly	Fernandes 1990
Same	Resistance, behavior	Beetles	Ohgushi 1992
Same	Resistance	Spittlebugs	Karban 1986b
Same	Behavior	Beetles	Olson and Roseland 1991
Same	Hypersensitive	Stem galler	Anderson et al. 1989
Subsequent	Resistance	Mammals	Seldal, Andersen, et al. 1994
Subsequent	Behavior	Mammals	Seldal, Andersen, et al. 1994
Same	Silicification	Mammals	McNaughton and Tarrants 1983
Same	Silicification	Mammals	McNaughton and Tarrants 1983
Same	Silicification	Mammals	McNaughton and Tarrants 1983
Subsequent	Resistance	Stem-boring caterpillars	Mook and van der Toorn 1985
Same	Resistance	Aphids	Morse et al. 1991

words, selection may favor annuals that produce many variable progeny in response to temporal variation and long-lived plants that are capable of changing phenotypes in response to the same variability. Mutikainen and Walls (1995) tested this hypothesis by comparing the responses of an annual and a perennial species of nettles (*Urtica* spp.) to artificial clipping; changes in trichome density were stronger in the perennial than in the annual, although this was based on only one species of each,

Table 4.2. Plants for which induced susceptibility has been reported

Family	Species	Growth Form	Relative Rate of Growth	Tissue Affected
Pinaceae	*Picea abies*	Evergreen	Fast	Leaves
	Picea sitchensis	Evergreen	Very fast	Leaves
Ochnaceae	*Ochna pulchra*	Woody perennial	Slow	Leaves
Malvaceae	*Gossypium hirsutum*	Woody perennial	Crop	Shoots
Brassicaceae	*Brassica oleracea*	Annual	Crop	Roots
Ebenaceae	*Euclea natalensis*	Evergreen	Slow	Leaves
Oleaceae	*Fraxinus excelsior*	Woody perennial	Slow-intermediate	Roots, leaves
Bignoniaceae	*Crescentia alata*	Semievergreen perennial	Slow	Leaves
Rosaceae	*Fragaria grandiflora*	Herbaceous perennial	Crop	Leaves
	Malus domesticus	Woody perennial	Crop	Leaves
	Prunus serotina	Woody perennial	Very fast	Leaves
Leguminosae	*Acacia karroo*	Deciduous, woody perennial	Fast	Shoots
	Burkea africana	Woody perennial	Slow	Leaves
	Phaseolus vulgaris	Herbaceous annual	Crop	Leaves
Betulaceae	*Alnus rubra*	Woody perennial	Fast	Leaves
	Betula pendula	Woody perennial	Fast	Shoots
	Betula pubescens	Woody perennial	Slower	Shoots
Fagaceae	*Quercus emoryi*	Evergreen	Slow	Leaves
	Quercus garryana	Woody perennial	Slow	Leaves
	Quercus nigra	Semievergreen	Fast	Leaves
	Quercus robur	Woody perennial	Intermediate	Leaves
	Quercus velutina	Woody perennial	Intermediate	Leaves
Salicaceae	*Salix lasiolepis*	Woody perennial	Very fast	Shoots
Urticaceae	*Urtica dioica*	Herbaceous perennial	Intermediate	Shoots
Myrtaceae	*Eucalyptus blakeyi*	Evergreen	Fast	Shoots
	Eucalyptus spp.	Evergreen	Fast	Leaves
	Eucalyptus spp.	Evergreen	?	Leaves

Same/Subsequent Season	Mechanism/Type of Effect	Herbivore Affected	Generalist/Specialist	Reference
Same	Hypersensitive	Aphids	Specialist	Fisher 1987
Same	Hypersensitive	Aphids	Specialist	Fisher 1987
Subsequent	% damage	Insects	?	Bryant, Heitkonig, et al. 1991
Same	Resistance	Mites	Generalist	Karban and Niiho 1995
Same	Behavior	Root flies	Specialist	Baur et al. 1996
Subsequent	% damage	Insects	?	Bryant, Heitkonig, et al. 1991
Same	% damage	Caterpillars beetles	?	Foggo and Speight 1993; Foggo et al. 1994
Same season, next generation	Resistance	Beetles	Specialist	Rockwood 1974
Same	Behavior	Mites	Generalist	Kielkiewicz 1988
Subsequent	Resistance	Caterpillars	Generalist	Roland and Myers 1987
Subsequent	Resistance	Beetles	Generalist	Schultz and Allen 1977
Same season, next generation	Numbers	Psyllids	Specialist	Webb and Moran 1978
Subsequent	% damage	Insects	?	Bryant, Heitkonig, et al. 1991
Same	Resistance	Mites	Generalist	English-Loeb and Karban 1991
Subsequent	Resistance	Caterpillars	Generalist	Williams and Myers 1984
Subsequent	% eaten	Moose	Generalist	Danell et al. 1985
Subsequent	% eaten	Moose	Generalist	Danell et al. 1985
Same	Damage	Arthropods	?	Faeth 1992b
Subsequent	Resistance	Caterpillars	Generalist	Roland and Myers 1987
Same	Resistance	Caterpillars	Generalist	Auerbach and Simberloff 1984
Subsequent	Resistance	Caterpillars	Generalist	Hunter 1987
Subsequent	Resistance	Caterpillars	Generalist	Wallner and Walton 1979
Same season, next generation	Resistance	Sawflies	Specialist	Craig et al. 1986
Subsequent	Behavior	Caterpillars	Specialist	Pullin 1987
Subsequent	Resistance	Many	?	Landsberg 1990
Subsequent	Numbers, damage (no data)	Sawflies	Generalist	Carne 1965
Same season, next generation	Resistance	Caterpillars	Generalist	Wallace 1970

Table 4.2. continued

Family	Species	Growth Form	Relative Rate of Growth	Tissue Affected
Aquifoliaceae	*Ilex opaca*	Evergreen	Slow	Shoots
Anacardiaceae	*Rhus glabra*	Perennial shrub	Fast	Shoots
Asteraceae	*Helianthus annuus*	Annual	Fast	Leaves and shoots
	Senecio jacobaea	Herbaceous perennial	Fast	Leaves
Cucurbitaceae	*Cucumis sativus*	Perennial	Crop	Leaves
Poaceae	*Agropyron desertorum*	Graminoid		Leaves
	Elymus lanceolatus	Graminoid		Leaves
	Thinopyrum intermedium	Graminoid		Leaves
	Pseudoroegneria spicata	Graminoid		Leaves

and the investigators did not determine whether trichome density resulted in greater resistance or defense.

Induced responses that are not systemic may allow individuals (single genets) to respond to spatially variable environments with different phenotypes, each appropriate for particular locations. This may be particularly advantageous for clonal plants and larger trees made up of many modular units. For example, the bottom branches of a tree may be moderately likely to experience browsing by large mammals, but the top branches are far less likely to face this form of selection (Milewski et al. 1991; Myers and Bazely 1991). Branches on the edge of a vegetation patch may experience more herbivory than those in the middle. The localized induction of spines may allow each branch to present a phenotype appropriate for such spatially variable probabilities of attack. Following a similar line of reasoning, Harvell (1990a, 1990b) observed that induced resistance in marine invertebrates was generally found for clonal or colonial animals. She argued that localized induced responses were particularly advantageous for large colonies, allowing them to partition and localize their responses within the colony because selection by predators was often localized.

Only a very small fraction of the plants in table 4.1 are annuals; the vast majority are large and long-lived. Ecologists became interested in.

Same/ Subsequent Season	Mechanism/Type of Effect	Herbivore Affected	Generalist/ Specialist	Reference
Same	Resistance	Agromyzid fly	Specialist	Potter and Redmond 1989
Same season, spring-summer delay	Damage	Cerambycid beetles	Specialist	Strauss 1991
Same	Behavior	Grasshoppers	Generalist	Lewis 1984
Same season, next generation	Resistance	Caterpillars	Specialist	Wilcox and Crawley 1988
Same	Damage	Beetles	Specialist	A. A. Agrawal, personal communication
Same	Resistance	Aphids	Generalist	Messina et al. 1993
Same season, next generation	Resistance	Aphids	Generalist	Messina et al. 1993
Same season, next generation	Resistance	Aphids	Generalist	Messina et al. 1993
Same	Resistance	Aphids	Generalist	Messina et al. 1993

induced resistance initially as a means of explaining multiyear fluctuations in herbivore numbers. Perennial plants are much more likely to exhibit long-term induced resistance capable of producing such patterns. However, even if only short-term induced responses are considered, annual plants are dramatically underrepresented in table 4.1. This trend is also apparent if crop plants are considered; although many crops are derived from ancestors with annual life cycles, few of the crops that have been found to show induced resistance are annuals in nature.

A comparison of plant species for which induced resistance has been found (table 4.1) and those for which induced susceptibility has been found (table 4.2) suggests that the time frames involved in the two effects may be different. Most examples of induced resistance occurred in the same year as the damage. In contrast, many of the examples of induced susceptibility were delayed, occurring in the seasons following damage. Several of those showing induced susceptibility that occurred in the same season were only apparent for the herbivore generations that followed the inducers (delayed according to Haukioja's terminology [1990a]). Responses that resulted in induced susceptibility were delayed in many cases because they resulted from plants that produced very susceptible tissues at atypical times when herbivores were in a position to benefit. The process of regrowth necessarily introduces a delay.

Part of the conventional wisdom of workers in this field is that induced resistance is not found in evergreen trees. Therefore, it was surprising to note how well represented the evergreen habit in general, and the family Pinaceae in particular, are in table 4. 1. However, many of the examples that involve members of the Pinaceae relied on increased resin flow or qualitative changes in resin chemistry. Hypersensitive responses and abscission of infested tissues are two other mechanisms that were often described for evergreens. For example, Faeth et al. (1981) compared the propensity of three oak species in Florida to abscise leaves in response to attack by leaf miners. An evergreen species had the highest rate of abscission, followed by a semievergreen species, and a deciduous species had the lowest rate. If any real distinction exists between deciduous trees and evergreens, it is in the mechanisms of induced resistance. Evergreens exhibited changes in quality and quantity of resins, leaf abscission, and hypersensitivity; the mechanisms of induced resistance that are common for evergreens are different than those commonly reported for other plant-growth forms.

The number of examples of induced resistance for evergreens that had effects in subsequent seasons are few, although there are several exceptions to this pattern (Overhulser et al. 1972; Leather et al. 1987; Cook and Hain 1988; Karban 1990). This oft-cited "observation" that evergreens do not exhibit long-term induced resistance has been the basis of several of the theories attempting to explain a mechanism for the induced resistance observed in severely defoliated deciduous trees in the years following defoliation (Bryant et al 1983, 1988; Tuomi, Niemelä, Chapin, Bryant, and Sirén 1988). These theories assume that carbon-based defenses are responsible for the long-term induced resistance found in many deciduous trees (see section 3.3.3.1). According to the argument, deciduous trees store carbon in stems and roots, whereas evergreens store more carbon and nutrients in their leaves. When deciduous trees are defoliated by herbivores, relatively more nutrients than carbon are lost from leaves, producing a surplus of carbon, which the plant accumulates as carbon-based secondary compounds, resulting in induced resistance. When evergreens are defoliated, carbon reserves stored in the leaves are drastically reduced, and such trees have been hypothesized to have a limited capacity to accumulate carbon-based secondary chemicals. More recently, Bryant, Heitkonig, et al. (1991) have considered the induced responses of six tree species following severe defoliation. As predicted, the one evergreen species failed to show induced resistance during the following growing season. However, two deciduous species also failed to show induced resistance, suggesting that the distinction between the deciduous and evergreen habit may be less important than the plant's inherent growth rate in determining whether long-term induced resistance will be found (see section 4.2.2).

Induced susceptibility, like induced resistance, has been reported for very diverse species (table 4.2). The fact that many of the same species have shown induced susceptibility as well as induced resistance indicates the futility of generalizations based on taxonomic labels (Janzen 1979). Other factors (see below) clearly have the potential to switch the effects of damage for a given plant species from induced resistance to susceptibility or vice versa.

As was the case for induced resistance, most of the plants that show induced susceptibility are woody perennials (table 4.2). There appears to be a higher overall percentage of evergreen species showing induced susceptibility than showing induced resistance. We have resisted the temptation to treat the species listed in these tables as independent samples and to conduct statistical analyses on these data. Such analyses include implicit assumptions that samples were selected independently and randomly from the population, assumptions that were probably violated by these data.

4.2.2 Plant Growth Rate

Plant growth rate figures prominently in current theories about when and where plant defenses against herbivores will be found (Coley et al. 1985). Overall commitments to defenses are predicted to be greater for slow-growing plants, which can less afford to regrow following herbivory. However, induced resistance is expected to be stronger in fast-growing plants for at least five reasons (see also Herms and Mattson 1992). The first two arguments are observations or proximal explanations, while the last three arguments are evolutionary rationales for why we might expect these patterns.

1. Many of the mechanisms that provide induced resistance are associated with growth processes. Many secondary chemicals implicated in plant resistance, such as phenolics, terpenes, PIs, and alkaloids, are synthesized at greater rates in fast-growing plants. Similarly, some of the specialized tissues that are involved in induced resistance, such as spines, secretory glands, resin ducts, lactifers, and so forth, are differentiated following periods of cell growth and enlargement. This argument is similar to that presented for the effects of plant ontogeny on induced resistance, developed in more detail in section 4.2.3. The essence of this argument is that slow-growing plants don't have the machinery required to induce resistance.

2. Whether a given change results in successful resistance against her-

bivores probably depends on the speed and magnitude of the induced response. Slow-growing plants may have too slow a rate of metabolic activity to mount a response sufficiently quickly to outpace the attacker. This may be true for resistance involving many mechanisms, including de novo synthesis of secondary chemicals, accumulation of chemicals produced elsewhere in the plant, hypersensitive responses, some resin responses, callus formation, and so forth. The essence of this argument is that, even if slow-growing plants are capable of induced responses, these responses are too slow to provide effective induced resistance.

3. Plasticity of many traits (perhaps including induced resistance) is generally associated with fast-growing plants (Grime 1979; Grime et al. 1986). According to this argument, plants that have evolved in productive habitats grow rapidly. When exposed to a stress, they respond rapidly with potentially large morphological changes that allow them to continue to maximize resource capture and production. Such responses may have little value under conditions of more consistent stress that some slow-growing plants have adapted to. In stressful habitats and others where plants grow slowly, plants may have little opportunity for phenological or morphological responses. Differentiating tissue, which has the greatest ability for morphological plasticity, is uncommon in slow-growing plants (see argument 1). When plasticity is found in slow-growing plants, it most often involves rapid and reversible physiological changes. These arguments were developed by Grime to explain plasticity in response to stresses caused by local variation in mineral nutrients and light, although they may also apply to "herbivore stress." This argument may include reasons 1 and 2 above, although it was developed to explain a different set of empirical observations.

4. Selection for fast-growing plants may result in species that are highly competitive at capturing light and other resources but are generally poorly defended (Bryant et al. 1983; Coley et al. 1985). Fast-growing species allocate resources to growth rather than to constitutive defenses against herbivores. However, when attacked, fast-growing plants may be better able to regrow (Louda et al. 1990), which may provide resistance, or to respond in other ways that provide induced resistance (Coley 1987). This argument assumes that growth processes and defense compete for the resources that the plant can allocate (Coley et al. 1985; van der Meijden et al. 1988; Herms and Mattson 1992), although the validity of this assumption has been questioned (Rosenthal and Kotanen 1994). Induced defenses may represent a less costly alternative for fast-growing species, which we might expect to be under strong selection to allocate resources to growth. The essence of this argument is that induced resistance may allow fast-growing plants to pay less "opportunity cost" for

defense. In other words, induced resistance may allow plants to save the costs associated with defense when they are not needed and instead to allocate resources to growth in environments that place a great reward on rapid growth.

5. Slow-growing plants that already have high levels of constitutive defenses may gain little from inducible defenses (Mattson, Lawrence, et al. 1988; Herms and Mattson 1992). Inducible systems of defense may be redundant and hence provide little additional benefit to plants with constitutive defenses. Inducible defenses may also be relatively less effective against herbivores that already have evolved mechanisms to deal with the constitutive defenses of a particular plant species.

If this explanation is generally true, plants with high levels of constitutive resistance are not likely also to have high levels of induced resistance. The evidence for negative trade-offs between inducible and constitutive defenses is equivocal. For conifers, induced accumulations of monoterpenes is negatively correlated with constitutive levels (Lewinsohn et al. 1991; Raffa 1991; Lerdau et al. 1994). Those monoterpenes present in the largest quantities were induced the least, and the rarer monoterpenes were induced more (Lerdau et al. 1994). Similarly, those tree species with high levels of constitutive resistance (e.g. many pines) had smaller induced responses, whereas those with minimal constitutive resistance (e.g. many firs, spruces, hemlocks, and cedars) relied almost exclusively on induced responses (Lewinsohn et al. 1991; Raffa 1991). Inducible production and constitutive production of furanocoumarins are negatively correlated among the different tissues of wild parsnip plants (*Pastinaca sativa;* Zangerl and Rutledge 1996). Leaves and roots were the most inducible tissues, but they had the lowest constitutive titers. Fruits had high constitutive titers but were not very inducible.

However, observations of most other plants do not conform to the notion of negative correlation between constitutive and induced defenses. Lupine leaves with the highest initial levels of alkaloids were the ones with the greatest induced increases relative to undamaged controls (Johnson, Rigney, et al. 1989). Genetic correlations between levels of constitutive and induced furanocoumarins in wild parsnip individuals were positive rather than negative (Zangerl and Berenbaum 1990). Potato clones that displayed constitutive expression of PI genes also showed much stronger induced responses to wounding (Hildmann et al. 1992). Levels of constitutive resistance were positively associated with wound-induced increases in tannins for three species of African karoo shrubs (Stock et al. 1993). Families of *Cynoglossum officinale* showed no relationship between constitutive and inducible levels of pyrrolizidine alkaloids (van Dam and Vrieling 1994).

Constitutive and induced resistance were also found to be positively associated in bioassay studies of induced resistance. In a comparison of resistance of two soybean cultivars to spider mites, the variety with more constitutive resistance also showed a stronger hypersensitive response to mite damage (Hildebrand, Rodriguez, Brown, and Volden 1986). The more resistant of two strawberry cultivars showed a greater induced response to mite damage, measured in terms of mite-feeding preference (Kielkiewicz 1988). Induced resistance in creosote bushes appeared to be positively, rather than negatively, related to the levels of constitutive resistance (Ernest 1994). There was no evidence of negative correlations between constitutive and induced resistance to spider mites for varieties of cultivated cotton (*Gossypium hirsutum;* Brody and Karban 1992). In that study, constitutive resistance to spider mites was positively associated with induced resistance to verticillium wilt. When different species of *Gossypium* were considered, no evidence for a negative correlation between constitutive and induced resistance to spider mites was found (Thaler and Karban 1997). In summary, the evidence to date does not support the notion that selection has "discouraged" redundant capabilities for both constitutive and inducible resistance.

In an effort to examine possible relationships between plant growth rate and induced resistance, we have attempted to categorize the inherent growth rates of the plant species listed in tables 4.1 and 4. 2. To do this objectively, we had hoped that some independent source could be found that would provide an index of growth for many of the species we considered. Unfortunately, a single independent ranking could not be found. We have therefore used several different sources and classified plants as slow, intermediate, fast, and very fast. When the author of a particular study explicitly provided such a characterization, we used this estimate. For North American species we also consulted Dirr (1975), Hightshoe (1988), and Burns and Honkala (1990); Bradley et al. (1966) and Evans (1984) were used for British species. This process was less straightforward than we had imagined. For instance, both Danell et al. (1985) and Neuvonen et al. (1987) considered *Betula pubescens* to be slow-growing. They interpreted the preferences and performances of moose and caterpillars respectively on this species in terms of induced responses characteristic of a slow-growing plant species. In contrast, Bryant et al. (1988; Bryant, Danell, et al. 1991) considered *Betula pubescens* to be fast-growing and interpreted the long-term induced resistance that they documented for this species to be a prime example of how fast-growing deciduous trees respond. Characterization based on yield tables was also less straightforward than we had naively anticipated. Some species grow very rapidly when they are young; thereafter growth rate is

Table 4.3 Patterns of induced responses and plant growth rate

	Fast Growing	Slow Growing
Induced resistance	Common	Less common
Delayed	Common	Rare
Herbivore behavioral responses	Common	Common
Abscission, hypersensitive	Common	Common
Resin, callus	Common	Common
Presumed secondary chemicals	Common	Rare
Localized	Common	Common
Systemic	Common	Rare
Induced susceptibility	Common	Common
Delayed	Less common	Common

much reduced. Other species grow slowly for the first few years and then accelerate their growth. In cases such as these we attempted to consider the overall rates of growth until maturity. For crop plants we have not included a relative rate of growth, although we assume that most crops have undergone selection for very rapid growth, greatly exceeding their wild ancestors. Of course, any inferences drawn regarding relationships between plant growth rates and likelihood of induced resistance or susceptibility are limited by the reliability of our estimates of growth rates.

Most of the species for which induced resistance have been reported were fast growers, although we found many examples of intermediate and slow growers as well (table 4.1). For delayed induced resistance the bias toward fast-growing species was more marked (although there were several exceptions). Most of the examples of induced resistance reported for plants with slow and intermediate rates of growth were localized responses that involved behavioral choices by the herbivores, abscission of damaged plant parts, hypersensitive responses, accumulation of resin at the site of injury, and callus formation. Systemic responses and those involving the presumed induction of secondary chemicals that reduced herbivore performance were less commonly reported for slow-growing plants. These patterns are summarized in table 4.3, although our conclusions are preliminary. As noted above, categorizing the growth rates of species is often subjective. We also have no null hypothesis with which to compare our distribution of species that showed induced resistance. In other words, if we selected plant species at random, would most of them be classified as fast-growing? As such, these patterns should be regarded as suggestive, at best.

Comparisons between different species and studies are difficult to interpret because of the complications described above. Therefore, we sought individual studies that included several different species. These

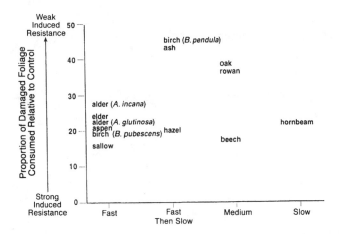

Fig. 4.1 The relationship between plant growth rate and the strength of induced resistance (data from Edwards et al. 1986). Plant species showing evidence of strong induced resistance lost little damaged foliage to herbivores relative to undamaged control foliage.

studies rely on bioassay results; they cannot tell us how much different plants respond but only how much those responses affected herbivores. Bryant, Heitkonig, et al. (1991) found evidence for induced resistance that affected herbivore performance in three species of fast-growing trees but not in three species of slow-growing trees. However, results of Neuvonen et al. (1987) do not conform to the generalization. Induced resistance that affected herbivore performance in a subsequent season was found to be strong for one slow-growing species and one fast-growing species; induced resistance was weaker for another slow-growing species and could not be detected for a second fast-growing tree.

Several studies that considered localized responses of various tree species did not find consistent differences in herbivore behavior associated with rate of plant growth. Edwards et al. (1986) examined the behavioral responses of a generalist caterpillar to leaves of thirteen tree species. They compared the consumption of induced foliage (damaged or adjacent) to consumption of undamaged foliage. We reanalyzed their data and found that plant growth rate was not particularly effective at explaining patterns of foliage consumption (figure 4.1).

In another study White (1981) compared the responses of fourteen species of trees to oviposition by periodical cicadas (*Magicicada* spp.). The cicada eggs remained within twigs for six to eight weeks before hatching. Some trees responded by producing callus or resin at the site of injury; both responses were effective at killing some of the cicada

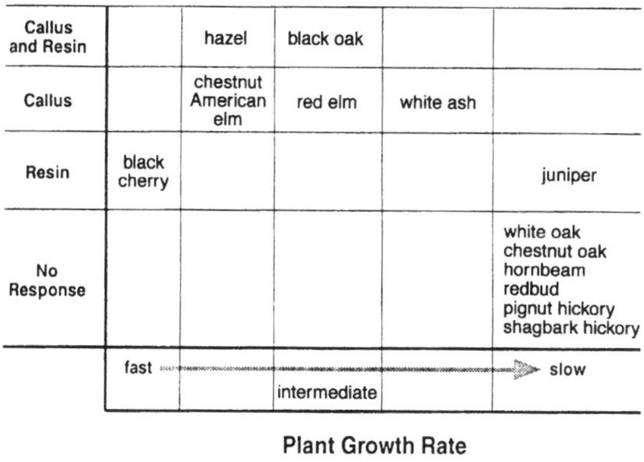

Callus and Resin		hazel	black oak		
Callus		chestnut American elm	red elm	white ash	
Resin	black cherry				juniper
No Response					white oak chestnut oak hornbeam redbud pignut hickory shagbark hickory
	fast ··⟫ slow				
			intermediate		

Plant Growth Rate

Fig 4.2 The relationship between plant growth rate and induced production of callus or resin in response to cicada oviposition (data from White 1981).

eggs. These data were consistent with the notion that fast-growing species are more likely to employ effective induced responses (figure 4.2). Since both resin production and callus formation kill cicada eggs only if the egg nests are surrounded during the six to eight weeks available, it stands to reason that these forms of resistance, in particular, are associated with rapid growth. Juniper, a slow-growing conifer with a preformed system of resin channels, is the exception to this generality.

In contrast, the list of species for which induced susceptibility has been reported includes roughly equal numbers of plants that grow slowly and rapidly (table 4.2). Induced susceptibility with effects that were not apparent until the subsequent herbivore generation (delayed) or growing season (long-term induction) tended to be found for species that grew slowly. Rapid effects were much less commonly reported for induced susceptibility (table 4.2) than for induced resistance (table 4.1; see discussion in section 4.2.1).

4.2.3 Plant Ontogeny

Plants, like animals, undergo a series of developmental changes as they mature. The selective environment (competitors, pathogens, herbivores, mutualists) for a seedling is very different than the selective regime that a mature plant experiences. The resources available and the specialized structures and tissues that could be used for defense against herbivores change as the plant develops. These changes occur on two scales, over the course of each growing season and over the life span of perennial plants.

Essentially nothing is known about ontogenic changes of relevance to induced resistance that occur very early in the plant's life as the seed develops and grows. During this interval the plant switches from being heterotrophic to being autotrophic. It is reasonable that ontogenic changes in how the plant allocates resources to induced resistance may be even greater during these early stages than those described below for more mature plants.

Induced responses may be constrained by the plant's developmental stage. Many of the secondary chemicals that are thought to be involved in induced resistance are synthesized in specialized tissues that are active only during periods of new growth. The correlation between young, actively growing tissue and the potential to produce secondary metabolites is widespread, although other factors and idiosyncrasies also affect production. For example, the ability to induce PIs in response to wounding of tomato leaves was found to peak while plants are young and to decrease as they mature (Wolfson and Murdock 1990). In this case the plants became less responsive to the same amount of signal or amount of wounding as they matured (Wolfson 1991; Jongsma et al. 1994). In tobacco, plants stop synthesizing new nicotine in response to wounding after they start flowering (Baldwin and Schmelz 1996). Instead, if they are wounded, they move the available nicotine to young leaves or reproductive parts (Baldwin et al. 1997; Ohnmeiss et al. n.d.). Similarly, induced phytoalexin responses were often associated with periods when new leaf tissue was being produced for 169 plant species found in a survey conducted in Brazil (Braga et al. 1986). In a more detailed study of induced phytoalexin production in cell cultures of soybean and cotton, Apostol et al. (1989) found that rates of synthesis are greatest during the exponential growth phase, intermediate during phases of intermediate growth, and negligible during periods of little cell growth. Extrapolating from patterns of inducible metabolism from tissue culture to whole plants is uncertain, blurring the interpretation of these results. Induced synthesis of phenolics appears to be age-dependent but with a complicated relationship. Damage to *Betula allegheniensis* caused decreases in phenolic content of the youngest, smallest leaves and increases in the phenolic content of older leaves (Baldwin and Schultz 1984). When seedlings and pole-sized trees of *Pinus ponderosa* had 50% of leaves removed artificially, concentrations of procyanidin (tannins) in the remaining needles increased for seedlings but decreased for trees (Wagner 1988).

Preformed induced responses have also been found to be more common in young tissues. Many plant species produce hydrogen cyanide in response to wounding (Jones 1972). The enzyme and substrate involved are compartmentalized separately in intact plant tissues; upon wounding, the enzyme and substrate come into contact, releasing hydrogen cyanide. For several species reviewed by Jones (1972), the ability to release cyanide is maximal at the seedling stage.

Conifers may be less constrained by ontogeny in their responses to bark beetles and other herbivores than the examples considered above (Lerdau et al. 1994). The storage organs for monoterpenes are produced early in ontogeny and are rapidly filled; growing leaves were found to have the highest concentrations of constitutive monoterpenes. During periods of growth in spring, however, monoterpene concentrations can increase in foliage that has already fully expanded. Lodgepole pine (*Pinus contorta*) and grand fir (*Abies grandis*) trees showed the highest capacity for defense while they were of intermediate age, with lower induced responses before maturity and after reproductive maturity (Raffa and Berryman 1982a, 1987). Young trees were poor hosts for other reasons, so that bark beetles often showed preferences for older or highly stressed individuals. In maritime pines (*Pinus pinaster*) the cells that release resin acids are differentiated early in development. These may be quiescent for some time, and following injury they reactivate, swelling, assuming the structural characteristics of active secretory cells, and increasing release of diterpene resins (Walter et al. 1989). In most plants, other than conifers, biosynthesis of terpenoids is restricted to specialized cells that are active only early in leaf development (Gershenzon and Croteau 1991). This developmental constraint may be responsible for the observation that terpenoids are less represented in cases of induced resistance and defense than would be expected considering their importance in constitutive defense (Gershenzon and Croteau 1991).

Induced resin production in birch trees appears to be more severely limited by plant ontogeny than similar processes in conifers. Resin production of *Betula pendula* is believed to be limited by the amount and size of the producing glands (Lapinjoki et al. 1991). These characteristics are determined at the time of tissue differentiation in the beginning of primary shoot growth. Density and size of glands constrain the ability of a shoot to change following damage. Excretion of resin was also found to peak coinciding with maximal apical growth of the juvenile shoot.

Many physical traits that could provide resistance cannot change once the tissue has matured (Myers and Bazely 1991). A shoot that has been browsed can produce thorns, spines, and hairs only on newly formed regrowth. Gray alders (*Alnus incana*) responded to folivory by beetles by producing new leaves that had high densities of trichomes, which beetles avoided (Baur et al. 1991). Increased density of trichomes was maximal for leaves that unfolded fourteen to twenty-one days after damage. No increase was possible for leaves that had already expanded prior to attack by beetles.

Selective abscission of leaves that contain insect galls or mines was also found to be higher for young leaves, early in the season. Leaves of *Salix lasiolepis* that had been attacked by leaf-mining caterpillars were more likely to be abscised early; this effect was much stronger (and statistically significant) for young leaves (Preszler and Price 1993). Similarly, leaves of narrowleaf cottonwoods (*Populus angustifolia*) with galls, containing aphids, were more likely to abscise prematurely; this effect was particularly apparent early in the season for young leaves (Williams and Whitham 1986).

Induced responses that involve reversion of mature plants to a juvenile stage are dependent on plant ontogeny, by definition. Reversion to the juvenile stage has been associated with both induced resistance and induced susceptibility. For example, browsing of several species of deciduous Alaskan trees returned them to the juvenile stage which afforded resistance to additional hare browsing (Bryant 1981). In contrast, heavy galling by sawflies maintained willow shoots at a juvenile stage, which was relatively susceptible to continued sawfly attack (Craig et al. 1986). Willow shoots (*Salix lasiolepis*) that were only lightly galled matured out of the juvenile stage and became far more resistant to sawflies.

As leaves mature, the relative concentrations of carbon and nitrogen may change. Tuomi et al. (1990) argued that, early in the growing season, young birch leaves contain relatively more of the plant's store of nitrogen and relatively less of the plant's carbon. Defoliation at this time is expected to produce a relative surplus of carbon and to result in induced resistance as the plant stores the excess carbon in phenolics that have deleterious effects on many herbivores. Later in the season, leaves and other plant parts have relatively equal amounts of carbon and nitrogen. Defoliation of mature leaves later in the season is not expected to cause a carbon surplus and hence no induced resistance. Observations of chemical patterns and responses of birch trees to defoliation were consistent with the hypothesis of Tuomi and coworkers.

Induced responses that are dependent on plant ontogeny should produce age-dependent patterns in induced resistance to herbivores. Indeed, induced resistance has been found to be strongly affected by plant ontogeny in many systems. Table 4.4 lists examples of induced resistance and induced susceptibility that were found to be dependent on plant or tissue age. Only examples in which effects on herbivores were documented are included in the table.

Several trends are apparent in the examples in table 4.4. When leaves were damaged, induced resistance tended to be stronger for young leaves, early in the growing season. This was true for the majority of studies, when the mechanism responsible was hypothesized to involve chemical changes induced by damage as well as for morphological responses such as premature leaf abscission or increased trichome density.

Several studies were not in accordance with this pattern. Death of leaf-mining caterpillars was greater on damaged leaves of *Quercus robur* later in the season (West 1985). This experiment was performed on only two trees, and the causes of herbivore death were not determined. Other studies describing damage that resulted in regrowth of foliage that was more vulnerable to herbivores because of ontogenic shifts were also not in accordance with the overall pattern of greater inducibility of young leaves early in the season. The mechanisms responsible for increased vulnerability of young leaves are probably different from the induced processes considered in chapter 3. Flea beetles much preferred young leaves of *Crescentia alata* trees to older leaves (Rockwood 1974). Hand removal of leaves caused branches to reflush younger leaves that were heavily attacked by the beetles and defoliated. This produced yet another cycle of young leaves and defoliation. Trees normally escaped this deleterious syndrome of repeated defoliation by producing new leaves only early in the season, when beetles were not present. Any event that caused the foliage to revert to a juvenile stage (experimental leaf removal in this case) stimulated repeated herbivory. Similar scenarios were described for *Eucalyptus* species (Carne 1965; Wallace 1970), *Quercus nigra* (Auerbach and Simberloff 1984), and *Ilex opaca* (Potter and Redmond 1989).

The shoot systems of woody plants proceed through a well-defined series of morphological and physiological changes as they mature, called phase changes (Poethig 1990). For example, cotyledonary leaves composed of embryonic tissue are different from leaves produced by nonreproductive seedlings, which differ in turn from leaves produced once the plant reaches reproductive maturity. Resistance to herbivores often changes dramatically as plants change from one phase to another. Any factors that cause plants to change phases have the potential to induce resistance or susceptibility. There are numerous examples of new, vigorous shoots (as opposed to leaves) being both more suitable and less suitable for herbivores. Obviously, this difference determines whether reversion to a juvenile state following herbivory induces resistance or susceptibility (table 4.4). Juvenile reversion probably results in leaves (as opposed to shoots) that are most often more susceptible. For example, in the study of herbivory on *Crescentia alata,* the new foliage was more suitable for herbivory than more mature leaves were (Rockwood 1974). Reversion to the juvenile stage may induce resistance to some herbivores

Table 4.4 Induced resistance/susceptibility dependent on plant ontogeny

Plant Species	Tissue Damaged	Possible Mechanism[a]	Same/Subsequent Season	Ontogenetic Relationship	Reference
Alnus glutinosa	Leaves	Chemical induction	Same	Resistance (preference) stronger early in season	Edwards et al. 1991
Betula pubescens	Leaves	Chemical induction	Same	Resistance stronger early in season	Haukioja and Niemelä 1979
Betula pubescens, B. pendula	Leaves	Chemical induction	Same	Resistance (preference) stronger early in season	Wratten et al. 1984
Citrus	Leaves	Chemical induction	Same	Resistance stronger for young leaves	Henderson and Holloway 1942
Crataegus monogyna	Leaves	Chemical induction	Same	Resistance (preference) stronger early in season	Edwards et al. 1991
Salix babylonica	Leaves	Chemical induction	Same	Resistance stronger for young leaves	Raupp and Sadof 1991
Vitis vinifera	Leaves	Chemical induction	Same	Resistance stronger for young leaves	Hougen-Eitzman and Karban 1995
Pinus contorta	Leaves	Chemical induction	Subsequent	Resistance stronger for leaves of young trees	Leather et al. 1987
Populus angustifolia	Leaves	Abscission	Same	Resistance stronger for small leaves	Williams and Whitham 1986
Salix lasiolepis	Leaves	Abscission	Same	Resistance stronger for young leaves	Preszler and Price 1993
Alnus incana	Leaves	Trichomes	Same	Resistance stronger for young leaves	Baur et al. 1991
Quercus robur	Leaves	?	Same	Resistance stronger for older leaves	West 1985
Crescentia alata	Leaves	Juvenile reversion	Same	Susceptibility stronger for young leaves	Rockwood 1974
Ilex opaca	Leaves	Juvenile reversion	Same	Susceptibility stronger for young leaves	Potter and Redmond 1989

Alnus crispa	Shoots	Juvenile reversion	Subsequent	Resistance stronger for young shoots	Bryant et al. 1983
Betula papyrifera	Shoots	Juvenile reversion	Subsequent	Resistance stronger for young shoots	Bryant et al. 1983
Populus tremuloides	Shoots	Juvenile reversion	Subsequent	Resistance stronger for young shoots	Bryant et al. 1983
Rhus glabra	Shoots	Juvenile reversion	Subsequent	Resistance to deer stronger for young shoots	Strauss 1991
Salix alaxensis	Shoots	Juvenile reversion	Subsequent	Resistance stronger for young shoots	Bryant et al. 1983
Betula pubescens	Buds	Juvenile reversion	Same	Susceptibility stronger for young buds	Haukioja et al. 1990
Rhus glabra	Shoots	Juvenile reversion	Same	Susceptibility to beetles stronger for young shoots	Strauss 1991
Salix lasiolepis	Shoots	Juvenile reversion	Same	Susceptibility stronger for young shoots	Craig et al. 1986
Urtica dioica	Shoots	Juvenile reversion	Subsequent	Susceptibility stronger for leaves of young shoots	Pullin 1987
Pinus contorta	Buds	Resin	Same	Resistance stronger for older buds	Harris 1960
Pinus nigra	Buds	Resin	Same	Resistance stronger for older buds	Harris 1960
Pinus sylvestris	Buds	Resin	Same	Resistance stronger for older buds	Harris 1960
Abies grandis	Phloem	Resin and hypersensitive response	Same	Resistance stronger for intermediate-aged trees	Raffa and Berryman 1987
Pinus contorta	Phloem	Resin and hypersensitive response	Same	Resistance stronger for intermediate-aged trees	Raffa and Berryman 1987

Note: Only examples in which effects on herbivores were documented are included.
[a] Chemical induction is the default when other mechanisms were not proposed.

and susceptibility to others within the same plant. For example, juvenile-form stump sprouts that regenerated after heavy grazing were less preferred by hares but more preferred by moose (Bryant, Danell, et al. 1991).

Although many specialized tissues, structures, and metabolic processes required for induced resistance were found in young, growing tissues, there were also a few counterexamples of older shoots and stems that are better equipped to induce resistance. These counterexamples all involve resin production by conifers. Buds of several species of pines develop larger resin canals as they mature. The greater induced resistance to shoot moths observed in older buds was attributed to this more mature bud morphology (Harris 1960). Induced resistance in conifers to bark beetles that attack the phloem is weak for young and old trees and peaks for healthy, intermediate-aged individuals (Raffa and Berryman 1987; Raffa 1991).

Most studies that have considered plant ontogeny have found that it exerts a large effect on induced plant responses. Experimental procedures and discussions of induced resistance must take the developmental stage of the plant into account.

4.2.4 Type and Extent of Damage

The type of damage that a plant suffers greatly affects the nature and strength of its response. Several examples of plant responses that differ depending on the specific attacking organism were discussed in section 3.2. Many other examples of responses that are idiosyncratic to particular types of damage (stimuli) could be described; the real challenge lies in trying to find patterns to the many examples.

The vast majority of reported cases of induced resistance followed damage to leaf tissue (roughly 70–80% of the examples listed in table 4.1). Shoots and stems are other plant tissues for which induced resistance was reported commonly (roughly 20–40% of the species listed in table 4.1). This bias probably reflects the interest that ecologists have had in folivory rather than in consumption of other organs such as roots or reproductive structures. To many ecologists, defoliation is almost synonymous with herbivory.

Induced resistance following damage to roots and to reproductive structures has been pretty much undescribed. It is ironic that induced resistance against cicadas has been found for stems of many plant species (White 1981; Karban 1983). This resistance involved resin and callus production of twigs in response to eggs; nothing is known about the relationships between cicada nymphs and root tissues despite the fact that cicadas spend the vast majority of their lives parasitizing roots, not stems.

Many of the plant systems for which induced chemical responses have been best characterized (i.e. cucurbitacin production in *Cucurbita* spp. and nicotine production in *Nicotiana* spp.) give hints that induced resistance may be very important in roots and reproductive organs. Concentrations of cucurbitacins in fruits and roots exceeded those found in leaves of wild species; nonetheless, it is leaf damage that has been examined (Tallamy and McCloud 1991). Similarly, farmers damage the flowering tops of tobacco plants (*Nicotiana tobacum*) to induce the greatest foliar concentrations of nicotine, and removal of flowering stalks induced the highest concentrations of alkaloids in a wild species of tobacco (Baldwin and Ohnmeiss 1993). Damage to roots of cultivated tobacco plants also increased the nicotine content of the leaves (Hanounik and Osborne 1977). Nonetheless, the effects of induced responses are known for only the folivores of tobacco (Baldwin 1988a, 1988b, 1988c). Given that nicotine biosynthesis is located in the roots, we suspect that a plant's ability to respond would rapidly diminish with increasing amounts of root herbivory.

There is limited evidence that induced resistance is possible against herbivores that attack roots and reproductive tissues of at least some plant species. Grape roots (*Vitis vinifera*) responded to the attack of *Phylloxera vastatrix* aphids by producing a local necrotic zone or "corky layer" surrounding each aphid (studies by Borner described in Muller 1959; Boubals 1966). These responses effectively isolated the *Phylloxera* from the rest of the healthy root tissue and prevented the insect from forming a root gall. The induction site was subsequently abscised.

Both localized and systemic responses to root-parasitic nematodes have been observed for numerous plant species (Sijmons et al. 1994). These responses often involve hypersensitive reactions, although increased gene expression, deposition of lignin, and synthesis of phytoalexins have also been reported in roots.

Damage to roots has also been found to make plants more susceptible to subsequent above-ground herbivory. For example, experimental root damage that was intended to mimic agricultural manipulations (ditching, close plowing) increased levels of damage caused by ash budmoth caterpillars (*Prays fraxinella*) to buds and weevil larvae (*Stereonychus fraxini*) to leaves of ash saplings (*Fraxinus excelsior;* Foggo and Speight 1993; Foggo et al. 1994).

Herbivores that attack plant reproductive structures may also experience induced resistance. For example, fruits of one-seed juniper (*Juniperus monosperma*) that were infested by seed-feeding insects were far more likely to be abscised than uninfested fruits (Fernandes and Whitham 1989). Mortality of larvae in abscised fruits exceeded that of larvae in fruits still attached to the tree. It may be plausible to argue that induced resistance against root feeders has not been reported because they are hidden underground where they remain inconspicuous and difficult to work with. However, this argument seems much less plausible for herbivory of flowers, fruits, and seeds, which are generally quite conspicuous. Perhaps examples of induced resistance involving these structures are underrepresented in the literature because they are truly rare in nature.

The defenses of reproductive parts appear to be less inducible over the short-term than foliar defenses. However, defenses of reproductive tissues are not completely fixed; they may be able to change developmental set points throughout the reproductive process and in response to damage to vegetative tissues (Baldwin and Karb 1995). Removing flowers did not increase nicotine concentrations of subsequently produced tobacco flowers. Similarly, leaf damage did not cause increases in nicotine in recently opened flowers or in the next three flowers produced. For flowers produced later, however, amounts of nicotine for plants that had experienced leaf damage were greater than for undamaged controls. In general, allocation of nicotine to reproductive parts was more strongly influenced by damage to vegetative tissues rather than reproductive tissues, and reproductive parts were generally less inducible than vegetative tissues.

Several related observations may help shed some light on the rarity of studies involving herbivory of reproductive parts. Coincidental with the differences in inducibility of plant parts are differences in the ability of plants to replace those parts. Generally, plants can compensate for losses of leaves more easily than losses of other tissues, especially reproductive organs (Inouye 1982; Whitham et al. 1991; Trumble et al. 1993). Compensation may reduce the influence of losses of leaf tissue on plant fitness relative to losses of other tissues. In addition, several reviews have concluded that plants have a much lower capacity to compensate for losses of any tissues to herbivores during the onset of reproduction (Bardner and Fletcher 1974; Trumble et al. 1993).

Perhaps most plants have not evolved substantial induced responses to damage to reproductive parts because such structures are (1) so valuable that they receive as much constitutive protection as the plant can afford, (2) so valuable that the plant allocates all available resources to their production, leaving little for induced defense, or (3) so ephemeral and easily destroyed that induced defenses are not effective. For example, floral parts of wild parsnip (*Pastinaca sativa*) contain three to thirty times the concentration of furanocoumarins as leaf tissue (Zangerl and Berenbaum 1990). Damage to leaves, but not flowers, induces production of additional furanocoumarins in leaves. When herbivores removed reproductive tissues, wild parsnip produced additional flowers (Hendrix and Trapp 1981), although it did not alter its allocation to defense of flowers. The fitness consequences of floral herbivory were much more severe than those of leaf damage for wild parsnip (Nitao and Zangerl 1987). In general, induced resistance is effective only if the tissue that is removed does not drastically reduce plant fitness.

The majority of studies that found induced susceptibility involved damage to leaf tissue (70–80% of species in table 4.2). Damage to shoots involved roughly 20–40% of the species in table 4.2. When all species were considered, knowledge of the tissue that was damaged provided no indication about whether resistance or susceptibility was likely to result.

Haukioja (1990a, 1990b) was puzzled when he compared his results that caterpillar feeding induced resistance in mountain birch with those of Swedish ecologists who found that moose feeding induced susceptibility in the same species. Simulated or real moose browsing made trees better food for moose. Browsed trees supported larger populations of aphids, psyllids, leaf miners, leaf gallers, other leaf feeders, and ants (Danell et al. 1985; Danell and Huss-Danell 1985). Caterpillars feed on leaf lamina, moose on buds and shoots. This suggested that the type of tissue that is removed affects the outcome of the induced response. Haukioja and associates (1990) tested this notion by artificially clipping twigs in winter, much as a moose would, and found that caterpillars grew more rapidly on leaves from those damaged shoots during the following season. Removal of apical buds was necessary and sufficient to improve plant quality for caterpillars. Removal of the same number of basal buds produced no effect when compared to undamaged controls. Haukioja et al. (1990) interpreted these results in terms of the effects of damage on apical dominance. Removal of apical buds altered the plant's hormonal control of growth; removal of basal buds had little effect on hormonal regulation. A similar experiment was conducted with cotton (Karban and Niiho 1995). Seedlings were damaged either by wounding the cotyledonary leaf surfaces or by pinching the apical bud. Damage to the cotyledonary leaves accelerated abscission of the cotyledons and made the plants more resistant to mites; removal of the apical bud retarded abscission of the cotyledons and made the plants more susceptible to mites. Similar experiments conducted on more species are required to determine the generality of this observation.

Haukioja and associates have recently developed a more complete and general model that attempts to reconcile why induced responses to different plant parts can produce such divergent results (Haukioja and Honkanen 1994; Honkanen et al. 1994; Senn and Haukioja 1994). The essence of the model is that plants become more resistant when damage removes strong sources of nutrients for next year's leaves (or other organs that herbivores use). Early in each growing season, leaves are strong sources that normally provision nearby buds and leaf primordia for next year's growth. With these sources removed, next year's leaf primordia are weakened and are unable to produce strong sinks to draw on reserves in the stems. This results in small leaves, with low concentrations of nitrogen and high concentrations of phenolics, qualities that make them poor hosts for herbivores. Haukioja et al. hypothesized that other kinds of damage that similarly remove important sources with this phenology should also produce delayed induced resistance. If leaves are damaged later in the growing season, after they have stopped provisioning nearby buds, but are provisioning stem and root reserves instead, Haukioja and associates hypothesized that damage will have little effect on induced resistance. If apical buds (or young leaves that are still sinks on the plant) are damaged, then basal buds are released from the normal suppression that they experience. These basal buds are able to essentially overexploit the reserves of the stems and produce big leaves, with high concentration of N, and low concentrations of phenolics; qualities that make them good hosts for herbivores. They hypothesized that the phenological shift from provisioning bud reserves to provisioning stem and root reserves probably occurs very rapidly; ignoring this critical distinction has led to much of the unexplained variation in defoliation experiments. Indeed, testing this hypothesis depends on the ease with which workers can unambiguously identify this phenological shift.

"Induction of resistance" implies that the plant has two possible states, induced and not induced. This simplistic model may be inaccurate for several reasons: (1) the response may not be a threshold or on/off process but may have different gradations depending on the type or strength of the cue and (2) the response may involve many different qualitative changes rather than a single response to all stimuli that varies only in intensity. Many immune responses of animals have been found to be graded reactions, increasing as the extent of damage increases (Harvell 1990a, 1990b). This process is referred to as amplification by immunologists.

Characterizing induced plant responses to different levels of damage is relatively straightforward although it requires experiments that vary the extent of damage and therefore have many treatments. When several levels of damage are applied, we can get a picture of the shape of the plant's response. Such response curves are often described as linear or nonlinear with respect to damage. The designations "linear" or "nonlinear" are rather arbitrary since they critically depend on what has been measured. For example, Lin et al. (1990) found that behavioral responses of beetles depends on the number of wounded cells that are in contact with healthy cells and not on the amount of plant tissue that is removed; measurement of one provided a linear response, whereas measurement of the other provided no evidence of a relationship. Whether the relationship between damage and response appears to be linear or characterized by thresholds or plateaus also depends on whether the data have been transformed (logarithms, arcsine, etc.). Finally, the range of damage levels that are considered can influence the shape of the relationship between damage and plant response (English-Loeb 1989, 1990). Therefore, a less arbitrary characterization divides the responses into those that increase monotonically as a function of damage versus those with no apparent relationship or those that increase over some range of damage and decrease over other ranges. For example, increasing spider-mite damage to soybean leaves increased induced resistance up to a level of ten mites per leaf (Brown et al. 1991). Higher levels of damage reduced the strength of the induced resistance.

Many plant species became more resistant following removal of only minute amounts of tissue (e.g. cotton, Karban 1986; tobacco, Baldwin 1988; tomato, Broadway et al. 1986; several species of deciduous trees, Edwards et al. 1986). These examples have been used to argue that large alterations in the plant's carbon and nutrient reserves are not necessary for induced resistance (Schultz 1988). These examples in which minute quantities of leaf removal resulted in induced resistance all occurred rapidly (within hours or days) and probably involved shifts in secondary chemicals that influenced herbivore behavior and/or performance. For three of the plant species mentioned above, several damage levels were examined (cotton [*Gossypium hirsutum*], Karban 1987; tobacco [*Nicotiana sylvestris*], Baldwin and Schmelz 1994; tomato [*Lycopersicon esculentum*], Wolfson and Murdock 1987). For all of these, responses became more effective as the level of damage increased.

Other examples of induced resistance became apparent only after severe defoliation events (larch [*Larix decidua*], Baltensweiler et al. 1977; paper birch [*Betula papyrifera*], Werner 1979; aspen [*Populus tremuloides*], Clausen et al. 1991; black oak [*Quercus velutina*], Wallner and Walton 1979; red alder [*Alnus rubra*], Williams and Myers 1984). In these examples trees had been almost completely defoliated for several seasons in a row. Resistance became stronger in proportion to the severity of defoliation and accumulated over several seasons of successive leaf removals. These examples have been used to argue that defoliation results in the progressive decline in food value and alteration of the plant's carbon and nutrient reserves (Williams and Myers 1984; Bryant et al. 1988).

The observations that some induced resistance can be stimulated by removal of only minute quantities of leaf tissue while others require massive tissue removal should not be viewed as contradictory. Different kinds of induced responses (mechanisms) are possible and will show different time courses, responses to stimuli, and so forth. It is interesting that both types of responses showed stronger induced resistance as damage increased.

Most of the studies that considered several different levels of damage found that the strength of induced resistance increased as the level of damage increased (table 4.5). Studies that considered different numbers of herbivores or amounts of damage during one season also included measurements of effects of induction within days or weeks after the damage occurred. The majority of these found that the strength of the induced effect increased monotonically as damage increased. The ability to induce resistance in most plant species appeared to be buffered against even high levels of leaf damage. For example, tobacco increased its nicotine production in proportion to the extent of damage until 88% of its leaf area had been lost (Baldwin and Schmelz 1994). Even then, concentrations of nicotine (rather than whole-plant pools) continued to increase as a function of damage. However, several studies reported that, at the very highest levels of damage, the strength of the induced resistance was reduced (e.g. Henderson and Holloway 1942; Brown et al. 1991). Presumably, most species are unable to respond when damage becomes so great as to compromise plant functioning. For some conifers, like *Pinus resinosa,* induced resistance was greatest following moderate levels of defoliation (25–40%), and plants were more susceptible at lower or higher levels of damage (Krause and Raffa 1995).

For some plants, the relationship between induced resistance and damage levels is inverse. The ability of pine trees to respond to attack by bark beetles and their symbiotic fungi decreases monotonically as the level of attack increases (Raffa and Berryman 1987). In this example, healthy individuals were best able to induce resistance.

Most of the studies that considered induced resistance that was measured during the years following the initial attack compared the cumulative effects of repeated defoliations. The minimum level of damage considered was generally quite great, often severe defoliation during a single season. Most of these studies found that the effects of repeated defoliations are cumulative. In other words, plants defoliated twice show stronger induced resistance than those defoliated only a single time.

Table 4.5 Relationship between extent of damage and induced resistance

Plant Species	Response in Same/Subsequent Season	Extent of Damage before Response Noted	Relationship between Extent of Damage and Induced Resistance	Possible Mechanism	Reference
Abies grandis	Same	1 needle puncture	Monotonic increase	Monoterpene accumulation	Lewinsohn et al. 1991
Alnus incana	Same	?	Monotonic increase	Trichomes	Baur et al. 1991
Alnus rubra	Subsequent	3 yr defoliation	Effects cumulative over years	?	Williams and Myers 1984
Betula papyrifera	Subsequent	1 yr defoliation	Effects cumulative over years	Chemical induction?	Werner 1979
Betula populifolia	Subsequent	1 yr defoliation	Effects cumulative over years	Chemical induction?	Wallner and Walton 1979
	Subsequent	No consistent effect	No consistent relationship	—	Valentine et al. 1983
Betula pubescens	Subsequent	15% leaf removal	Effects cumulative over years	Chemical induction?	Neuvonen et al. 1987; Haukioja and Neuvonen 1987
Citrus sinensis	Same	Slight injury	Monotonic increase until severe damage	Lowered nutrition?	Henderson and Holloway 1942
Glycine max	Same	20% defoliation	Monotonic increase	Phytoalexin or other chemicals?	Kogan and Fischer 1991
	Same	Slight damage	Maximum induction at intermediate damage	Chemical induction?	Brown et al. 1991
Gossypium hirsutum	Same	Slight damage	Monotonic increase	Chemical induction?	Karban 1987

Table 4.5 continued

Plant Species	Response in Same/ Subsequent Season	Extent of Damage before Response Noted	Relationship between Extent of Damage and Induced Resistance	Possible Mechanism	Reference
Larix decidua	Subsequent	1 yr defoliation	Monotonic increase	Leaf toughness?	Baltensweiler et al. 1977; Fischlin and Baltensweiler 1979
Lycopersicon esculetum	Same	Slight damage	Monotonic increase	Proteinase inhibitors	Broadway et al. 1986; Wolfson and Murdock 1987
Malus domesticus	Same	10% defoliation	Monotonic increase	?	Roland and Myers 1987
	Subsequent	10% defoliation	Minimum resistance at intermediate damage	?	Roland and Myers 1987
Nicotiana sylvestris	Same	Slight	Monotonic increase	Alkaloid accumulation	Baldwin and Schmelz 1994
Omphalea diandra	Subsequent	3 defoliations	Effects cumulative	Chemical induction?	Smith 1983
Pinus contorta	Same	1 attack	Monotonic decrease: resistance strongest at low damage	Monoterpene accumulation	Raffa and Berryman 1987
Pinus taeda	Subsequent	1 attack	Effects cumulative	Monoterpene concentration	Cook and Hain 1988
Phaseolus vulgaris	Same	Moderate damage	Monotonic increase	Lowered nutrition?	Bernstein 1984
	Same	Slight damage	Decrease	Increased nutrition?	English-Loeb and Karban 1991
Populus angustifolia	Same	1 gall	Monotonic increase	Abscission	Williams and Whitham 1986

Species	Timing	Damage	Response	Mechanism	Reference
Populus fremontii	Same	1 gall	Monotonic increase	Abscission	Williams and Whitham 1986
Populus tremuloides	Same and subsequent	50% defoliation	Effects cumulative over years	Lowered nutrition?	Clausen et al. 1991
Prunus persica	Same	Slight damage	Monotonic increase	?	Foott 1963
Prunus serotina	Same	Moderate damage	Monotonic increase	Resin	Karban 1983
Quercus emoryi	Same	1 attack	Monotonic increase	Abscission	Faeth 1992b
Quercus garryana	Same	10% defoliation	Monotonic increase	?	Roland and Myers 1987
	Subsequent	10% defoliation	Monotonic decrease: resistance strongest at low damage	?	Roland and Myers 1987
Quercus geminata	Same	1 attack	Monotonic increase	Abscission	Simberloff and Stiling 1987
Quercus rubra	Same	Slight damage	Monotonic increase	Chemical induction?	Rossiter et al. 1988
Quercus velutina	Subsequent	1 yr defoliation	Mortality cumulative, pupal weight not cumulative	Lowered nutrition?	Wallner and Walton 1979
	Subsequent	1 yr defoliation	Pupal weight cumulative	Lowered nutrition?	Valentine et al. 1983
Sinapis alba	Same	Slight damage	No relationship	?	Palaniswamy and Lamb 1993
Solanum tuberosum	Same	Slight damage	Monotonic increase	?	Tomlin and Sears 1992

There were several exceptions to this general pattern, however (*Quercus velutina*, Wallner and Walton 1979; *Betula populifolia*, Valentine et al. 1983; *Quercus garryana*, Roland and Myers 1987). If amplification of resistance to increasing damage is considered to be an important criterion for an immune response (Harvell 1990a, 1990b), then in general both rapid and long-term induced responses satisfy this condition.

4.2.5 Evolutionary History of Interaction

Induced resistance involves traits that are plastic and vary in response to herbivory. Although the traits themselves are not fixed genetically, the ability to respond may be selected (see section 5.1). Plants that have evolved with intermittent herbivory may be expected to show induced responses, while those that have evolved without herbivores may be expected not to show responses that have been shaped specifically by herbivory. Plants that have evolved with constant herbivory may be expected to evolve constitutive resistance or greater tolerance of herbivory rather than induced resistance.

Several workers have developed this argument and attempted to test it. McNaughton and Tarrants (1983) reported that three species of African grasses accumulated silica in their leaves following grazing. Grasses from four populations were tested, including those from short grasslands with a history of heavy grazing and those from tall grasslands with a history of only light grazing. Those plants exposed historically to higher levels of herbivory (short-grass populations) had higher silica contents than those from locations with lower herbivore exposure. Unfortunately, the argument is not as convincing as it could be, because the interaction between origin of the grasses and the effects of experimental clipping is not statistically significant.

Damage to the leaves of wild parsnip (*Pastinaca sativa*) induced production of furanocoumarins and made plants poorer hosts for generalist herbivores (Zangerl 1990). Two populations were compared, one that had experienced more herbivory (about 40% chance of being damaged) and one that had experienced less herbivory (about 5% chance). The population that experienced higher herbivory had both higher constitutive levels of furanocoumarins and greater heritability for inducibility. Zangerl and Berenbaum (1990) argued that these results are consistent with the hypothesis that the evolutionary history of attack affected the level of inducibility of these presumed defensive compounds, although the relationship between high heritability of inducibility and high inducible resistance (or defense) was unclear. They also examined patterns of constitutive and inducible furanocoumarin production within plants (Zangerl and Rutledge 1996). Fruits of wild parsnip had the highest rates of attack and had high constitutive levels that were not altered by herbivore attack. Leaves and roots had lower, inducible levels and suffered lower rates of attack. These correlations are consistent with the hypothesis that plant parts with negligible risk of herbivory should be relatively undefended, parts with the greatest risk of herbivory should be defended constitutively, and parts experiencing variable or low risk should have inducible defenses. However, interpreting these correlations is difficult; perhaps the type of defense causes the level of herbivory rather than the other way around. In other words, perhaps leaves have lower risk of herbivory than reproductive parts because leaves have relatively effective induced defenses rather than constitutive defenses.

Nettles from southern Finland are subjected to herbivory caused by insects, whereas plants from northern Finland experience herbivory primarily by reindeer (Mutikainen and Walls 1995). These different potential evolutionary pressures match the responses on the southern and northern subspecies of *Urtica dioica;* those in the north increase trichome density following apical excision, whereas those in the south increase trichome density following leaf clipping.

Two studies failed to find any relationship between the evolutionary history of attack and the strength of induced resistance. Bryant, Heitkonig, et al. (1991) found that three African tree species with no history of defoliation showed induced resistance to artificial leaf removal while two species with such evolutionary histories failed to induce resistance. Haukioja and Hanhimaki (1985) compared birch trees from within the range where periodic outbreaks of defoliating moths have occurred for as long as records have been kept and from an area where no outbreaks have been recorded. Trees with both histories responded to mechanical defoliation with equal induced resistance.

All of these experiments were very weak tests of the hypothesis that evolutionary history has affected induced resistance. Indeed, the hypothesis may be difficult to test rigorously. For one thing, comparisons between populations within an outbreak area and outside of it have no true replication. Many other differences between the two places could have given rise to any differences or lack of differences that we observe today. With only one independent replicate (sample) of each treatment (history of herbivory or not), we cannot separate these other differences from our hypothesis about history of attack.

A first step toward a more rigorous analysis has been taken by Matson and Hain (1987), who have considered patterns of induced resistance found in the pines of North America. The defenses that are important for pines correlate with the type of beetle attack they receive. One group relies most heavily on constitutive, preformed defenses such as the primary resin system. Trees employing this primarily constitutive defense are subject to two to ten complete and asynchronous generations of bark beetles per year and may be attacked synchronously by up to five species. Other pine species rely more heavily on induced defenses such as the hypersensitive response to beetles and their symbiotic fungi. Trees employing this primarily induced defense may be attacked by one to three synchronous generations of beetles per year. Matson and Hain (1987) argued that the difference in the constancy of attack might favor different defensive systems. A constitutive defense may be favored for pines that are attacked by beetles emerging continuously throughout the year. Pines that respond to a reduced number of generations of beetles may be able to rely on short-lived induced responses.

These hypotheses are not easy to test because it is difficult to separate adaptive explanations for induced resistance based on the history of herbivory from those based on effects of shared plant phylogeny. To clarify this problem, imagine, as Harvey and Pagel (1991) did, that we are interested in determining the conditions favoring egg laying versus live births. We would note that many feathered species (birds) tend to lay eggs and many hairy species (mammals) tend to have live births. However, it would be premature to conclude that egg laying is an adaptation associated with conditions that favored feathers and giving birth to live young is an adaptation associated with conditions that favor hair. If each species were considered an independent sample, a chi-square analysis would indicate that these two features are significantly associated. However, a phylogeny would reveal that the events are not independent and that the many cases of egg laying and of live births each evolved only one time. The species are similar, not necessarily because they evolved to solve a common problem in a similar way, but perhaps because they are closely related.

Future studies of plant responses to herbivory should examine correlations between induced resistance and history of attack, as Matson and Hain (1987) did. In addition, much more may be learned about the evolution of induced resistance by using a phylogenetic approach (Ridley 1983; Sillen-Tullberg 1988, 1993; Donoghue 1989; Harvey and Pagel 1991; Miles and Dunham 1993). This approach uses existing plant phylogenies to count the number of times that induced resistance is found. It allows one to ask the question, Is the occurrence of induced resistance better explained by history of attack or by plant phylogeny? For example, was the correlation between history of beetle attack and induced responses that Matson and Hain (1987) observed due to all of the species in the southeast that rely on constitutive resistance sharing one common ancestor and all of the species in the west that rely on induced resistance sharing another, or did the correlation reflect adaptive responses to the frequency of attack? The correlations alone cannot allow us to differentiate between phyletic heritage and adaptation to herbivory, although phylogenetic analysis possibly could. An example of the kind of information that such an investigation could reveal is provided by Sillen-Tullberg (1988, 1993), who used a phylogenetic approach to examine the relationship between gregarious behavior, unpalatability, and aposematic coloration in butterfly larvae. She found that unpalatability always preceded gregariousness and concluded that unpalatability predisposes the evolution of gregarious behavior. Similar studies would add greatly to our understanding of the relationship between induced resistance and history of herbivory.

Several recent studies have attempted to examine the historical origins of induced resistance. Induced resistance against mites in several grape species was found not to be clustered among clades that shared common ancestors (English-Loeb et al. n.d.). Plant phylogeny is not a good predictor of whether a plant species would show evidence of induced resistance against mites. In this study, induced resistance appeared to be more evolutionarily labile than was constitutive resistance against mites. For cotton species, induced resistance against spider mites appears to be a trait that has been more constrained by phylogeny (Thaler and Karban 1997). Of twenty-one species of *Gossypium* that were examined, induced resistance was found to have originated four times. Induced resistance was less common than constitutive resistance, although both are derived rather than ancestral traits. Induced resistance arose in clades in Australia and tropical America, although not all clades from these two geographic regions are inducible. Species from islands, which presumably have reduced pressure from herbivores, were no less likely than mainland species to have high levels of either constitutive or induced resistance. There was no evidence that induced resistance is less likely to be found in species that also have high levels of constitutive resistance.

Phylogenetic analyses can potentially provide insights into how often and under what conditions induced resistance arose. Few phylogenetic analyses of plant defenses have been attempted, although the techniques are well established and require only reliable plant phylogenies and assays of which taxa express resistance. In the future, as these last two requirements become available for more plant taxa, we will probably get a much clearer picture of the evolution of induced resistance.

4.2.6 What Herbivores Are Affected by Induced Responses?

Studies of induced resistance and susceptibility have most often involved effects on insect herbivores, although there are also many examples involving phytophagous mammals and mites (tables 4.1 and 4.2). Most of these herbivores feed on plants by chewing (chewing caterpillars, beetles, mites, sawflies); sucking insects and gall formers are also well represented. Caterpillars are dramatically overrepresented in these studies in comparison to their natural commonness and diversity.

Several mechanisms of induced resistance may be particularly effective against certain insect guilds. Induced chemical and morphological responses that reduce plant attractiveness are believed to be effective against many mobile herbivores that can leave plants or plant parts that they find less suitable. Such responses may be less effective against immobile insects that cannot be deterred. In cases involving relatively immobile insects, induced responses are effective only if they kill these individual herbivores before they have had a chance to cause much loss of plant fitness (see section 5.1).

Hypersensitive plant responses and abscission of infected plant parts might be expected to provide more effective defense against sedentary rather than mobile herbivores. Hypersensitive plant responses were found to be most often elicited by immobile insects that form relatively long-lasting intimate relationships with their hosts (table 4.1; also see review by Fernandes 1990). Similarly, hypersensitive responses were most often found to affect sedentary, endoparasitic nematodes (Giebel 1982). Sessile insects are also more vulnerable to induced early abscission of plant tissues such as leaves (Faeth et al. 1981). Williams and Whitham (1986) suggested that the relatively high vulnerability of galling aphids to induced gall abscission may explain why species that form galls are 3.5 times as likely to have host-alternating life cycles as nongalling aphids.

Rhoades (1979) proposed that induced responses may have different effects on specialist and generalist herbivores. According to his argument, rapid induced responses should involve "qualitative defenses" that are effective against generalists but not specialists. Long-term induced responses should involve "quantitative defenses" and should be effective against both specialists and generalists. Van Dam et al. (1993) developed a similar argument for the relative effectiveness of rapid induced responses against generalists and specialists. They reasoned that constitutive defenses are more effective against those generalist herbivores that are often present, and inducible defenses are advantageous for those plants that experience low and unpredictable pressure from generalist herbivores. They argued that specialist herbivores generally are not greatly affected by the plant's defenses (constitutive or induced) so plants should not respond facultatively to specialists.

Several workers who have considered the effects of induced resistance have found support for this hypothesis, particularly the expectation that rapid induced responses will be more effective against generalists than specialists. For example, rapidly induced soybean phytoalexins are effective deterrents against two generalist species, the southern corn rootworm (*Diabrotica undecimpunctata*) and the Mexican bean beetle (*Epilachna varivestis*), but not against the bean leaf beetle (*Cerotoma trifurcata*), a specialist on soybeans (Kogan and Fischer 1991). Field experiments confirmed that damage induced more resistance to generalist Mexican bean beetles than to specialist soybean loopers (*Pseudoplusia includens*). However, results that were not consistent with Rhoades's hypothesis (1979) are also common. For example, Wheeler and Slansky (1991) assayed the effectiveness of induced resistance against several other caterpillars that feed on soybean. They found that rapid induced resistance was effective against both a generalist and a soybean specialist. Several other species of generalist caterpillars were not affected.

It would probably be informative to examine all of the examples in table 4.1 to determine if rapid induced responses are more commonly reported to provide resistance to generalists rather than to specialists. However, we did not attempt this because most authors of the studies in table 4.1 did not include information about the degree of dietary specialization of the herbivores they studied. When this information is available, there are many cases involving both specialists and generalists, as indicated by the examples above for soybean. It is our subjective impression that the data do not lend strong support for Rhoades's argument.

One pattern that impressed us was the number of cases of induced susceptibility that involved herbivores that were specialists, almost half of the examples in table 4. 2. This observation has no null hypothesis for comparison; it is unclear how many examples would involve specialists if a random sample could be selected. It does make good intuitive sense that specialists might be less adversely affected, and might even benefit, by responses that they induced in their host plants.

4.3 DOES INDUCED RESISTANCE AFFECT HERBIVORE POPULATIONS?

There is considerable and incontrovertible evidence that damage-induced changes affect the performance of herbivores to their detriment (table 4.1) and sometimes to their benefit (table 4.2). It makes good sense that these changes in the survival, growth rate, fecundity, and behaviors of herbivores should be reflected in the sizes and dynamics of herbivore populations. However, several workers have argued that damage may have little or no effect on herbivore populations, regardless of how intuitively appealing the connection between herbivore performance and populations might appear (Fowler and Lawton 1985; Myers 1988). These authors argued that many of the early bioassay results were analyzed incorrectly, using pseudoreplication that inflated the degrees of freedom in statistical comparisons (see Fowler and Lawton 1985; Neuvonen and Haukioja 1985; for a general discussion of this problem see Hurlbert 1984). These statistical problems should not invalidate studies but do make them subject to other interpretations. Most of the more recent studies were analyzed correctly, at least with regard to this problem.

Second, many of the reported effects of damage on performance of herbivores were relatively small effects (Fowler and Lawton 1985). Relatively small increases in mortality associated with feeding on damaged foliage or relatively small changes in behavior may be swamped by other factors. As a result these small changes in performance may exert no effect on population dynamics. This problem is amplified because most of the studies demonstrated effects for only a small portion of the herbivore's life cycle (most often the larval stage for Lepidoptera). Differences in one developmental stage may be compensated for by factors in other stages, so that the net effects on populations are minimal.

Compensatory mortality and behaviors that can negate small effects of induced resistance and susceptibility may take many forms. Often only a small proportion of the total available plant tissue or individuals becomes induced. For instance, induced gum production by wild cherry trees (*Prunus serotina*) increased mortality of cicada eggs (Karban 1983). However, only a small fraction of the forest was composed of cherry trees, so that induced resistance of this one plant species had a negligible impact on overall populations of cicadas. In a slightly different example, early-season feeding by arctiid caterpillars (*Platyprepia virginalis*) was found to induce changes in lupine bushes (*Lupinus arboreus*) that made them poorer hosts for later-feeding tussock moth caterpillars (*Orgyia vetusta*), lupine specialists (Harrison and Karban 1986). However, only a small fraction of the available foliage was induced and the effects of induction were small relative to other factors, so that induced resistance had negligible effects on populations of tussock moths. Induced resistance was insufficient to reduce populations of tussock moths, and bushes were repeatedly defoliated by high numbers of tussock moth larvae for at least ten years (Harrison 1995). In a detailed study of leaf-miner populations, Faeth (1992b) found that increasing levels of attack on oak trees (*Quercus emoryi*) led to increasing rates of leaf-miner (*Cameraria* sp.) mortality due to early leaf abscission. However, this mortality was compensated for by decreased death from bacterial and fungal pathogens for miners on heavily attacked branches. The number of co-occurring larvae exerted the most important influence on leaf-miner survival. In the context of these other sources of mortality, induced resistance had no detectable effect on populations of this leaf miner.

Behavioral responses by herbivores tend to negate the detrimental effects of induced resistance on herbivore populations in some instances. Bioassay results that rely on caged herbivores or individuals whose normal behaviors are limited by the experimental protocol may overestimate potential effects on populations. Under normal conditions herbivores may choose not to encounter damaged tissue, an option that may not be available to caged herbivores. For example, larvae of pine beauty moths (*Panolis flammea*) that were forced to develop on damaged trees (*Pinus contorta*) had reduced rates of growth and survival compared to larvae on undamaged trees (Leather et al. 1987). However, ovipositing female moths were very sensitive to induced changes in monoterpene concentration of resin and avoided damaged trees. When undamaged trees were available, this behavioral discrimination reduced the consequences of induced resistance on populations of pine beauty moths. There are surprisingly few studies that demonstrate that herbivore behaviors reduce effects of induced resistance on herbivore populations, although this phenomenon may be widespread in nature. More attention to herbivore behavior in future studies of induced resistance will surely repay the effort.

Many of the damage-induced changes in herbivore performance do not affect herbivore populations. In addition, changes in herbivore performance are much easier to assay than are effects on populations. For both of these reasons fewer studies report effects of induced resistance and susceptibility on herbivore populations (table 4.6) than on herbivore performance (tables 4.1 and 4.2). The studies listed in table 4.6 considered herbivore numbers for at least one complete generation or parts of at least two generations. Not surprisingly, the herbivores reported in these studies were mites, small homopterans, and leaf miners.

Table 4.6 Experiments reporting effects of induced resistance and susceptibility on populations of herbivorous arthropods

	Herbivore Affected		
Order	Species	Common Name	Field/Lab
Induced resistance			
Acari	*Oligonychus punicae*	Avocado brown mite	Lab
	Oligonychus subnudus	Pine spider mite	Field
	Panonychus ulmi	European red mite	Lab and field
	Tetranychus pacificus	Pacific spider mite	Field
			Lab and Field
	Tetranychus urticae	2-spotted mite	Field
			Lab
			Lab
			Lab
		2-spotted mite	Lab
Homoptera	*Rhopalosiphum padi*	Bird cherry-oat aphid	Field
	Empoasca fabae	Potato leafhopper	Lab
	Philaenus spumarius	Meadow spittlebug	Field
Lepidoptera	*Archanara geminipuncta*	Stem borer	Field
	Bucculatrix thurberiella	Cotton-leaf perforator	Field
	Panolis flammea	Pine beauty moth	Lab
	Phyllonorycter spp.	Leaf miner	Field

No. of Generations Studied	Comments	Reference
1+	Populations reduced on leaves that had been heavily damaged (competition?)	McMurtry 1970
Several	Populations reduced on pine trees damaged in previous year	Karban 1990
Several	Populations reduced on apple leaves previously damaged by many organisms	Cutright 1963; Kuenen 1948; Foott 1963
Several	Populations reduced throughout season on cotton plants damaged earlier	Karban 1986a
Several	Populations reduced on grapevines damaged earlier	English-Loeb and Karban 1988; Hougen-Eitzman and Karban 1995
Several	Populations reduced on apple trees with European red mites (competition?)	Lienk and Chapman 1951
1+	Populations reduced on cotton plants damaged earlier	Karban and Carey 1984
1+	Populations reduced on soybean plants damaged earlier	Hildebrand, Rodriguez,Brown, and Volden 1986; Brown et al. 1991
Several	Populations reduced on strawberry plants damaged earlier	Shanks and Doss 1989
1+	Populations reduced on cucumber plants damaged earlier	A. A. Agrawal, personal communication
1+	Colonization and number of eggs reduced on cherry trees damaged earlier	Leather 1993
1	Populations of nymphs reduced on potato plants damaged earlier	Tomlin and Sears 1992
1+	Populations reduced on plants damaged earlier, refuges destroyed	Karban 1986b
Several	Reduced survival and populations on narrow reed shoot that followed herbivory.	Mook and van der Toorn 1985
Several	Cumulative population over season reduced on plants damaged earlier	Karban 1993b
1	Populations of eggs reduced on trees defoliated previously	Leather et al. 1987
2	Local extinction on most damaged leaves	West 1985

Table 4.6 continued

| Order | Herbivore Affected | | Field/Lab |
	Species	Common Name	
	Zeiraphera diniana	Larch budmoth	Field
	Stilbosis juvantis	Leaf miner	Field
Diptera	*Liriomyza trifolii*	Leaf miner	Lab
Induced susceptibility Homoptera	*Acizzia russellae*	Psyllid	Field
Coleoptera	*Oedionychus* sp.	Flea beetle	Field
	Deporaus betulae	Curculionid	Field
Lepidoptera	*Stilbosis juvantis*	Leaf miner	Field
Lepidoptera?		Leaf miners and other leaf eaters	Field
Hymenoptera	*Euura lasiolepis*	Tenthredinid sawfly	Field

These species lend themselves to estimation of their demographic rates, and they are the same herbivores whose population dynamics have been most thoroughly investigated (e.g. examples in Price 1984). These species are all small, with short generation times and low mobility. The reason these species were chosen for study is obvious, although studies involving less convenient herbivores are required to assess the generality of these results.

Roughly equal numbers of studies finding effects of induced resistance on herbivore populations were done in the lab and in the field. Several studies also reported effects of induced susceptibility on herbivore populations. These were field studies and generally were unexpected results of experiments designed to answer other questions.

Critics of this field have argued that induced responses have not been shown to affect herbivore populations. The studies in table 4.6 suggest that this criticism is not warranted for at least some herbivore species. However, the species represented are a small subset; no mammals have

No. of Generations Studied	Comments	Reference
Several	Populations reduced following defoliation, corresponding to experimental bioassays	Baltensweiler et al. 1977; Baltensweiler 1985
1	Local extinction on trees defoliated late in season	Faeth n.d., 1987
1	Populations of flies reduced on chrysanthamum leaves damaged earlier	Van de Vrie et al. 1988
Several	Populations increased on leaves that regrew 1 yr following defoliation	Webb and Moran 1978
Several	Populations increased on leaves that regrew following artificial defoliation	Rockwood 1974
Several?	Higher populations observed on birches following artificial defoliation	Neuvonen et al. 1988
1	Higher densities on trees defoliated early in season	Faeth n.d., 1987
Several?	Populations increased on birches following defoliation	Danell and Huss-Danell 1985
3	Population density increased on previously damaged shoots	Craig et al. 1986

been shown convincingly to be affected, for example. In no case has a definitive connection been made between a particular mechanism (induced response) and resistance that affected herbivore populations. In other words, the mechanisms responsible for the effects listed in table 4.6 remain unknown.

Demonstrating an effect on herbivore populations (statistically significant or otherwise) is much less informative than comparing the importance of induced resistance relative to other ecological factors that can also affect herbivore populations. An effect can be statistically significant but ecologically weak if a sufficiently large sample size was examined. Few studies have attempted to compare the strength of induced resistance versus other ecological factors. As the discussion of compensatory mortality indicated (see above), induced resistance has often been found to be less important than other mortality factors. For example, wild cotton plants (*Gossypium thurberi*) that were experimentally damaged early in the season supported fewer leaf miners (*Bucculatrix thur-*

beriella) over the rest of the season (Karban 1993b). However, effects of induced resistance were weaker than negative effects of plant crowding on cumulative herbivore populations.

Do induced responses affect populations of herbivores? The answer to this question depends on your a priori expectations. If your expectation was that induced resistance does not and generally cannot affect herbivore populations in either natural or agricultural systems, then the examples in table 4.6 should convince you that you were wrong. However, if your expectation was that induced resistance was "the factor" that controlled herbivore populations, then the evidence argues against this simplistic view. The truth appears to lie somewhere between. For at least several species, induced resistance has been shown to produce large effects on herbivore numbers, although in no case has induced resistance been shown to "regulate" populations.

4.4 Does Induced Resistance Drive Cycles of Herbivore Outbreaks?

Populations of some herbivores exhibit spectacular fluctuations. Some of these fluctuations exhibit predictable periodicity, peaking every four years or every ten years; such periodic fluctuations are called cycles. Controversy exists about whether some of these herbivore outbreaks are truly cycles, but everyone agrees that numbers of fluctuating populations can span many orders of magnitude. During years of high abundance millions of individuals may defoliate their host plants. During years of low abundance individuals may be so rare as to escape detection by all but the most thorough searches.

The oft-cited dogma of this field is that large population cycles are more characteristic of Arctic herbivores. The best-known herbivore population fluctuations include ten-year cycles of hares (*Lepus americanus;* Keith 1963) and autumnal moths (*Epirrita autumnata;* Haukioja et al. 1983), both from the far north. Populations of larch budmoths (*Zeiraphera diniana*) exhibit eight- to ten-year cycles in the higher elevations of the European Alps (Baltensweiler and Fischlin 1988). Although many cycles have a periodicity of approximately ten years, other periodicities are known; populations of voles tend to have outbreaks every three or four years (Krebs and Myers 1974). A survey of examples showed that cycles are found in temperate and tropical environments as well. Krebs and Myers (1974) concluded that all microtine populations for which data had been collected for at least four consecutive generations show the characteristic population peaks and troughs with a three- or four-year periodicity. However, rodent populations from the Arctic have more pronounced cycles than populations farther south, at least in Europe (Hansson and Henttonen 1985; Stenseth and Ims 1993). Gallinaceous birds, particularly grouse, exhibit population fluctuations of both four and ten years (Lack 1954a, 1954b). These may be linked to the population cycles of rodents with which the grouse are sympatric. The numbers of predators that feed on both rodents and grouse are believed to follow the ups and downs in rodent abundance. These predators may then impose the four- or ten-year pattern on populations of grouse. Alternatively, food and other resources that grouse require may be periodically devastated by fluctuating rodent populations (Lack 1954a, 1954b).

There are many well-documented cases of insect herbivores that exhibit dramatic population fluctuations (Myers 1988). These may have periodicities of ten years, although other periods have also been reported. For example, spruce-budworm (*Choristoneura fumiferana*) populations peak approximately every thirty-five years. Like the microtine cycles, cycles of insect outbreaks have been reported from both Arctic and "less extreme" habitats. For example, populations of day-flying *Urania* moths fluctuate greatly throughout the New World tropics (Smith 1982). Large outbreaks occur approximately every eight years and smaller ones every four years. These are the exceptions rather than the rule, and certainly most insect populations show no evidence of population fluctuations or cycles (Mason 1987).

4.4.1 Models of Herbivore Population Cycles Driven by Induced Resistance

Induced resistance first caught the imaginations of ecologists as a possible mechanism to explain herbivore population cycles. First, verbal models were proposed by Benz (1974, 1977) and Haukioja and Hakala (1975). According to their schemes, plants that are damaged react by producing "less nutritious" tissues that contain high concentrations of "toxic" chemicals. When this occurs over a wide area, populations of herbivores may decrease abruptly. Some years following the decline in herbivore numbers, plants return to the "less defended" state. This allows herbivore populations to increase again to levels that damage their host plants and induce resistance, causing another population crash. This model was developed to explain the eight- to ten-year cycles of larch budmoths (*Zeiraphera diniana*) observed in the Alps and the ten-year cycles in abundance of autumnal moths (*Epirrita autumnata*) throughout northern Scandinavia.

A similar scenario was proposed a few years later by Bryant (1981) and Fox and Bryant (1984) to explain the ten-year cycles in abundance of hares in Alaska and northern Canada. When hares are very abundant, they overbrowse or girdle many of the shoots of their preferred host species. This results in the resprouting of adventitious, juvenile-form shoots, which are much less palatable. The shortage of preferred food causes hare populations to crash. Low numbers of hares may lead to the production of mature growth–form twigs after several years, which in turn allows hare populations to increase again, renewing the cycle. Low levels of browsing may even result in an increased production of high-quality food for hares relative to no browsing. Increased production caused by light browsing may delay the onset of deterioration in plant quality, allowing the hare population to become very large.

Earlier models of microtine population cycles involved reduction in the availability of food though not necessarily changes in food quality. For example, Lack (1954a, 1954b) argued that cycles were due to overexploitation of habitat and destruction of food and cover. Freeland (1974) modified this notion when he suggested that vole populations increase when their most preferred and least "toxic" foods were plentiful and declined when they had exhausted these and were forced to feed on more "toxic" plants or plant parts. These verbal models do not contain the hypothesis that plant quality changes as the result of feeding, merely that the best stuff gets used up.

These first models were followed by several more explicit mathematical treatments of induced resistance as a possible driving force for herbivore population cycles. These later models did incorporate induced changes in plant quality. Fischlin and Baltensweiler (1979) developed a systems model that included only local interactions between host plants (larch trees) and herbivores (larch budmoths). Their intent was to generate several plausible models and to compare the herbivore population dynamics generated by these models with real data. The model of induced resistance that they developed included the empirical observation that raw fiber content of larch needles correlated well with budmoth larval mortality. Defoliation caused new leaves to be produced, which had higher raw fiber content, another empirical result. When herbivore numbers were low, levels of foliar raw fiber decreased, although few field data existed to evaluate this assumption. This simple model produced herbivore population fluctuations that were similar in period and amplitude to those observed in the field. Simulation runs were conducted to determine effects of altering the model's parameters. In general, the model was quite robust to various parameter changes, although the rate at which plants recover after being damaged had a large effect on the cycles that were produced. The rate of recovery must be slow (seven years) to produce cycles with a ten-year period and the tremendous amplitude that have been observed in nature. Unfortunately, rate of recovery is one of the most difficult and time-consuming parameters to estimate empirically, and relatively little information is available about whether the assumption that plant recovery takes many years is reasonable (see below).

Edelstein-Keshet and Rausher (1989) presented a simple analytical model with two differential equations to represent dynamics of plant quality and herbivore density. The herbivores were assumed to be mobile but not selective, feeding without regard to plant quality. Damage to plants caused food quality to decline, and this reduced herbivore population growth. Plants improved in quality when herbivory was diminished. The model was solved for several different cases. When induced resistance was the only density-dependent factor capable of regulating the herbivore population, stable solutions were found as long as the half-life of the defenses was longer than the induction time. When other regulatory factors were included in the model (e.g. predation), induced resistance and the other factors generally depressed herbivore populations to a steady equilibrium. However, induced resistance showed no particular tendency to cause herbivore populations to fluctuate or cycle. Lundberg et al. (1994) modified this basic approach, modeling herbivory occurring as discrete events (rather than as accumulated nibbles) and modeling relaxation to occur completely after some lag time. Their model predicted oscillating herbivore populations, although they did not get stable limit cycles. Several other factors could be added to the model to achieve this behavior (see also Adler and Karban 1994). In fact, cycling behavior can be produced in models by so many different factors that some authors have wondered why more cycles are not observed in nature (Finerty 1980).

Models of cycling populations almost universally highlight the importance of time lags: the lengths of time between when damage occurs and when resistance is induced and between when resistance is induced and when it is relaxed so that plants are again high-quality hosts. These time lags are highlighted because they determine whether the induced resistance is likely to stabilize or destabilize the herbivore population (May 1973, 1981; Haukioja 1982, 1990a, 1990b; Berryman et al. 1987). Induced resistance with very short time lags (that are induced rapidly and decay rapidly) may affect the individual herbivores that cause the response. These responses have the potential to provide a regulating influence on the herbivore population and reduce fluctuations. Induced resistance with longer time lags may affect only individual herbivores after the initial attackers have left. If relaxation times are long, the plant may remain highly resistant long after herbivore populations have declined. Negative factors with time delays have the potential to destabilize herbivore populations and may cause herbivore populations to oscillate. Conversely, if low levels of herbivory increase the production of high-quality food, then this positive factor may permit the herbivore population to continue to increase and may delay the negative effects of over-browsing (Fox and Bryant 1984). If such positive factors with time delays exist, they are also likely to destabilize herbivore populations.

Models have been developed for several eruptive species of forest Lepidoptera that highlight induced plant responses, viral pathogens of the herbivores, and especially the interactions between induced plant chemistry and pathogenicity as potentially important forces in driving herbivore population dynamics. Models of the Douglas-fir tussock moth (*Orgyia pseudotsugata*) and its nuclear polyhedrosis virus failed to produce the seven- to ten-year cycles that are observed in nature (Vezina and Peterman 1985). Foster et al. (1992) constructed a similar model of nuclear polyhedrosis virus and gypsy-moth (*Lymantria dispar*) hosts. They included inducible levels of phenolics in their model; gypsy-moth defoliation is known to lead to increased levels of tannins, which in turn reduce the susceptibility of gypsy-moth larvae to virus. Simulations that included effects of induced deterioration of food quality on larvae, mortality caused by virus, and reduced susceptibility of larvae to virus when tannin levels were induced could generate the fluctuating densities with peaks every five to ten years that are observed in nature.

These models are useful in helping to identify whether induced resistance could potentially cause herbivore cycles and under what conditions cycles would result. However, model results indicate only what is possible, not whether such processes actually occur in nature. For example, several different groups have modeled the population cycles of larch budmoths (*Zeiraphera diniana*). Van den Bos and Rabbinge (1976, cited in Baltensweiler et al. 1977; Haukioja et al. 1983) reported that three trophic levels are necessary to produce the observed patterns. Fischlin and Baltensweiler (1979) found that induced resistance alone is sufficient to produce cycles with the proper amplitude and periodicity. Anderson and May (1979) found that diseases are also sufficient to produce the observed cycles. Models can suggest possible conditions and outcomes, but observations and experiments are clearly necessary to determine which natural processes work in reality.

4.4.2 Induced Resistance and Cycles: Empirical Evidence

For induced responses to drive the observed cycles in herbivore populations, several predictions should follow (modified from Karban and

Myers 1989). First, herbivore performance should be influenced by the history of attack. Survival or fecundity of herbivores must decline following increasing levels of damage. The evidence reviewed above (tables 4.1 and 4.5) is generally consistent with this prediction, although exceptions certainly are common.

Second, for induced resistance to contribute to fluctuations in herbivore populations, the relaxation of induced resistance should take several years. Turchin (1990) reviewed the population patterns observed for fourteen species of forest insects for which ten consecutive population estimates had been collected. Eight of these fourteen species showed evidence of delayed density dependence with a time lag. All eight species also showed oscillating population patterns. The two species with the most regular cyclic fluctuations in density, the larch budmoth (*Zeiraphera diniana*) and the autumnal moth (*Epirrita autumnata*), were also the two that showed the most convincing evidence of induced resistance with relaxation times of several years. For the larch budmoth, mortality was high immediately following an outbreak and remained high for several years (Baltensweiler 1985). High mortality and dispersal were associated with smaller larch needles (Benz 1974, 1977). Needles remained small for four to five years following an outbreak (Baltensweiler 1985). For the autumnal moth the largest effect of induced resistance was found in reduced pupal weights (Haukioja 1980). The reduction in fecundity was greatest one year after artificial defoliation but remained significant (roughly 10–20% reduction) four years after damage (Haukioja 1982; Haukioja and Neuvonen 1987). Similarly, stump sprouts of several tree species eaten by snowshoe hares remained less palatable to hares for three to four years following browsing (Bryant, Danell, et al. 1991). Time lags of this length are consistent with the hypothesis that induced resistance may keep populations from immediately increasing after an outbreak.

Third, induced resistance should translate into reductions in herbivore population densities if it is a factor that affects fluctuations. The examples in table 4.6 provide evidence that induced resistance can reduce populations of herbivorous arthropods, although none of the herbivore species listed in the table have been found to cycle. The existing data for herbivore species with cyclical fluctuations does not convincingly implicate induced resistance as a factor that affected or drove the cycles.

Many hypotheses have been proposed to explain the population cycles of microtine rodents. Most of these make rather similar predictions that are consistent with observed patterns of demography, growth, and behavior. However, Seldal (1994) argued that only hypotheses related to changes in food quantity or quality also predict the observed changes in physiology and pathology of lemmings. Of particular interest are observations that lemmings from the decline phase showed enlarged pancreases and spleens, as well as undigested dietary protein in their feces (Seldal, Andersen, et al. 1994). In addition, only hypotheses relating to induced host responses can explain observed patterns in plants during a cycle of microtines (Seldal 1994), although this last argument may be circular. These considerations led Seldal (1994) to advocate grazing-induced changes in food-plant quality as the most plausible hypothesis to explain microtine cycles. The primary problem with this advocacy is that it was based on correlations and does little to establish cause and effect. For example, while it may be true that only a hypothesis that includes interactions between herbivores and vegetation can predict changes observed for plant populations, this indicates that the herbivores are likely to have caused the plant responses but not that the plant responses necessarily caused changes in numbers of rodents. Manipulative experiments are much more suited to establish causality than are correlations, even when the correlations are expressed as a priori predictions.

Although manipulative experiments are difficult to perform, several different experimental approaches have been attempted. Several workers have introduced herbivores from areas of low herbivore populations into areas that have recently experienced outbreaks. If induced resistance is responsible for causing declines following outbreaks, such introduced herbivores should do poorly in areas with histories of recent outbreaks. Experiments and observations that compare different herbivore populations can provide valuable information about the role of induced resistance in cycles, although these studies are all limited by inadequate replication. In these experiments each population rather than each herbivore is an independent sample. With such low replication other factors not considered by the investigator may explain the results.

Auer et al. (1981) introduced a large number of larch budmoth pupae into an area that was in the decline phase. Numbers of larch budmoths did not increase in this population, consistent with the hypothesis that induced resistance (or any other site-specific factor such as disease or predation) was limiting the herbivore population. In contrast, population dynamics of western tent caterpillars (*Malacosoma californicum*) did not support the hypothesis that induced resistance produced the observed population cycles. History of previous attack did not affect the synchronous decline of several subpopulations (data from Wellington analyzed by Myers 1988). Myers (1988) argued that populations of cycling forest insects (tussock moths, western tent caterpillars, larch budmoths, spruce budworms) have been found to decline synchronously, independently of the maximum densities achieved or the extent of defoliation that they caused. She reasoned that this pattern was more likely to have been caused by disease epidemics than by induced changes in foliage quality. Similarly, populations of voles brought from a low-density area were not found to decline after being placed in enclosures that had been heavily depleted of vegetation (Krebs and Myers 1974). Krebs and Myers concluded that food quality does not drive the observed cycles of vole populations, although several authors have questioned whether their small enclosures were sufficient to capture the processes that occur in natural situations (Wiens 1989; Stenseth and Ims 1993).

Fourth, experimentally reducing damage to plants should prolong the outbreak. Similarly, experimentally introducing herbivores from outbreak areas to suitable host plants should produce an outbreak out of synchrony with natural populations. There is no evidence that these predictions have been met for any herbivores. On two occasions foresters were successful in artificially reducing populations of larch budmoths early in the outbreak cycle so that they did not cause defoliation (Baltensweiler et al. 1977). In both instances, however, populations of budmoths from treated and control areas declined subsequently, independent of the history of defoliation. Similarly, a population of western tent caterpillars was brought from an outbreak area to an isolated island where undamaged host plants were abundant (Myers 1981). Contrary to expectations, this population did not remain large, prolonging the outbreak, when other mainland populations crashed. Myers (1990) repeated this approach on a larger scale by introducing egg masses of western tent caterpillars, collected during the population increase or peak, to seven sites that had received little previous damage. The population decline was delayed at some of these sites with little prior damage. However, all populations declined to very low levels within two years of the start of the decline phase, regardless of the history of damage at the site. Myers reasoned that the resilience of the cyclic pattern in which no introduced population was more than one year out of phase argues against the notion that induced foliage quality drives cycles of tent caterpillars.

Several experiments have also been conducted in which microtine rodents were introduced to areas where food quality was experimentally enhanced. Early experiments suggested that a lack of recent defoliation is not sufficient to cause outbreaks of any of the herbivore species examined (Krebs and Myers 1974). These enclosure experiments were difficult to evaluate because they were probably not large enough to capture important natural processes (Wiens 1989; Stenseth and Ims 1993). More recently, Krebs and coworkers have conducted a larger-scale factorial experiment in which they provided extra food, excluded predators, and added nutrient fertilizers to plots of 1 km² in the Yukon (Krebs et al. 1995). Predator exclosure doubled the density of hares, food addition tripled hare numbers, and the combined treatment of predator removal and food addition increased density elevenfold. Addition of nutrient fertilizers had little effect on hare densities. In this and other studies, providing extra food sometimes increased individual growth rate, survival, or reproduction as well as population densities attained, although in no case did extra food prevent a decline completely (Ford and Pitelka 1984; Akcakaya 1992; Krebs et al. 1995). Similarly, attempts to experimentally improve food quality failed to affect population cycles.

Ecologists have long been fascinated with periodic cycles of herbivore populations, and records of outbreaks are found in the Old Testament, in Aristotle's writings, and in various accounts of European history. Certainly since Elton (1924) many ecologists have been looking for the single factor that drives these dramatic cycles. The exceptions to the four predictions described above make it quite clear that induced resistance cannot provide the single factor that is necessary and sufficient to produce cyclic dynamics. Unfortunately, it has also become clear that no other single mechanism considered to date explains the dynamics of the herbivore cycles either (Stenseth and Ims 1993). What is much less clear is whether it makes sense to expect any single factor to apply to all herbivore populations, all the time.

Perhaps ecologists could make more progress by considering explanations that are pluralistic and involve several different processes. The experimental approach of Krebs et al. (1995) is particularly encouraging because it suggests that predation and food availability are both involved and that these two kinds of factors act synergistically to produce effects much larger than the effects of either one, considered alone. Time-series analyses of hare populations support the notion that synergisms of this sort are important (Stenseth 1995).

Even pluralistic hypotheses may have to be conditional on the species involved and on the ecological situation, although such an approach is far less elegant than a single-factor explanation (Lidicker 1973; Haukioja et al. 1983). Many ecologists have been reluctant to give up the notion that a single factor would explain all dynamics of cycling herbivore populations, because clear, testable predictions can be generated from single factors. In contrast, multiple-factor explanations potentially can be modified a posteriori to explain any desired result, so that they become untestable (Krebs and Myers 1974; Tamarin 1978; Gaines et al. 1991; Stenseth and Ims 1993; Seldal 1994). Any multiple-factor explanations that involve induced resistance must be capable of making unambiguous and specific predictions. Viewed in this context, it still remains unclear what role induced resistance plays in driving the population dynamics of herbivore species.

4.5 FUTURE DIRECTIONS: EXTENDING OUR KNOWLEDGE TO THE POPULATION AND COMMUNITY LEVELS

Most ecologists now acknowledge that induced resistance is a common and, in some systems, important phenomenon. This reflects a change during the past ten or fifteen years that is based on the many examples listed in table 4.1. The rate at which that table is growing has decreased, as ecologists no longer consider another example of damage affecting herbivore performance to be exciting. Almost all of these examples involve leaves and shoots, however; we still know next to nothing about induced resistance of roots and reproductive parts. We have not yet begun to explore how induced changes may affect the way plants are perceived by their herbivores, their pollinators, and their seed dispersers. Similarly, we have few clues about the role that induced resistance might play in affecting guilds and interactions other than folivory. The effects of induced responses on organisms at other trophic levels is just beginning to be explored; chemical changes in internal and volatile chemicals may have effects on herbivores as well as on the predators and parasites of those herbivores.

With a few exceptions, what we know about induced resistance stops at the level of herbivore performance. Induced resistance can affect some herbivore populations (table 4.6), but the generality of this interaction is still in question. Demonstrating an effect of induced resistance on populations of herbivores (statistically significant or otherwise) does not give us an appreciation for the ecological importance of induced resistance. Such a demonstration is dependent on the sample size involved in the experiment and the skill of the experimenter at reducing other sources of variation. Far more useful are experiments (or observations) that allow us to compare the strength of induced resistance relative to other ecological factors and processes that also affect population dynamics. For example, experimental manipulations of both early-season damage and plant density revealed that induced resistance affects populations of herbivores on wild cotton but that plant density exerts an even larger influence (Karban 1993b).

The task of understanding herbivore population dynamics is even more difficult than assessing relative effects of induced resistance on herbivore populations at any given point in time. Experiments must continue for many years, and even then, interpreting the causes of dynamic behavior is anything but straightforward. Mathematical models provide one useful avenue of research into this problem. Models of population dynamics resulting from induced responses are underdeveloped. Models incorporating any of a variety of factors might conceivably provide new insights; these models include spatially explicit dynamics, more than one plant species, plants that vary in their responses, different spatial or temporal scales, herbivores with different life histories, herbivores with different behaviors, and so forth.

Many ecologists became interested in induced resistance because of its potential to drive herbivore cycles. The role induced responses play in producing these cycles is still unresolved, although induced resistance is not sufficient to explain all of the effect. However, this long-standing problem is amenable to large-scale, multiple-factor experiments of the sort that Krebs and coworkers (1995) have conducted in the Yukon. An understanding of the processes that produce these dynamics will almost certainly involve several factors that likely interact nonadditively. Such explanations are not elegant, and revealing them requires large experiments, but they are more likely to illuminate the true nature of cyclic dynamics than is continued advocacy for a single, favored causal factor.

5   Induced Defense and the Evolution of Induced Resistance

5.1 EVOLUTIONARY PROCESSES AND INDUCED RESPONSES

Induced responses are, by definition, traits that change in response to the plant's environment (see section 1.2.1). Although the response itself and the character states expressed by the plant are facultative, the response system is constitutive and has evolved by normal, classical evolutionary processes (Sultan 1987; Bradshaw and Hardwick 1989; Via et al. 1995). Selection can be envisioned to act on the plant's ability to change and the range of phenotypes that the plant can assume. An individual's phenotypic repertoire is called its norm of reaction (Johannsen 1911; Schmalhausen 1949). Each genotype is assumed to have its own reaction norm that describes the phenotypes formed under different environmental conditions (van Noordwijk 1989). For traits that change monotonically and predictably in response to damage, the norm of reaction may reasonably be described by specifying the level of the trait with no damage, the maximum level of the trait following damage, and the rate of change in the trait as damage increases (figure 5.1). Some traits may exhibit more complicated responses to the environment, and more complicated norms of reactions are required to describe them (Via et al. 1995). In summary, the way in which a plant responds to herbivory may be viewed as a characteristic that is subject to selection much as fixed morphological or physiological characters are.

It is difficult to determine the evolutionary origin of a defensive trait (Endler 1986; Rausher 1992); in many cases the best we can do is to understand something about the evolutionary factors that maintain it (Berenbaum and Zangerl 1995). Traits originate by mutation (or perhaps recombination or insertion of a transposable element), and these are random events. However, herbivores may have been responsible for evolutionary changes in the trait: maintaining it, shaping it, or eliminating it. For example, as large fleshy fruits ripen, changes occur that seem hard to explain except as adaptations to increase the probability of consumption by frugivores. The changes that are associated with fruit ripening make fruits more attractive to frugivores, and these changes are frequently the mirror image of responses found after herbivore attack

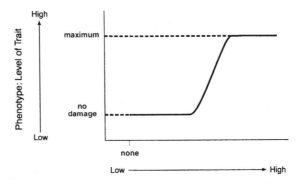

Fig. 5.1 A norm of reaction showing a plant's phenotypic response to a change in its environment. In this case the norm of reaction is monotonically increasing; the plant expresses the low phenotype when herbivory is low and the level of the trait increases as herbivory increases, up to a maximum level. Adapted from Harvell 1990a.

on leaves; tannins polymerize to lose their protein-binding activity, PIs and alkaloids are catabolized, cell walls are weakened enzymatically (Goldstein and Swain 1963; Heftmann and Schwimmer 1972; McKey 1974; Pearce et al. 1988). When these suites of traits increase in response to herbivory, it is plausible to assume that they are selected responses.

Stronger evidence for the evolution of traits is desirable, however, and it is within our abilities to obtain more convincing evidence. For an induced-response system, or any other trait, to evolve by natural selection three requisites must be met: (1) variability in the trait must exist in the population, (2) some of the variability must be heritable, and (3) possessing the trait must be associated with differential Darwinian fitness. The evidence for each of these requisites for systems of induced responses to herbivory will be considered below.

Evidence that these three requisites are met supports the hypothesis that an induced response could have been shaped by evolution, influenced in part by herbivory. This information relates only to what is now plausible, however, rather than what actually occurred in the past or the reasons that it occurred. For example, an induced response that increases a plant's fitness in contemporary environments that include herbivores is valuable in these environments. It may or may not have been valuable in the past. Such a trait may have originated as the result of selection by herbivores, or it may have been molded primarily by other agents.

It is virtually impossible to determine unambiguously the evolutionary forces that gave rise to particular defensive traits. Plant pathologists have been fascinated for nearly a century with the possibilities of coevolution between pathogens and plants (Farrer 1898; Flor 1971). Despite efforts to understand the evolutionary origins of resistance to pathogens, we have no clear demonstrations that any of the traits that provide resistance to plant diseases evolved in response to selection specifically by pathogens. It is not surprising that the much briefer efforts of workers interested in responses of plants to herbivores have not demonstrated convincingly that these facultative traits arose in response to selection specifically by herbivory.

5.1.1 Variability in Induction

Variability in a constitutive character is relatively easy to observe by growing different individuals in a common environment. However, variability in an induced trait requires that individuals be observed in at least two environments, with herbivores and without. Most studies of induced responses have not included information about whether some individuals or genotypes have consistently produced different responses to damage. For the seven plant species listed in table 5.1, however, different genotypes were found to differ in their induced responses (norms of reaction) when grown in a common environment and subjected to herbivory. This suggests that genetic variability exists for induced responses in some species, a result that should not be very surprising.

Some authors (e.g. Rauscher 1992) consider variability in complex traits like induced responses to be a given. However, this variability may not necessarily include phenotypes that are "optimal" for the plant, in terms of defense against herbivores. Variability in traits that affect defense may also be affected by other selective factors that are difficult to appreciate when only specific, limited defensive functions are considered. For example, Ehrlen and Eriksson (1993) considered various adaptive hypotheses to explain why fleshy fruits are toxic to some potential frugivores, a constitutive trait. They concluded that none of their hypotheses regarding fruit-frugivore interactions were particularly good at explaining the patterns that they observed. Instead, they found that plants that had leaves with high concentrations of secondary chemicals had fruits that also contained some of these same chemicals. In other words, variability in the defensive phenotypes of fruits may be driven by selection directed primarily by traits expressed in the leaf tissues. Variability in induced responses may similarly be affected by correlations

Table 5.1 Studies reporting a genetic basis to induced responses

Plant	Herbivore	Induced Response	Variability
Betula pubescens	*Epirrita autumnata*	Reduced growth rate, pupal weight	Length of response differed with geographic provence.
Brassica nigra	*Pieris* spp.	Hypersensitive response	Some individuals responded, some not.
Cynoglossum officinale	Artificial damage	Pyrrolizidine alkaloid accumulation	Variation between families; some increased, others decreased.
Gossypium hirsutum	*Tetranychus urticae*	Reduced population growth	Different varieties showed different strength.
Pastinaca sativa	*Depressaria pastivacella*	Furanocoumarin accumulation	Variation between individuals, populations and within individuals.
Pinus contorta	*Dendroctonus ponderosae*	Monoterpene accumulation	Variation in extent and rate of response.
Solidago altissima	*Eurosta solidaginis*	Hypersensitive response	Clones differed in resistance.

with other plant functions that are difficult to predict or understand. Parker (1992) argued that plant resistance may not always evolve to include the "optimal solution" because any of several factors may constrain the variability that selection can act upon. He suggested that time lags, population bottlenecks, constraints on gene flow, and plant mating systems may limit plant resistance. This is a problem that will be difficult to resolve, since we are limited to observing the variability that is now present and drawing inferences about the variability that may or may not have been present during evolutionary time.

5.1.2 Heritability of Induced-Response Systems

By examining a population of plants, an investigator can get some notion about how variable a particular trait, like an induced-response system, may be. However, not all of the variation observed may be subject to selection. Some of the variation may reflect a historical event, independent of the plant's response. For an extreme example, consider the resistance to insects of a plant that has been treated previously with a persistent synthetic insecticide. Such a plant may have become highly resistant to herbivores although this characteristic will not be passed on to its descendants.

The additive genetic variance is that portion of the phenotypic variance that is subject to natural selection; it can be estimated by several

Heritability	How Measured	Reference
Not measured	Common garden	Haukioja and Hanhimaki 1985
Not measured	Common environment	Shapiro and DeVay 1987
0.35 for shoots, 0.36 for roots, 0.36 for whole plants	Full-sib analysis	Van Dam and Vrieling 1994
Not measured	Common environment	Brody and Karban 1992
High, as much as 94% of variation	Half-sib analysis	Zangerl and Berenbaum 1990
Moderate to high	Unclear?	Raffa and Berryman 1982a, 1982b; Raffa 1991
High: 1.12 ± .22; 0.92 ± .33	Common garden, parent-offspring regression, sib analysis	Anderson et al. 1989; Maddox and Root 1987

approaches. Using a common garden to keep environmental conditions as constant as possible is one way to try to ensure that much of the variation observed is probably heritable. However, a common garden cannot provide as reliable an estimate of the additive genetic variance as parent-offspring regressions, sib analyses, or controlled breeding experiments. Heritability of a trait is equal to the ratio of the additive genetic variance to the overall phenotypic variance. Heritability is difficult to estimate, and rigorous estimates have been made for only a few plant defense traits that are phenotypically plastic (Schlicting 1986; Bradshaw and Hardwick 1989). Most estimates have very high variances and are meaningful only under limited environmental conditions.

Studies of the genetic basis for variation in induced responses of three plant-insect systems are particularly noteworthy. Clones of tall goldenrod (*Solidago altissima*), varied in their resistance to the ball gallmaker (*Eurosta solidaginis;* Anderson et al. 1989). Approximately 96% of larvae that developed within meristems of susceptible ramets survived, but only 27% of those that developed within meristems of resistant ramets survived. Much of the difference was due to a localized hypersensitive response of the plant to the gallmaker. Maddox and Root (1987) found that 24% of the variance in abundance of the gallmaker on plants in a common garden was attributable to plant genotype. Nearly all of this resistance was heritable, estimated by both parent offspring regression ($h^2 = 1.12 + 0.22$) and by sib correlation ($h^2 = 0.92 + 0.33$).

In a study of wild parsnip (*Pastinaca sativa*), accumulation of furano-coumarins after damage was found to vary among tissues within an individual, among individuals within a population, and among populations (Zangerl and Berenbaum 1990). Half-sib analysis was used to determine the heritability of this variation. A considerable fraction of the variation in plant response was explained by knowing the family that a particular plant came from: as much as 94% of the phenotypic variation in induced accumulation of furanocoumarins was attributable to additive genetic effects.

Families of hound's tongue (*Cynoglossum officinale*), grown in an experimental room, differed in their constitutive and induced concentrations of pyrrolizidine alkaloids (van Dam and Vrieling 1994). The families differed in the direction of induction, some increasing and some decreasing in response to damage. Differences in the magnitude of the induced responses were less than differences in constitutive variation in pyrrolizidine alkaloids among the same families. Approximately a third of the phenotypic variance in inducibility of pyrrolizidine alkaloid concentration was found to be heritable ($h^2 = 0.35$ for shoots, $h^2 = 0.36$ for roots, and $h^2 = 0.36$ for whole plants). Although pyrrolizidine alkaloids have been found to be deterrent and antibiotic to herbivores, van Dam and Vrieling (1994) argued that the inconsistency in the direction of the induced response makes it unlikely that the response could have evolved as a defense against herbivores. Their assumption that such variability is inconsistent with defense will be considered in section 5.4.1.

Estimates of heritability are only as meaningful as the information on which they are based. Good measures of a simple, monotonic induced response should include at least minimum and maximum levels of the trait, as well as the rate of change (see figure 5.1). Unfortunately, these parameters are not known for the systems described above. This makes interpretation of the high heritability measures less straightforward; high genetic variation is only one of several possible interpretations of these values. Nonetheless, the heritable genetic variation for inducibility suggested in these species probably indicates that selection can act on inducible plant systems, in other words, that inducibility can evolve. However, the scarcity of studies that have examined this issue precludes generalizations. Induced resistance that results from several chemical or physical traits interacting with each other and with several environmental conditions may have lower heritability than resistance caused by a few relatively simple traits. If this is true, estimates of heritability based on relatively simple traits, such as induction of furanocoumarins, may reflect an upper limit rather than a value that applies generally.

5.1.3 Differential Fitness: Do Induced Responses Defend Plants?

Heritable variation provides the raw material for evolution. However, induced responses will be favored as defensive systems by selection only if plants that respond under the appropriate circumstances experience higher Darwinian fitness than plants that lack such responses. Fitness should be determined ideally by measuring the relative representation of different phenotypes in the subsequent plant generation (Endler 1986). This is exceedingly difficult to do. Ecologists are often forced to measure correlates of Darwinian fitness such as plant survival, seed production, and pollen or flower production. In many cases the amount of leaf tissue consumed or plant biomass removed is used as an estimate of plant performance. Since plants are often able to compensate for lost tissue (Trumble et al. 1993), the amount or percentage of leaf tissue removed is a convenient measure to estimate, but it is a particularly difficult correlate of fitness to evaluate. Because measures of plant fitness in the discussion of induced defenses that follows are so crude, the evidence should be regarded with a large degree of skepticism.

Many studies have attempted to measure the benefits of induced responses in environments that include herbivory. Most of the studies listed in table 5.2 compared the effects of a subsequent challenge (feeding) by herbivores on plants that were either previously damaged (induced) or not. In order to more accurately estimate only fitness benefits of the induced defense, it would be desirable to remove potential fitness costs of the response. This can be accomplished by including two additional treatments: induced plants and undamaged plants, all of which are subsequently protected from all further herbivory (e.g. Karban 1993b). Despite the large list in table 5.2, the evidence that phenotypes that express induced responses experience higher fitness when herbivores are present than those that do not is weak. The entries in the table represent plant species for which any fitness benefit has been reported, although the benefits are inconsistent in many cases. For example, the best-studied species are the birches, *Betula pendula* and *B. pubescens* (table 5.3). The variable and inconsistent results from these studies of the effects of induction on subsequent consumption of birch foliage underscore the point that generalizations cannot be made, based solely on knowledge of a particular plant species. In general, information based on field tests is likely to be more reliable than lab results. The laboratory assays remove differences in recruitment of herbivores that could be important in determining actual amounts of damage to plants. Laboratory assays also exclude much of the external environment that could deter-

Table 5.2 Induced responses that provide a benefit to the plant

Plant	Herbivore Affected	Lab/Field	Fitness Measure Affected	Relationship	Mechanism Hypothesized	Reference
Abies grandis	Bark beetles	Field	Tree survivorship	Positive correlation: induction and survival	Monoterpene accumulation	Raffa and Berryman 1982
Acacia seyal	Giraffes	Field	% branches eaten, # shoots, branch growth	Dethorned branches eaten more and grew less	Thorns	Milewski et al. 1991
Alnus crispa	Hares	Field	% consumption of bioassay twigs	Resins decreased biomass removed	Resins	Bryant 1981
Alnus glutinosa	Generalist caterpillars	Lab	Leaf area consumed	Less damaged foliage removed	?	Edwards et al. 1986
Alnus incana	Generalist caterpillars	Lab	Leaf area consumed	Less damaged foliage removed	?	Edwards et al. 1986
	Hares	Field	% biomass consumed	Less consumption of regrowth shoots	Juvenile reversion	Bryant, Danell, et al. 1991
Asclepias syriaca	Generalist caterpillars	Lab	Leaves fed upon	Leaves with latex escaped damage	Latex	Dussourd and Eisner 1987
Betula glandulosa	Hares	Field	% biomass consumed	Less consumption of regrowth shoots	Juvenile reversion	Bryant, Danell, et al. 1991
Betula pendula	?	Field	Leaf area consumed	Less damaged foliage removed	?	Silkstone 1987
	Hares	Field	% biomass consumed	Less consumption of regrowth shoots	Juvenile reversion	Bryant, Danell, et al. 1991

Plant species	Herbivore	Field/Lab	Measurement	Result	Mechanism	Reference
Betula pubescens	?	Field	Leaf area consumed	Less damaged foliage removed	?	Silkstone 1987
Betula pubescens	Hares	Field	% biomass consumed	Less consumption of regrowth shoots	Juvenile reversion	Bryant, Danell, et al. 1991
Betula resinifera	Hares	Field	% biomass consumed	Less consumption of regrowth shoots	Juvenile reversion	Bryant, Danell, et al. 1991
Carpinus betulus	Generalist caterpillars	Lab	Leaf area consumed	Less damaged foliage removed	?	Edwards et al. 1986
Corylus avellana	Generalist caterpillars	Lab	Leaf area consumed	Less damaged foliage removed	?	Edwards et al. 1986
Crataegus monogyna	Generalist caterpillars	Lab	Leaf area consumed	Less damaged foliage removed	?	Edwards et al. 1986
Fagus sylvatica	Generalist caterpillars	Lab	Leaf area consumed	Less damaged foliage removed	?	Edwards et al. 1986
Fucus distichus	Snails	Field	Leaf area consumed	Less damaged biomass consumed	?	Van Alstyne 1988
Larrea tridentata	Jackrabbits	Field	% damage to bioassay shrubs	Less new damage to previously damaged plants	?	Ernest 1994
Lepidium virginicum	Generalist caterpillars	Lab	Leaf area consumed	Less damaged foliage removed	?	A. A. Agrawal, personal communication
Lycopersicon esculentum	Generalist caterpillars	Lab	Leaf area consumed	Less damaged foliage removed	?	Edwards et al. 1985
Populus balsamifera	Hares	Field	% consumption of bioassay twigs	Resins decreased biomass removed	Resins	Bryant 1981

(continues)

175

Table 5.2 continued

Plant	Herbivore Affected	Lab/Field	Fitness Measure Affected	Relationship	Mechanism Hypothesized	Reference
Populus tremuloides	Hares	Field	% consumption of bioassay twigs	Resins decreased biomass removed	Resins	Bryant 1981
Quercus robur	Generalist caterpillars	Lab	Leaf area consumed	Less damaged foliage removed	?	Edwards et al. 1986
Raphanus raphanistrum	Grasshoppers	Field	Leaf area consumed	Less damaged foliage removed	?	A. A. Agrawal, personal communication
Salix spp.	Leaf beetles	Field	Leaf area consumed	Less foliage removed from damaged shoots	?	Raupp and Sadof 1991
Salix alaxensis	Hares	Field	% biomass consumed	Less consumption of regrowth shoots	Juvenile reversion	Bryant, Danell, et al. 1991
Salix arbusculoides	Hares	Field	% biomass consumed	Less consumption of regrowth shoots	Juvenile reversion	Bryant, Danell, et al. 1991
Salix bebbiana	Hares	Field	% biomass consumed	Less consumption of regrowth shoots	Juvenile reversion	Bryant, Danell, et al. 1991
Salix caprea	Hares	Field	% biomass consumed	Less consumption of regrowth shoots	Juvenile reversion	Bryant, Danell, et al. 1991
Salix cinerea	Generalist caterpillars	Lab	Leaf area consumed	Less damaged foliage removed	?	Edwards et al. 1986
Salix glauca	Hares	Field	% biomass consumed	Less consumption of regrowth shoots	Juvenile reversion	Bryant, Danell, et al. 1991

176

Species	Herbivore		Measure	Result	Mechanism	Reference
Salix nigricaus	Hares	Field	% biomass consumed	Less consumption of regrowth shoots	Juvenile reversion	Bryant, Danell, et al. 1991
Salix pentandra	Hares	Field	% biomass consumed	Less consumption of regrowth shoots	Juvenile reversion	Bryant, Danell, et al. 1991
Salix phylicifolia	Hares	Field	% biomass consumed	Less consumption of regrowth shoots	Juvenile reversion	Bryant, Danell, et al. 1991
Sambucus nigra	Generalist caterpillars	Lab	Leaf area consumed	Less damaged foliage removed	?	Edwards et al. 1986
Sinapis alba	Flea beetles	?	Leaf area consumed	Less damaged foliage removed	?	Palaniswamy and Lamb 1993
Sorbus aucuparia	Generalist caterpillars	Lab	Leaf area consumed	Less damaged foliage removed	?	Edwards et al. 1986
Trifolium repens	Snails	Field	Leaf area consumed, seedling survival	Less damaged foliage removed, positive correlation: induction and survival	Cyanogenesis	Dirzo and Harper 1982a, 1982b; Kakes 1989
Vitis spp.	Phylloxera	Field	Plant survivorship	Positive correlation: induction and survival	Abscission	Borner, cited in Muller 1959; Boubals 1966

177

Table 5.3 Studies of effects of induction on subsequent leaf area removed for birch trees

Birch Species	Source of Initial Damage	Herbivore Used in Bioassay of Subsequent Damage	Lab/Field
Pubescens	Artificial	Snail, *Helix aspersa*	Lab
Pubescens	Artificial	Generalist caterpillar, *Spodoptera littoralis*	Lab
Pubescens	Artificial	Generalist caterpillar, *Orgyia antiqua*	Lab
Pubescens	Artificial	*Spodoptera littoralis*	Lab
Pubescens	Artificial	?	Field
Pubescens	Artificial	?	Field
Pubescens	Artificial	?	Field
Pubescens	Artificial	Caterpillar, *Apocheima pilosaria*	Field
Pendula	Artificial	Generalist caterpillar, *Spodoptera littoralis*	Lab
Pendula	Artificial	Generalist caterpillar, *Orgyia antiqua*	Lab
Pendula	Artificial	*Spodoptera littoralis*	Lab
Pendula	Artificial	Caterpillar, *Orthosia stabilis*	Field
Pendula	Artificial	?	Field
Pendula	Artificial	*Apocheima pilosaria*	Lab
Pendula	Artificial	*Epirrita dilutata*	Lab
Pendula	Artificial	*Erannis defolaria*	Lab
Pendula	Artificial	*Euproctis similis*	Lab
Pendula	Natural mining	*Apocheima pilosaria*	Lab
Pendula	Natural mining	*Epirrita dilutata*	Lab
Pendula	Natural mining	*Erannis defolaria*	Lab
Pendula	Natural mining	*Euproctis similis*	Lab
Pendula	Natural chewing	*Apocheima pilosaria*	Lab
Pendula	Natural chewing	*Epirrita dilutata*	Lab
Pendula	Natural chewing	*Erannis defolaria*	Lab
Pendula	Natural mining	Caterpillars	Field
Pendula	Artificial	*Coleophora serratella*	Field

Local/Systemic	Result	Reference
Local	Snail ate more of undamaged leaf disk.	Edwards and Wratten 1982
Systemic to branch	Caterpillars ate more undamaged foliage.	Wratten et al. 1984
Local only	Caterpillars ate more undamaged foliage.	Wratten et al. 1984
Damaged and adjacent leaf	Caterpillars ate more undamaged foliage.	Edwards et al. 1986
Local	More undamaged foliage removed.	Edwards and Wratten 1985
Local and systemic to tree	Damaged foliage more likely to escape further damage.	Silkstone 1987
Systemic to tree	No differences.	Fowler and Lawton 1985
Systemic to branch	No differences.	Fowler and MacGarvin 1986
Systemic to branch	Caterpillars ate more undamaged foliage.	Wratten et al. 1984
Systemic to branch	Caterpillars ate more undamaged foliage.	Wratten et al. 1984
Local	Preference for undamaged not statistically significant.	Edwards et al. 1986
Local and systemic to shoot	Caterpillars ate more undamaged foliage.	Edwards et al. 1991
Local and systemic to shoot	Damaged foliage more likely to escape further damage.	Silkstone 1987
Local	No difference.	Hartley and Lawton 1987
Local	Caterpillars ate more damaged foliage.	Hartley and Lawton 1987
Local	Caterpillars ate more damaged foliage.	Hartley and Lawton 1987
Local	No difference.	Hartley and Lawton 1987
Local	Caterpillars ate more undamaged foliage.	Hartley and Lawton 1987
Local	Caterpillars ate more undamaged foliage.	Hartley and Lawton 1987
Local	Caterpillars ate more undamaged foliage.	Hartley and Lawton 1987
Local	Caterpillars ate more undamaged foliage.	Hartley and Lawton 1987
Local	No difference.	Hartley and Lawton 1987
Local	No difference.	Hartley and Lawton 1987
Local	Caterpillars ate more undamaged foliage.	Hartley and Lawton 1987
Local	No difference.	Hartley and Lawton 1987
Local	Caterpillars ate more undamaged foliage, but damage was clumped.	Bergelson et al. 1986

mine the effects of induced responses on plant fitness. For example, the consequences of induced responses may be affected strongly by plant density in nature (Karban 1993a, 1993b). Studies of potted plants in the lab or greenhouse would likely have presented an unrealistic picture in this case. Laboratory assays are also particularly prone to methodological biases that could influence the results (Risch 1985; Hartley and Lawton 1991).

A serious limitation of most of the studies in table 5.2 is that they measure only leaf area or biomass removed. The relationship between leaf area removed and plant fitness is unclear. Survival is a much better correlate of fitness than leaf area or biomass removed. Only three studies in table 5.2 (Borner's and Boubal's work on grapes [*Vitis* spp.], Dirzo and Harper's studies of clover seedlings [*Trifolium repens*], and Raffa and Berryman's work on grand fir trees [*Abies grandis*]) reported that plants showing induced responses are more likely to survive than those showing no response. In the case of clovers the induced response was correlated with the distribution of plants in the field, providing further evidence that the induced response increased plant fitness in environments with herbivory (Dirzo and Harper 1982a, 1982b). In areas where mollusks were abundant, plants with the ability to release cyanogenic glycosides in response to damage were overrepresented. Conversely, where mollusks were rare, plants that lacked this ability were more common.

Only two published studies that we are aware of have looked for differences in reproductive benefits for plants with induced response when herbivores were present. Cultivated cotton plants that were damaged as seedlings supported smaller populations of spider mites throughout the growing season than undamaged control plants in field trials (Karban 1986a). Despite this effect, vegetative growth, number of bolls, and weight of seeds produced per plant were no higher for plants with induced responses. Similarly, for wild cotton, *Gossypium thurberi,* induction reduced the cumulative number of leaf miners (the most abundant herbivores), but no benefits in terms of numbers of bolls, seeds per boll, or mean seed mass were detected (Karban 1993b). This negative evidence may indicate that induction really does not increase plant fitness or may reflect an inability to detect such an effect. Since *Gossypium* spp. are woody shrubs that set seeds numerous times during their lives, possible benefits of induction may become apparent only in long-term studies. Future studies of annual plants may provide more definitive estimates of benefits of induced responses on lifetime reproduction.

In one such study exogenous application of jasmonic acid to the roots of wild tobacco (*Nicotiana attenuata*), an annual of the Great Basin, induced plants to increase their nicotine pools (see section 3.3.3.4).

Treated plants produced approximately 20% more viable seeds during their lifetime than uninduced control plants when both treatments were subjected to naturally occurring herbivory in a field experiment (I. T. Baldwin, personal communication).

Hypothetically, plants may benefit from induced responses by several indirect means. First, induced responses may affect the spatial distribution of the damage that a plant receives without affecting the total amount of biomass removed. Different distributions of damage may have very different effects on plant fitness. Second, induced responses may make the plant more attractive to the predators and parasites of herbivores. Increased numbers of herbivore enemies may reduce the negative effects of herbivory, without changing the actual performance of the herbivores (growth rate, fecundity, etc.). Evidence for these scenarios is presented in sections 5.1.4 and 5.1.5.

5.1.4 Induced Responses and the Distribution of Damage within a Plant

Induced responses may provide a benefit by redistributing the damage caused by herbivores without actually reducing the amount of biomass removed (Janzen 1979; Edwards et al. 1991). This hypothesis requires that the following conditions are met: (1) Induced responses in plants are local and either are not systemic or are at least diminished away from the site of attack. (2) Herbivores move away from the site of damage. (3) This movement has the effect of benefiting the plant if evenly distributed loss of tissue is less detrimental than highly clumped damage, if herbivores remove biomass of less value to the plant when damage is dispersed, or if plants can more easily compensate for dispersed damage. These conditions are discussed below.

The distributions of induced responses within a plant have been relatively poorly studied. Many workers have reported that chemical changes or effects on bioassay herbivores could be detected away from the site of attack, but detailed maps of induced responses have not been constructed. Two studies of distributions of induced responses are particularly noteworthy. Induced responses of damaged birch trees were much stronger within the branch that was damaged compared to undamaged branches (Tuomi, Niemelä, Rousi, Sirén, and Vuorisalo 1988). In a beautifully designed experiment, Jones et al. (1993) showed that induced responses of cottonwood trees were affected by the architecture of the plant's plumbing (figure 5.2). Every fifth leaf on a cottonwood shoot is attached to a common vascular connection (Davis et al. 1991). Damage to a particular leaf induced resistance to leaf-chewing beetles in the

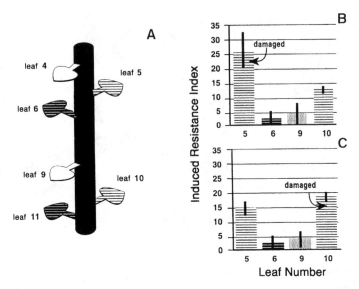

Fig. 5.2 Induced resistance and the plumbing of cottonwood trees (redrawn from Jones et al. 1993). *A*, every fifth leaf shares a common vascular connection (and symbol). For example, leaves 4 and 9 are connected, as are leaves 5 and 10, and leaves 6 and 11. *B*, following damage to leaf 5, leaves 5 and 10 (connected leaves) exhibit increased resistance relative to leaves 6 and 9. *C*, the signal for induced resistance can move up or down the shoot. Following damage to leaf 10, leaf 10 and leaf 5 (connected leaves) exhibit increased resistance relative to leaves 6 and 9.

leaves above and below the damaged leaf, but only for those leaves that shared a vascular connection with the damaged leaf. Adjacent leaves that had different vascular connections responded far less. Interestingly, these two studies were conducted on plants that had been assumed to exhibit systemic induction (see Haukioja and Niemelä 1977, 1979, and Haukioja et al. 1985 for *Betula pubescens;* Clausen et al. 1991 for *Populus* spp.).

Several workers found that herbivores moved away from or avoided damaged foliage, which resulted in a more evenly distributed pattern of damage than expected. Leaf miners were found to prefer undamaged oak leaves to previously damaged ones (Faeth 1986). However, densities of miners were not different between control and experimental trees, suggesting that females preferred undamaged leaves but not undamaged trees. Stem-galling sawflies rarely oviposited in a previously attacked node of arroyo willow, producing a distribution that was more uniform than expected by random attack (Craig et al. 1986). Casebearer moth larvae moved away from artificially damaged birch foliage (Bergelson et al. 1986). They didn't move far, however, and damage was found to be aggregated rather than evenly dispersed. Artificially damaged birch leaves received less subsequent overall tissue loss throughout the season and were more likely to escape all further damage (Silkstone 1987). Finally, caterpillars of the common quaker moth (*Orthosia stabilis*) moved more often, tried more different leaves, and removed less area per leaf when feeding on artificially damaged birch leaves than on undamaged control leaves (Edwards et al. 1991). Only Craig et al. (1986) demonstrated that the pattern of damage that they observed was statistically more uniform than expected if attack had been random, although most of the other authors assumed that such a pattern was probable for their systems.

Unfortunately, the consequences of more evenly distributed loss of tissue were not evaluated for these studies in which herbivores moved away from damaged foliage. However, several other workers have compared the effects on plant fitness of damage that was either evenly dispersed throughout the plant or concentrated. Seedlings of coachwood, *Ceratopetalum apetalum*, had either $x\%$ leaf tissue removed from every leaf or complete removal of $x\%$ of entire leaves (Lowman 1982). Removal of entire leaves (concentrated damage) was more detrimental than removal of portions of leaves (dispersed damage). Similarly, 10% of the leaf area of the tropical shrub *Piper arieianum* was either removed from a single branch or spread throughout the entire canopy (Marquis 1992). Plants with concentrated herbivory grew less overall and produced fewer viable seeds than those with more diffuse damage. The same result was also found for wild radish (*Raphanus sativus*) when subjected to four different distributional patterns of leaf removal (Mauricio et al. 1993). More dispersed damage was less detrimental in terms of numbers of flowers produced, reproductive biomass, and total biomass.

Plants may benefit by redistributing damage, by spreading it evenly throughout the canopy as discussed above, or by directing damage away from the most valuable and productive tissues. Krischik and Denno (1983) reviewed studies that considered the effects of removal of leaves at different shoot positions. Most of the evidence came from studies of conifers or agricultural crops. Removal of terminal and subterminal foliage had greater negative effects on conifer growth and survival than removal of older basal needles. Similarly, defoliation of the upper parts of sorghum, potato, and sunflower plants was more detrimental than removal of leaves from their respective lower parts. Edwards and Wratten (1987) developed the argument that, for many plants (trees) in productive habitats, light becomes the limiting resource. Plants that are competing or "foraging" for light do so primarily with their outermost leaves.

Under these conditions, induced responses that direct herbivory away from the outermost leaves may increase plant fitness without reducing the total amount of leaf area consumed. Similar conclusions were reached when the nonuniform distribution of photosynthetic capacity throughout a plant's canopy was considered (Stockhoff 1994). This mechanism involves changes in competitive abilities of plants resulting from induction; it seems plausible, although it has not been tested directly.

Plants may be able to compensate more fully for evenly distributed damage than for concentrated damage. Watson and Casper (1984) argued that, if plants are made up of autonomous units, such as branches, with only limited exchange of resources, then damage to foliage that was concentrated within one of these units may severely reduce reproduction. According to this model the plant does not shift resources to average out, or spread, the effects of its losses (see also Sprugel et al. 1991). If the same amount of damage is spread evenly over the plant, the remaining leaves may more easily compensate for the losses by increasing photosynthesis, and there may be no discernible decrease in reproduction. There is considerable evidence that many plants are unable to compensate for complete defoliation of entire branches (reviewed by Marquis 1992). The converse, that plants can more easily compensate for evenly distributed damage, is less well documented.

5.1.5 Induced Responses and Increased Predation on Herbivores

Induced responses may benefit plants indirectly by making herbivores more vulnerable to predation, parasitism, and disease. This mechanism was considered in section 3.2.3. There is limited evidence that herbivores feeding on plants of constitutively low quality suffer greater rates of predation than individuals on higher-quality hosts (Haggstrom and Larsson 1995). Willow leaf beetles (*Galerucella lineola*) on poor hosts grew more slowly, resulting in longer exposure to predators and increased daily rates of predation. In addition, several authors have argued that induced responses can benefit the damaged plants by attracting "bodyguards." This anthropomorphic terminology can be confusing; the suggestion is not that plants benefit by being damaged but rather that predators and other natural enemies of herbivores may reduce plant losses if they are more effective on damaged plants.

Induced responses could conceivably benefit damaged plants via natural enemies in several ways (Schultz 1983; Price 1986; Faeth 1991). (1) Natural enemies may use induced responses as cues to locate herbivores and therefore reduce the damage caused by herbivory. (2) Induced responses may make plants more heterogeneous, concentrating the distribution of herbivores (contrary to the argument in section 5.1.4) and making density-dependent enemies more effective. (3) Induced responses may cause herbivores to move more frequently or farther (see supporting evidence in section 5.1.4), increasing their risk of contacting natural enemies and reducing the damage they inflict. (4) Induced responses may prolong development of herbivores, making them more likely to be discovered by natural enemies and decreasing the damage they cause.

The evidence that any of these mechanisms operate in nature is rather limited, and even when the mechanisms do occur, the consequence to plant fitness has rarely been considered. For example, Dicke and Sabelis (1989) found that predatory mites are attracted to induced cues released by spider mites feeding on the foliage of lima beans. They argued that this interaction must necessarily benefit the plants because spider mites can be devastating pests, and predatory mites can destroy spider-mite populations before damage has occurred. These arguments were based on lab results, however, and Dicke and Sabelis have not compared the relative fitness of plants that release cues and attract predators with plants that do not.

Until recently, predator attraction to cues released by damaged plants was a phenomenon restricted to laboratory wind tunnels and olfactometers. However, studies involving anthocorid bugs that are attracted to pear trees (*Pyrus communis*) experiencing attacks by herbivorous psyllids (*Psylla* spp.) offer exciting, though preliminary, field results supporting this hypothesis (Drukker et al. 1995). When pear trees are infested with psyllids, the concentration of polyphenols increases, as does the amount of volatiles released (Scutareanu et al. 1996). Two of the volatiles released, *E, E-a*-farnesene and methyl-salicylate, in Y-tube olfactometer studies are attractive to the anthocorid predator *Anthocoris nemoralis*. In a pear orchard Drukker et al. (1995) manipulated psyllid densities with screened cages that contained the psyllid herbivores but allowed the putative volatiles to pass through. They found more bugs surrounding caged pear trees with high infestations of psyllids than surrounding caged trees with low infestations. They replaced the screens with plastic sheets that prevented airflow from one of the high infestation trees; the aggregation of predators disappeared. When the plastic sheet was replaced with screen again, the predator aggregation reappeared. These field experiments are on the right track, although they suffer from problems of experimental design; the cages may have multiple effects in addition to those intended, the experiments were poorly replicated, and the few replicates of each treatment were not interspersed (Hurlbert 1984). These results suggest that predators are attracted over long distances by some cues released by trees with high densities of psyllids; they do not demonstrate which of the many potential cues are being used by predators for this aggregation response. For example, predators could be using sound, or odors released by the herbivores themselves, rather than synomones released by the damaged plants. The story will become more compelling once these workers from the Netherlands have isolated the attractive volatiles, conducted "perfume" experiments in the field, and demonstrated that the predators reduce psyllid populations and benefit host trees.

We are aware of only one study that considered the effects of increased attack by natural enemies from the plant's perspective. Faeth (1990) found increased rates of attack by natural enemies and decreased survival of leaf miners on damaged leaves of emory oaks (*Quercus emoryi*). However, experimentally damaged trees did not produce significantly more leaves (a correlate of plant fitness?) than did undamaged trees (Faeth 1992b). Unfortunately, these negative results were particularly difficult to interpret because, in the experiment that considered leaf production (Faeth 1992b), leaf damage did not influence attack by hymenopteran parasitoids or vertebrate and invertebrate predators, contrary to earlier studies (Faeth 1986, 1990).

The net effect of indirect interactions can be either positive or negative for the plant. On the one hand, induced responses might increase the probability of exposing an herbivore to predation, parasitism, or disease and thereby function as a plant defense. However, the induced response might have a greater negative impact on the predators or diseases of the herbivore than on the herbivore itself and, as such, represent a liability for the plant. For example, leaf trichomes sometimes hinder the parasites of herbivores more effectively than the herbivores themselves (Norris and Kogan 1980). Similarly, dietary phenolics, particularly hydrolyzable tannins, which are known to be inducible in oak trees (Schultz and Baldwin 1982; Rossiter et al. 1988), inhibit the ability of a pathogenic nuclear polyhedrosis virus to infect gypsy-moth caterpillars (*Lymantria dispar;* Keating et al. 1990; Schultz and Keating 1991), thereby decreasing an important mortality factor for gypsy moths. In addition, induced chemical responses may prove to be a liability for a plant when it is faced with an herbivore that sequesters those "defensive" chemicals for its own uses. Hence, all induced plant traits, even those that have proven defensive in laboratory assays, need to be tested in the field before they can be considered as playing a defensive role in nature. The complexity of ecological interactions often wreaks havoc with even the best-laid plans and working hypotheses.

In general, possible benefits of induced responses to plants, due to all of the indirect mechanisms considered (evenness of damage, plant compensation, interactions with natural enemies), have been virtually unexplored. Certainly, examinations of the consequences of induced responses from the plant's perspective would repay further study. Information about the situations in which induced defenses benefit plants and the mechanisms by which this occurs is essential before an understanding of the evolution and maintenance of induced responses is possible.

5.2 Induced Defenses as Incidental Effects of Defoliation or Evolved Responses to Herbivory: A Bogus Dichotomy?

It is naive to assume that traits that provide defense against herbivory necessarily evolved to "fulfill this function." They may or may not have been maintained in the recent past because they help to fulfill this hypothesized function. For example, traits we presume to have been preserved because they provide defense may have been pulled along by pleiotropy, in which one gene affects many different traits, or by linkage, in which genes are bound together and not inherited as distinct entities. We must be willing to consider and develop other plausible evolutionary scenarios, what Kitcher (1985) called rival Darwinian histories, to explain why we find induced defenses. We must also be willing to accept that some current traits may not result in the highest fitness but may persist because of lack of variation, as a consequence of genetic drift, or because of selection on another correlated trait.

The argument that induced responses are adaptations to herbivory attempts to provide an ultimate explanation for the characteristics we observe. It addresses questions such as, why are induced responses maintained? The observation that induced responses are affected by the resources that a plant receives during the course of an experiment helps to elucidate possible physiological mechanisms of induced defense. This is a proximate explanation and addresses "how" questions such as, how does induced defense work? Understanding the proximate mechanisms is vitally important in its own right and will lead to formulation of better ultimate explanations. Proximate explanations cannot replace ultimate explanations, however, and the two should not be compared as competing, mutually exclusive hypotheses, as they have often been in the past.

The notion that induced resistance is the incidental result of defoliation stemmed from observations that carbon-based phenolics accumulate following extensive removal of leaves (see C/N balance in section 3.3.3.1). Tuomi et al. (1990) argued convincingly that patterns of delayed inducible responses are affected by proximate mechanisms and by the functional organization of the plant. Those plants with storage systems that leave them with a relative excess of carbon following severe defoliation are the ones that accumulate high concentrations of phenolic compounds. The way that different plants store nutrients may constrain the evolution of this particular form of induced response. In some instances, understanding mechanisms may provide considerable insight into the whens and wheres of induced resistance (Tuomi et al. 1984, 1990, 1991; Bryant, Danell, et al. 1991; Bryant, Heitkonig, et al. 1991; Bryant et al. 1993). In addition, understanding the physical consequences of removing tissues that are relatively rich in one nutrient and poor in another may explain why defoliated plants respond as they do to simulated herbivory. In this case physical laws may affect why we see the particular "atomized trait" that we are calling induced resistance. We may be trying to come up with an explanation for induced responses, independent of the rest of the plant, when in fact induced responses are the result of larger-scale plant processes. However, such a consideration cannot substitute for evolutionary hypotheses that try to address plant processes, including induced resistance. To explain why we observe induced defenses may require a credible evolutionary scenario explaining why different plants have different primary metabolisms or storage systems in the first place. Furthermore, the existence of physiological constraints on plant metabolites that could potentially produce induced defenses does not preclude the possibility that these "incidentally" produced chemicals have been tailored to serve defensive functions (see Baldwin 1994).

Credible hypotheses include those that attempt to explain the commonness of induced responses because they increase plant fitness when pathogens or other stresses are present. In addition, induced responses may have been favored because they increase plant fitness in some "civilian" rather than "defensive" roles. For example, Hendry (1986) suggested that preformed defenses that are accumulated and isolated in plant vacuoles might have been acquired by plants as an adaptive solution to avoid autotoxicity (a civilian role) and now provide a secondary benefit as defenses against herbivores. Hypotheses such as these have not been developed to any extent for most induced responses of plants, and development of rigorous alternative hypotheses would be of great value to the field.

Some authors have argued that induced responses are incidental rather than evolved because the responses serve other functions in the plant. For example, resins, tannins, and toughened leaves may contribute to drought tolerance. Secondary chemicals that are deterrent or antibiotic to herbivores may provide benefits in reducing germination or growth of other plant competitors. Secondary chemicals may also benefit plants by reducing attacks by pathogens and other plant parasites. Traits with multiple functions may be adaptations against herbivory if they increase the plant's fitness in environments that include herbivores (following Reeve and Sherman 1993). Natural selection preserves traits that increase fitness without regard to where those traits came from originally, or why. Janzen (1979, 334) foresaw this confusion when he argued that "a change in secondary chemistry following damage may be the result of a rearrangement of internal priorities rather than an explicit attempt to produce facultative defenses." Indeed, there are no such explicit attempts in the process of evolution. Traits that arise by mutation or as "solutions to other problems" may be preserved and molded by natural selection if they increase plant fitness in response to herbivory. It seems quite clear that most, if not all, of the plant traits that provide defense against herbivores produce many other consequences as well. Selection probably acts on any character from many different directions, due to the varied consequences associated with the character. Thinking about traits as adaptations molded solely by herbivory is unrealistic. Arguing that herbivory played no role simply because the traits serve multiple purposes is equally unrealistic. At some level almost all precursors to current traits evolved to serve functions other than their current uses (Gould and Vrba 1982). However, even if we could determine why a trait originally arose (and we rarely can for plant defensive traits), this does not preclude the hypothesis that the trait has been maintained and tailored because it defended the plant during more recent times (Reeve and Sherman 1993).

Some authors have argued that induced responses that can be stimulated by diverse cues, in addition to herbivores, probably are not adaptations to herbivory. This argument is not convincing. Much as adaptations to herbivory may have multiple functions, so too the number of different stimuli that provoke a response does not provide strong evidence about why the plant responds now or did so in the past.

In summary, physiological hypotheses such as C/N balance are extremely useful and interesting. Elucidating the physiological mechanisms of induced defense will improve our understanding of all aspects of induced responses, including their evolution. However, proximal and ultimate explanations answer different questions. In no way are they mutually exclusive; they cannot be compared as competing hypotheses. Ultimate explanations for induced responses resulting from selective forces other than herbivores should be developed (see section 5.4.3). These and the notion that induced defenses were shaped by herbivory should be considered as plural, rather than mutually exclusive, hypotheses.

The debate about whether induced defenses are the result of selection exerted by herbivory mirrors the larger debate that has raged in evolutionary biology during the past fifteen years about whether particular traits can be considered as adaptive or optimal solutions to evolutionary problems. Gould and Lewontin (1979) challenged the conventional adaptationist practice of attempting to explain "atomized traits" as structures optimally designed by natural selection for their functions. They argued that some traits such as the spandrels or triangular supports to a domed ceiling cannot be explained, by themselves, as optimal solutions. Instead, triangular supports are necessary consequences of mounting domes on rounded arches. The shape of the supports is constrained by physical laws, once the other architectural components are specified. In another example, they argued that we cannot explain why blood is red, even if it turns out that blushing is affected by sexual selection. Red blood is constrained by historical, biological, and chemical properties that are far more useful in explaining the color of blood than is sexual selection. Their argument boils down to separating cause and effect. A trait may well be useful currently for a particular function without its having evolved (or even been maintained) for that function. In summary, it is important to recognize that physical laws, selection against traits that disrupt other processes, and lack of variation may constrain adaptations (Maynard Smith et al. 1985; Reeve and Sherman 1993). However, recognition of such constraints cannot replace attempts to understand and critically evaluate ultimate, evolutionary explanations.

5.3 Costly Defenses and Evolutionary Hypotheses to Explain Induced Defenses

5.3.1 Optimal Allocation Models

Traits that provide resistance to herbivory are generally assumed also to benefit plants that possess them. Although this assumption need not hold in all circumstances, it probably is true in most. Induced responses were associated with reduced loss of leaf biomass in many examples (table 5.2); this effect is probably reflected in improved fitness for plants with induced responses to herbivory, although the real evidence supporting this hypothetical relationship is surprisingly weak (section 5.1.3). If induced traits provide defense and benefit plants when herbivores are present, why doesn't natural selection fix them in the population as constitutive traits?

Traits that defend plants may be costly. Plants have a finite pool of resources that they can allocate to defense or to current reproduction or to growth (future reproduction). Natural selection may eliminate plants that allocate resources to defense at the expense of current or future reproduction. If so, the kinds and amounts of resources that are available to a plant will affect the costs of different defenses.

Induced defenses may be favored because they may allow a plant to erect defenses when they are needed and to dismantle them when they are not needed. Plants with induced defenses may accrue the benefits of defense when herbivores are present and not pay the costs of being defended when herbivores are absent. In theory such a facultative system should allow plants to allocate limited resources so as to maximize fitness in response to current, unpredictable conditions.

If defenses are truly costly, plants should allocate resources to defenses so as to optimize their investments, measured as the ratio of benefits to costs of defense (Feeny 1976; McKey 1979; Rhoades 1979; Fagerstrom et al. 1987). Plants or plant tissues that are more likely to be attacked over evolutionary time should invest more resources in defense than those that are more likely to escape attack. Plant tissues that are more valuable to the plant should be better defended than tissues of lesser value or those that can be readily replaced.

In the absence of a cost, it has been difficult to imagine what could favor intermittent induced defenses rather than permanent constitutive defenses (but see section 5.4). Most of the verbal models of the evolution of induced defenses were based on the notion that plants reduce costs by employing facultative responses triggered by herbivory (e.g. Haukioja and Hakala 1975; Rhoades 1979; Harvell 1986, 1990a, 1990b; Karban and Myers 1989; Baldwin et al. 1990).

Similarly, all of the ten formal attempts to model the conditions that allow plants with inducible defense systems to invade a population without inducible defense systems considered the cost savings of defenses that are employed only when beneficial (table 5.4). Several of the models used relatively simple comparisons between benefits and costs and concluded that induced or conditional defenses are favored whenever benefits associated with reduced herbivory exceed costs of the defense (Clark and Harvell 1992; Riessen 1992; Till-Bottraud and Gouyon 1992; Frank 1993). Clark and Harvell (1992) constructed their model with induced morphological defenses of aquatic invertebrates in mind and argued that costs of defense become more important late in the season (or in the lifetime of a perennial organism), when compensation for costs becomes more difficult. Riessen (1992) designed his model using actual demographic data from *Daphnia* populations and argued that the cost of defense was more likely to be increased development time for induced animals rather than allocation of energy or nutrients. Till-

Table 5.4 Mathematical models of the conditions favoring conditional (induced) defense versus constitutive defense

Reference	Relationship between Fitness and defense cost
Levins 1968	Semiquantitative relationship; benefit: cost ratio maximized by comparison of fitness sets.
Lloyd 1984	Conditional and multiple strategies only stable when costs are not too high.
Lively 1986a	Conditional strategy stable under more conditions when costs are intermediate.
Clark and Harvell 1992	Allocation model assumes cost of defense in terms of reduced growth and reproduction.
Riessen 1992	Reduced fecundity is cost of defense. Riessen suggests that cost is increased development time, not allocation of energy or nutrients.
Till-Bottraud and Gouyon 1992	Cost of defense maintains polymorphism for induced phenotype that is modeled as constitutive trait. Cost may be allocation or autotoxicity associated with frost damage.
Frank 1993	Inducible strategy favored when benefits exceed costs. High densities of host and parasite favor inducible strategy.
Adler and Karban 1994	Optimal inducible strategy favored at intermediate levels of cost. At high cost, best strategy is no defense. Cost not required for induced defense to be favored.
Astrom and Lundberg 1994	Reduced fecundity is cost of defense.
Padilla and Adolph 1996	Inducibility is costly in terms of lifetime fitness, although saving costs is not essential for induced responses to be favored. Costs change the quantitative but not qualitative predictions about induction.

Bottraud and Gouyon (1992) modeled the frequency of cyanogenic and acyanogenic leaves of clover (*Trifolium repens*) and trefoil (*Lotus corniculatus*) and suggested that costs in their system might result from allocating nutrients to the inducible cyanide system or from autotoxicity associated with release of cyanide caused by frost damage. Several of the models

Caveats and Special Features	Effect of Reliability of Cue as Predictor of Future Environments
	How good current environment is at predicting future environment is important; frequency of environmental change and lag in response time are considered.
Conditional and multiple strategies observed to be poorly represented in flora of New Zealand and elsewhere	Conditional strategies are only favored when probability of making the correct choice > .5.
Whether conditional strategy can invade is more sensitive to probability of making right choice than to costs	Conditional strategies can only invade when probability of switching correctly > .5.
Induced defenses are never favored at end of the season (lifetime) when compensation for costs of defense are not possible.	Induced defenses are not favored when timing of attack is certain or when probability of attack is very low.
Demographic model with estimates of costs and benefits from actual data for *Daphnia*.	Behaviors that reduce spatial overlap between predator and prey reduce benefit of induced defense and make it less advantageous.
Model developed for cyanogenic/acyanogenic leaves of clover and trefoil predicts their frequency, observed in different environments.	—
Optimality model that maximizes host population growth, modifying Lotka-Volterra model.	Model assumes that host can accurately and instantly assess the probability of attack.
Considers more than one kind of herbivore. If induced defense not universally effective, optimal induced defense not favored.	Optimal inducible strategy favored by high reliability of cue. If reliability is low, moving target model of induced defense is favored.
The relationship between the value of a defense is a complicated function that depends on risk of herbivory, stochastic variation, plant life history, the form of the defense function.	Induced resistance becomes favored over constitutive resistance generally as the risk of herbivory becomes more variable and less predictable.
Optimally model that highlights the importance of time lags between environmental change and plant response.	Induced resistance is favored only when the lag time for induction is low, i.e. when the current environment predicts future environments.

(Lloyd 1984; Lively 1986a; Adler and Karban 1994) concluded that induced defenses are favored only when costs are intermediate. If costs of defense become too high, then a "constitutive" strategy of no cost and no defense would exclude those with even an intermittent deployment of a very expensive defense. Models have also been developed that allow an induced strategy to invade even in the absence of costs of allocation to defense (see section 5.4). In several of the models (Adler and Karban 1994; Padilla and Adolph 1996), the important costs were not allocations to defense versus growth or reproduction but rather the possibility that the plant would not be well matched to its environment. Inducible, plastic phenotypes are favored because they reduce the probability of such mismatches.

Almost all of the models assumed that reliability of cues as predictors of future environments plays a major role in determining when and where induced defenses would be favored. In the early models of Levins (1968), reliability of the current environment in predicting future conditions reflected the frequency of environmental changes and the length of time required for the plant to respond to changed conditions. Lloyd (1984) and Lively (1986a) found that conditional strategies could invade only when the probability of the plant's switching to the correct state exceeds .5; that is, plants have to be capable of making the correct choice, on average. Whether conditional strategies could invade was more sensitive to reliability of predictions than it was to costs in Lively's model. Lloyd suggested that the rarity of induced strategies observed in the angiosperm flora of New Zealand and elsewhere results from an inability of plants to assess the environment with sufficient accuracy to determine which alternative would have higher fitness. Clark and Harvell (1992) found that induced defenses are not favored over constitutive strategies when the environment (timing of attack) is completely predictable. At the other extreme, when the frequency of attack is very low, induced strategies are also not favored; most signals contain too little information of interest to worry about. Clark and Harvell argued that predator persistence enhances the value of an inducible defense. In their model, predators accumulate over time; once a host (plant) is attacked, it is certain to be attacked again. Astrom and Lundberg (1994) developed the notion that predictability of attack is an important determinant of the relative success of induced resistance versus constitutive resistance. As the risk of herbivory becomes more variable, they predicted that induced resistance becomes relatively more favored compared to constitutive resistance. In the demographic model of Riessen (1992), behaviors that reduce spatial overlap between predator and prey reduce benefits of induced defenses and make them less advantageous. In such cases current cues of predation are not good predictors of future risk. Frank (1993) assumed that the host can accurately and instantly assess the probability of attack. Biological factors that alter this assumption affect the model predictions. In the models of Adler and Karban (1994) the optimal inducible strategy was favored only when cues were highly reliable. Similarly, in the model developed by Padilla and Adolph (1996) time lags were critically important because they determined how well the conditions that the plant responded to would predict future conditions that the plant would experience.

Almost all of these models highlight the importance of reliability or predictive power of information. Therefore, it is surprising that almost nothing is currently known about the reliability of current damage or other cues (see section 2.2) in predicting future damage in plant-herbivore systems. This assumption that damage early in the season provides information about the likelihood of attacks later in the season has been examined for three populations of wild cotton (*Gossypium thurberi*) that are attacked by leaf-mining caterpillars (Karban and Adler 1996). For one population, damage early in the season was correlated with the number of new mines that would be initiated throughout the season. This result was neither evidence for nor a consequence of induced resistance, which affects survival of miners but not the initiation of new mines. The only population where frequency of early-season attacks correlated with late-season attack was also the only one that provided evidence for induced resistance. No evidence was found that the correlation between early-season damage and later damage degraded as the season progressed. Mines, which were produced only by the miner *Bucculatrix thurberiella*, a specialist, provided more information about subsequent attacks by *B. thurberiella* than did other forms of damage. Plants responded more strongly to chewing and rasping damage, however, even though it provided less information than mining damage. Other empirical measures of serial correlation of risk of attack would be well worth study.

An interesting variant of the argument that allocation costs favor inducible defenses rather than constitutive ones has been developed by van der Meijden and coworkers (1988). They regarded regrowth and plant compensation as an alternative strategy to defense. Regrowth is an inducible strategy, since it occurs only following damage. Their rationale for the evolution of a strategy that relies on regrowth and plant compensation is that it represents a cheap alternative to constant deployment of expensive and unnecessary defenses. They argued that defense and regrowth are redundant strategies and made three predictions: (1) Plants should invest heavily in either regrowth or defense, but not both. (2) Herbivory and regrowth should be positively correlated, since plants investing little in defense should suffer high levels of herbivory and invest more in regrowth. (3) Root-shoot ratios and above-ground herbivory should be positively correlated, since root-shoot ratios provide an index of the potential for regrowth. These predictions have received little consideration apart from the dune system where they were first conceived (van der Meijden et al. 1988; Prins et al. 1989). The first prediction that regrowth and defense should be negatively correlated is consistent with results from six inbred lines of wild morning glory, *Ipomoea purpurea* (Fineblum and Rausher 1995). Those lines that were most resistant to damage were least tolerant of the damage they received (i.e. a given level of artificial damage had most impact on fitness).

5.3.2 Measuring Costs of Defense

Developing a means of measuring costs of plant defenses is critical because reducing costs is perceived as the driving force responsible for the evolution of induced defenses. Costs of defenses have been estimated using three methods: (1) estimation of the resources that are required to construct and maintain the defense, (2) correlations between levels of defensive traits and plant fitness in environments without herbivores, and (3) correlations between defenses and levels of herbivores across geographic gradients.

There have been numerous attempts to calculate the costs of plant defenses in units of energy or materials. The basic notion behind these attempts is that energy, carbon, or nitrogen can become limiting to plants and that the amounts of these limiting resources required for construction and maintenance of defenses can be calculated (Chew and Rodman 1979; Mooney and Gulmon 1982). This method has several limitations.

First, it is rarely known whether the currency that is being used (energy, carbon, nitrogen) is really limiting to the plants in question. Many workers have calculated the energetic or material costs of producing particular defenses and argued that these calculations indicated either substantial or minimal costs. However, investments involving relatively small caloric expenditures may or may not have relatively large fitness consequences. For example, the caloric and nutrient investments of cyanogenic morphs of clover differ little from those of acyanogenic morphs (Kakes 1989), perhaps suggesting that cyanogenesis is relatively inexpensive. However, these measurements were not reflected in calculations that were more representative of plant fitness. The Ac gene that regulates the presence of cyanogenic glucosides was associated with a reduction in flower-head production of approximately 50% (Kakes 1989). Presumably, flower production is limited by some currency other than calories or elemental nutrients.

Evaluating whether the resource in question is limiting to a plant is extremely difficult because we cannot easily assess the resources that the plant takes in (Mole 1994). Compared to animals, plants absorb resources through many "mouths," including stomata, roots, and root-associated fungi. Because direct measurement of resource absorption is difficult, plant biologists have generally assumed that levels of resources that are in the environment are good indicators of what the plant absorbs and has available for allocation to various functions. Ecologists generally assume implicitly that "consumption of resources" is proportional to external environmental levels. However, this assumption has not been supported in studies of animals where actual consumption can be measured (Mole 1994).

Second, defenses vary in the frequency with which they must be replaced; some defenses outlast the rest of the plant, others turn over and must be replenished frequently. Estimating the level of materials tied up in a defense at any one time provides little information about turnover, a process that has been exceedingly difficult to measure accurately (Mihaliak et al. 1991).

Third, there may be other costs associated with defense besides the construction and maintenance of the chemicals or structures involved (see section 5.3.3). An accounting of calories and materials required to produce the defense misses these other costs.

Fourth, resources allocated to defense do not necessarily result in reduced allocations for growth or reproduction. To understand how this counterintuitive situation can come about, consider a Y-shaped model of resource allocation (figure 5.3). The total pool of resources available (R) may be allocated to either defense (D) or to growth and reproduction (G). The assumption is often made (e.g. Coley et al. 1985) that $D + G = R$, so that resources allocated to defense cannot be used for growth and reproduction. However, several real-world complications may render this model misleading (Mole 1994). First, plants may be able to adjust R, the total amount of resources available, in response to herbivory, particularly if the resource in question is not limiting. Second, when specific "defensive" traits are being considered, they may use only a small fraction of the total resource budget. In other words, if the traits considered don't account for the plant's total budget for R, then altering some fraction of R allocated to a defensive trait does not necessarily produce a predictable trade-off between reproduction and growth. Indeed, if the fraction of R allocated to the defensive trait is small relative to the plant's total budget and we consider that the plant allocates to functions other than D and G, then the change in allocation to the defensive trait may affect other components (competitiveness, tolerance to stress, etc.) and have no direct effect on allocation to current and future reproduction (Herms and Mattson 1992; Mole 1994).

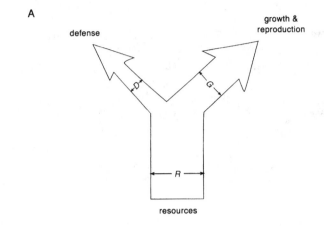

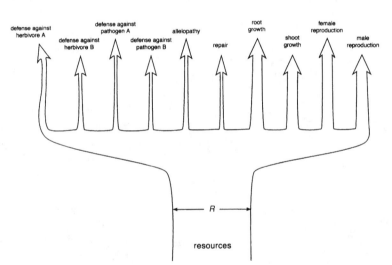

Fig. 5.3 Models of plant allocation to various functions. Resources are acquired at the bottom of the diagram and allocated to functions at the top. The width of each arrow represents the amount of resources allocated to any particular function (redrawn from Mole 1994, modified from deJong and van Noordwijk 1992). *A,* the y-shaped model assumes that the total pool of resources (*R*) that the plant takes in can be allocated to either defense (*D*) or to growth and reproduction *(G)* such that *D + G = R.* An increased investment in defense necessarily results in a corresponding decrease in growth and reproduction. *B,* a more complicated model of plant functions. Here the total pool of resources (*R*) that a plant takes in can be allocated to many diverse functions. An increase in defense against herbivore A (*leftmost arrow*) need not necessarily produce a corresponding decrease in any particular component of growth or reproduction. In fact, attack by herbivore A may not only increase defense against that herbivore but is likely to have direct and indirect effects on many of the arrows.

A second, and for our purposes more useful, method of calculating costs involves estimating differential plant fitness associated with the defense. In environments in which herbivores have been excluded, there should be no net benefits of the presumed defense against herbivory. Any observed differences in fitness between plants that have the induced defense and those that do not should be due to the costs of the induced response plus whatever damage was required to achieve induction.

This method gives a more reliable estimate of the costs of induction if the damage required to achieve induction can be kept to a minimum. For an extreme example of the problems resulting from a failure to minimize the damage of induction, consider the attacks of bark beetles. Each attack by beetles reduces the ability of pine trees to resist future attacks (Raffa 1991). Trees that have been depleted by past inductions are less likely to survive a current attack. However, it makes little sense to consider this increased mortality risk to be a cost of defense rather than a consequence of damage. In this case the damage needed to induce resistance has not been kept to a minimum. In summary, the ability of the experimenter to reduce other costs of damage, independent of the induced response of interest, limits the usefulness of this technique.

This problem may be solved by using chemical elicitors to induce resistance, without actual damage. The only limitation of this technique is that the chemical elicitors may produce other confounding effects on the plant. Excluding herbivores to create an environment without herbivory may also produce unwanted experimental side effects or biases. Therefore, it is preferable to use several methods to remove herbivores (cages, pesticides, etc.) so that the magnitude of this potential problem can be assessed. Finally, this method is most convincing if lifetime plant fitness can be estimated. Reductions in one correlate of fitness (i.e. growth) may result in increases in another correlate (i.e. seed production). Estimates that include many correlates of fitness are preferable; estimates of lifetime fitness are preferred to those that include only a portion of the plant's life.

A third method of estimating costs involves comparing plant populations that have evolved in environments with herbivores to those that have evolved in environments without herbivores. If defenses are costly, we would expect them to be missing in environments without herbivores, where defenses would accrue no benefits. However, the loss of defenses where herbivores are absent could result from genetic drift rather than from selection favoring plants that minimize costs (Rausher 1992). Results using this method are difficult to analyze with confidence (see discussions in Simms 1992 and Rausher 1992). For example, the persistence of spines on cacti in the Chihuahuan desert, thousands of years after the extinction of large mammals from this area, suggests that spines are not very costly (Myers and Bazely 1991). This argument is difficult to evaluate, however, because spines may provide benefits independent of herbivores, such as thermoregulation (Nobel 1978; Myers and Bazely 1991). Similarly, the very high diversity of mostly inactive secondary compounds suggests that they are not very costly to maintain (Jones and Firn 1991). This argument is also difficult to evaluate, since our understanding is so incomplete of the roles of secondary compounds that currently appear to be of little use to the plant. In summary, correlations between geographic occurrences of herbivores and of defensive traits are extremely important to the advancement of the field because they stimulate hypotheses about processes that may be important. They are less useful for providing evidence to test the hypotheses.

5.3.3 Types of Costs

Plants that construct and maintain defenses against herbivores may accrue costs for several reasons.

1. Resources that are allocated to providing defense cannot be allocated simultaneously to growth or reproduction. These costs resulting from allocated resources are called direct costs, physiological costs, or allocation costs. One assumption behind this notion is that traits conferring defense against herbivory do not serve other functions. Direct costs of secondary metabolites are generally small in terms of the plant's total resource budget (Gershenzon 1994). If, in fact, traits conferring defense also provide other benefits (make the plant more competitive against other plants, dissipate heat, provide plant structure, etc.), then the energy and materials associated with the trait may cost relatively little in terms of plant fitness even when herbivores are not present. Unfortunately, relatively little is known about the "primary" or "civilian" roles of most plant defenses (Seigler 1977; Baldwin 1994; Gershenzon 1994), and these roles have not been integrated into estimations of allocation costs. The direct costs of "defensive" compounds are also reduced if these defenses can be catabolized and the resources salvaged by the plant. Although data on such processes are limited, preliminary evidence suggests that this recycling may be important for some compounds (Gershenzon 1994). In general, however, recent evidence suggests that rates of turnover may be far lower than previously assumed; in other words, investments in defense may not be available for salvage (Mihaliak et al. 1991; Gershenzon et al. 1993; Baldwin, Karb, and Ohnmeiss 1994; Gershenzon 1994).

2. Investments that occur early in ontogeny cost the plant in terms of lost opportunities throughout its life if those investments cannot be reclaimed. Plants may be viewed as investment machines that live off the interest of their investments (Givnish 1986; Gulmon and Mooney 1986). If they invest early in functions that do not contribute to growth, they lose a lifetime of compound interest and hence incur a large opportunity cost. Resources that are allocated to defense may not become part of the machinery that allows the plant to acquire more energy and materials (see above). As a result, the plant may pay an opportunity cost if it diverts resources from productive tissues into less productive foliage, storage, rapid expansion of foliage, and so forth associated with defense. Such opportunity costs may be quite great in situations in which the competitive environment becomes progressively more severe (e.g. rapid early growth is essential to prevent severe shading later). Under these circumstances the missed opportunities associated with diverting resources may be much more important than direct allocation costs. Opportunity costs may be very important even in environments where plant competition is not strong, if investment of resources in defense cannot be recovered or occurs early in plant ontogeny.

3. Products that deter or harm herbivores may also be toxic to the plant producing them (see section 3.3.1); defended plants may pay a cost associated with self-toxicity. The effects of self-toxicity may increase if the defense is maintained for an extended period of time, as in a constitutively defended plant. For example, some phytoalexins are toxic to plants at concentrations that inhibit microbial pathogens (Kuc 1987). Repeated applications to bean leaves of fungus-derived elicitors of these phytoalexins caused severe necrosis and stunted growth. This self-toxicity is avoided if phytoalexins are produced only rarely, when needed. Many other compounds that are toxic both to parasites and to the host plant are normally stored as benign precursors in plant vacuoles (Duffey and Felton 1989). These precursors are activated only after vacuoles have been ruptured by feeding damage.

The costs associated with self-toxicity may be reduced if plants that increase synthesis of a chemical also increase tolerance of the same material. For example, damage-induced nicotine synthesis reduced photosynthetic capacity of *Nicotiana sylvestris* (Baldwin and Callahan 1993). Leaf damage also marginally increased the photosynthetic tolerance to nicotine in this species, however, suggesting that synthesis and tolerance may be coordinated responses.

4. Plants may incur indirect costs due to many ecological interactions. At a physiological scale new resistance traits may disrupt normal metabolic processes in some instances, especially when they first arise (see

section 5.3.4). At an ecological scale, traits that provide defense against one herbivore may make a plant more attractive or more vulnerable to others. Effects of induced responses that are idiosyncratic for different herbivores were considered in section 2.2.2. Even more dramatic examples have been described for substances that are generally repellent or antibiotic but have become strong attractants for specialist herbivores. For example, damaged squash leaves became more resistant to one species of beetle and more vulnerable to another species naturally co-occurring with the plant in southeastern Mexico (Carroll and Hoffman 1980). In other parts of the world, responses by cucurbits to damage have been found to deter some beetles but actively to attract other species (Tallamy and McCloud 1991). Similarly, for many crucifers glucosinolates or mustard oils are released enzymatically from precursors following damage to leaves (Harborne 1988; Bodnaryk 1992; Giamonstaris and Mithen 1995). Glucosinolates are repellent or toxic to most mammalian and insect herbivores. However, these same chemicals are powerful feeding stimulants for cabbage butterfly larvae (*Pieris rapae*). Caterpillars feed on an artificial diet only if it contains mustard oils, and females will oviposit on filter paper as long as it contains mustard oils. Cabbage aphids (*Brevicoryne brassicae*) are also attracted to glucosinolates. They reject very high concentrations found in young leaves and very low concentrations found in senescent leaves and prefer mature leaves with intermediate concentrations. In summary, deterrents against most herbivores may be costly if they provide information to specialists. The diverse consequences of a trait that seems to provide defense may be difficult to predict. For example, cyanogenesis in clovers deters slugs and snails but is associated with dramatically increased susceptibility to rust infections (Dirzo and Harper 1982a, 1982b).

Traits that defend plants against herbivores may make them less attractive to pollinators. Some secondary metabolites are found in floral nectar, and these decrease floral visitation under some circumstances (Baker and Baker 1975). Recent work with wild radish (*Raphanus sativus*) suggests that damaged plants received fewer visits by syrphid flies even when flower number and size were controlled (Lehtila and Strauss, personal communication). Increased attraction of specialist herbivores or reduced attraction of pollinators and other mutualists may be important indirect costs of defense, although the evidence for these costs is preliminary.

5. Plants may incur indirect evolutionary costs if they deploy defenses constantly. Herbivores may be more likely to evolve means of circumventing defenses that they constantly encounter. Such adaptation by herbivores may be less rapid or less common if defenses are not constant

and therefore selection on the herbivores is also intermittent (Whitham et al. 1984; Schultz 1988; Harvell 1990b). Induced defenses may allow plants to present inconsistent phenotypes to their herbivores, especially defenses that involve new mechanisms rather than amplified constitutive defenses. However, it seems likely that group selection must be invoked for plants to employ defenses only intermittently to avoid evolutionary tracking by their herbivores. Selection at the individual level allows this mechanism to work only if lineages of herbivores adapt to, or track, their individual host plants over evolutionary time (sensu Edmunds and Alstad 1978; Karban 1989). Otherwise, individual selection would probably favor plants that deployed defenses against herbivores for as long as those defenses were effective, even if such heavy deployment eventually rendered them useless as the result of counteradaptation by herbivores.

Dicke and Sabelis (1989) have suggested that plants advertise for bodyguards when they are damaged. Specifically, predatory mites are attracted to the volatiles released by plants attacked by herbivorous spider mites. They argued that plants that continuously release synomones to attract predators, even when they are not damaged, pay direct costs of producing the volatile materials and pay an indirect cost of reducing the information value of their "cry." However, it is unclear if this crying wolf could be prevented by selection at the individual level.

5.3.4 Empirical Evidence That Induced Defenses Are Costly

Because the question of costs of plant defenses has been central to evolutionary theories, there have been many attempts to measure costs of both constitutive and induced defenses. A variety of methods and arguments have been developed (see section 5.3.2). Many of these are extremely difficult to interpret, however, because various causal factors could have produced the patterns that authors suggested resulted from costs or a lack of costs. We argued above that the best method to evaluate costs of defenses is to induce plants with minimal damage and to compare the fitness of induced and uninduced plants when herbivores are absent; fitness differences under these conditions will reflect differences in costs of defenses (see section 5.3.2). Ideally these experiments should be conducted under natural conditions where opportunity costs due to competition or limited resources are more likely to be seen than under lab conditions, which provide unlimited resources or other unrealistic circumstances. For example, fitness costs associated with resistance to herbicides was much greater for jimsonweed plants (*Datura stramonium*) grown with neighbors compared to costs for plants without nearby competitors (Williams et al. 1995).

Experiments of this design have been conducted for several plant species (table 5.5). The most convincing demonstration of costs of induced defenses involves wild tobacco plants that were induced by artificially clipping leaves (Baldwin et al. 1990). Induced plants had increased concentrations of alkaloids and associated with this change was a reduction in both total plant mass and number of seeds produced. Since *Nicotiana sylvestris* is an annual, these reductions in growth and reproduction should provide relevant measures of lifetime fitness costs associated with induced production of alkaloids. A more elegant experimental approach to determining the fitness costs of induced nicotine production involves inducing nicotine synthesis with the wound signal, jasmonic acid (see section 3.3.3.4). This technique has the advantage that neither induced nor uninduced control plants must actually be wounded. *N. sylvestris* plants incurred no reduction in seed output associated with increased nicotine production, induced by application of jasmonic acid, when plants were grown in hydroponic culture in individual containers (Baldwin and Hamilton n.d.). When induced plants were forced to compete with uninduced plants for common resources in the hydroponic pools, however, induced plants produced significantly fewer seeds than did control plants. The fitness differences were principally opportunity costs, probably reflecting differences in the abilities of two treatments to acquire limiting nitrate. Demonstrations of costs of facultative resistance induced with jasmonic acid will be even more convincing when they are conducted under natural competitive conditions. *N. attenuata* plants that were induced by application of jasmonic acid and subsequently protected from herbivores produced 23.6% and 18.4% fewer total viable seeds than uninduced and subsequently protected controls in two field experiments (I. T. Baldwin, personal communication).

The costs of cyanogenesis involving mixing of preformed enzymes and substrates were evaluated for clovers (*Trifolium repens*) in England (Dirzo and Harper 1982a, 1982b) and Holland (Kakes 1989). In these experiments the costs of induction were estimated by comparing plants with different, genetically determined, abilities to respond. Both studies found that the phenotypes that contain substrates for cyanogenesis, cyanogenic glucosides, produce fewer flowers than phenotypes that lack these compounds and therefore lack the ability to release cyanide following damage. These costs may have represented resources that were allocated to defense rather than to reproduction, or they may have represented costs caused by autotoxicity when freezing ruptured plant vacuoles and released the cyanogenic compounds, damaging the plant's own cells.

Since there were relatively few examples of costs of induced responses against herbivores, we looked for similar effects in the plant-pathogen

Table 5.5 Fitness costs of induction

Plant Species	Inducer	Lab/Field	Effects on Plant Fitness Associated with Induction	Reference
Gossypium thurberi	Natural damage early in season	Field	No differences in survival, growth, reproduction.	Karban 1993a
Hordeum vulgare	Avirulent race of powdery mildew	Lab	Grain yield reduced by 7%; kernel weight reduced by 4%; grain protein reduced by 11%.	Smedegaard-Petersen and Stolen 1981
Lycopersicon esculentum	Chitin injections	Lab	No differences in survival, growth, reproduction.	Brown 1988
Nicotiana sylvestris	Artificial damage	Garden	Total plant mass reduced by 15%; number of fruits reduced by 38%.	Baldwin et al. 1990
Trifolium repens	Only ability to induce	Field	Frequency of flowering reduced by 10–30%.	Dirzo and Harper 1982a, 1982b
	Only ability to induce	Garden	Dry weight of flower heads reduced by 20–60%.	Kakes 1989

literature. Barley plants (*Hordeum vulgare*) can be inoculated with avirulent strains of powdery mildew and gain protection against more virulent strains. Plants with this ability are termed resistant varieties although the resistance must be induced. Induced resistance in barley plants was associated with increased biosynthetic activity and respiration rates (Smedegaard-Petersen and Stolen 1981). However, resistant plants remained entirely free of disease symptoms following inoculation with avirulent fungi. Comparisons between resistant plants infected with avirulent fungi and resistant plants that were not inoculated provided an estimate of the costs of resistance. Grain yield of induced plants was reduced by 7%, kernel weight by 4%, and grain protein by 11% relative to uninduced controls. These reductions probably reflected costs of induced resistance. However, several other factors may also have contributed to the differences observed. There may be other effects caused by inoculations that were not controlled for in this experiment. In addition, all of the induced plants were kept in one growth chamber and all of the uninduced controls in another. This design did not allow a separation of effects caused by different chambers from those caused by different induction treatments.

Several workers who have looked for costs of induction failed to detect any. Repeated injections of chitin induced the production of PIs in young tomato plants (*Lycopersicon esculentum;* Brown 1988). Induction of these chemicals had been found to reduce growth of beet armyworm caterpillars (*Spodoptera exigua*) when incorporated into artificial diets (Broadway et al. 1986). Similarly, growth of hornworm caterpillars (*Manduca sexta*) was retarded when they fed upon leaves of transgenic tobacco plants that expressed PIs (Johnson, Narvaez, et al. 1989). However, induction of PIs caused no measurable fitness decreases in plant growth or reproduction (Brown 1988). Because these experiments were conducted in the lab and plants were provided with unlimited resources, the possibility exists that real opportunity costs were missed in this experiment.

Low levels of early-season damage by caterpillars caused wild cotton plants (*Gossypium thurberi*) to become less suitable as hosts for caterpillars throughout the remainder of the season (Karban 1993b). When herbivores were absent, however, this induced resistance was not associated with reductions in the growth, survival, or seed production of induced plants relative to uninduced controls (Karban 1993a). The design of this field experiment was sufficiently powerful that it is unlikely that costs were present but not detected. Furthermore, plant competition was found to be costly in terms of growth and seed production in this same experiment. Growth, survival, and reproduction were measured for only two years; it is possible that costs would have been revealed if the lifetime fitness of plants had been monitored.

The evidence that defenses are costly and that induced defenses save costs from studies of plants and herbivores is equivocal. Just what this means is itself hard to evaluate. Rather small but consistent costs, below the threshold of existing analytical techniques, could be sufficient to direct evolution over long periods of time (Simms 1992; Karban 1993a). Induced defenses that were costly when they first arose may have been subjected to subsequent selection that has reduced their costs (Simms 1992). A similar progression has been documented for the reduction in costs associated with the evolution of resistance to pesticides in populations of Australian sheep blowfly (McKenzie et al. 1982). Recently, this process has been observed experimentally for *E. coli* strains that have become resistant to antibiotics (Schrag and Perrot 1996). When they first arose, resistant mutants had a cost of 14–19% lower fitness per generation compared to strains that were not resistant when both were grown free of antibiotics. Despite these costs, after 180 generations free of antibiotics, selection had not eliminated the resistant strains. Rather, they had evolved compensatory physiological functions so that the fitness costs originally associated with the resistance was gone. In other words, the traits conferring resistance persisted although the original fitness costs of resistance had disappeared.

The hypothesis that costs were involved in the evolution of induced defenses is practically unfalsifiable because costs that may have been present quickly become difficult to detect. Transgenic plants that have been recently "given" an inducible defense may shed some light on this question. Transgenic plants could be produced by inserting the genes for the inducible trait into lots of different parts of the genome to simulate the presumed early stages in the evolution of an inducible defense when it arises by some mutation in the promoter gene. This may minimize linkage problems and produce plants with inducible defenses before the presumed evolution of factors that subsequently reduce costs.

Costs of induced defenses have also been studied intensively for aquatic and marine invertebrates. Rapid progress testing cost-benefit models has been possible in these systems because the induced responses can be elicited experimentally with chemical cues from the predators, without damaging the prey. In plants, induced responses are thought to be elicited by endogenously produced signals that require wounding for their production. The requirement of wounding makes tests of the importance of any potentially defensive trait difficult to interpret. If herbivores are used to wound plants, they must be randomly assigned to test plants to avoid the potentially confounding influence of herbivore choice (Neuvonen and Haukioja 1985); the damaging herbivores may transmit diseases to herbivores that use the plant subsequently (Fowler and Lawton 1985); wounding may change the apparency of plants to herbivores (sensu Feeny 1976) and may cause a plethora of growth-related responses (Baldwin 1994; Baldwin and Ohnmeiss 1994) in addition to the induced defenses, which can influence a plant's resistance and fitness. The recent progress identifying the endogenous chemical cues that are responsible for signaling in plants should help researchers who study plants to make more reliable estimates of the costs and benefits of induced responses and to emulate the recent successes seen in aquatic systems.

Many aquatic and marine invertebrates produce spines, keels, wings, helmets, or thicker shells within hours after exposure to cues from their predators (Havel 1987; Dodson 1989; Harvell 1990a). Costs of these morphological structures were suggested by experiments in which induced individuals had reduced rates of growth (barnacles, bryozoans, protozoans, cladocerans, snails), survival (rotifers), or reproduction (barnacles) compared to uninduced individuals when predators were absent. Unfortunately, most of these studies with few exceptions (e.g. Lively's work [1986b] on barnacles and Harvell's work [1992] on bryozoans) have been conducted in the laboratory. Workers in this area who have worried about extrapolating from the laboratory to natural settings have found evidence that such leaps are likely to be misleading in many cases. For example, costs of defense in rotifers are very sensitive to environmental conditions, which casts doubt on the usefulness of lab results. Induced spines in rotifers are costly when food levels are high but not costly when levels are low, which they probably experience for much of the season (Stemberger 1988). Induced spine production was costly for bryozoans in the field, but this effect was much smaller than cost estimates based on short-term laboratory experiments (Harvell 1992).

Behavioral changes such as diel vertical migration by zooplankton have also been studied with regard to their costs (Lampert 1989; Loose and Dawidowicz 1994); daphnia move as much as 60 m down through the water column at dawn to avoid visually orienting predators and move back up at dusk to surface waters. The extent of diel vertical migration depends on the concentration of "predator factor" in the water (Loose and Dawidowicz 1994). The behavioral responses of the daphnia are specific depending on whether the predators are fish or invertebrates and the result is effective against each of these specific predators. However, enhanced vertical migration brings the animals into colder waters and decreases their growth and reproductive rates (Loose and Dawidowicz 1994). This cost results from ecological factors rather than from the energy allocation to swimming.

Like daphnia, ciliated protozoans exhibit morphological and behavioral responses to their predators (Kuhlmann and Heckmann 1985; Kusch 1993). These responses are specific for different predators (or cues from different predators) and make induced individuals more likely to be rejected or avoided by their predators (Kuhlmann and Heckmann 1994; Kusch 1995). When the ciliate *Euplotes octocarinatus* was placed in a laboratory flask with a predator, it quickly changed morphology and "grew wings"; this induced response reduced population growth rate of the ciliate by approximately 15% (Kusch and Kuhlmann 1994).

Predator cues not only induce morphological and behavioral responses for daphnia, but also change life-history patterns. Recent research suggests that these shifts in the timing of development and reproduction may be adaptive responses to unpredictable juvenile and adult mortality (Tollrian 1995). For example, predator cues derived from fish feeding on daphnia result in shorter generation times and smaller sizes at maturity. On the other hand, predator cues derived from *Chaoborus* larvae feeding on daphnia result in longer generation times and larger sizes at maturity (Tollrian 1995). Similar specificity in induced responses to different predators has also been documented for *Euplotes* spp. (Kusch 1995). These life-history shifts not only have consequences for population growth of the prey but also affect the vulnerability of prey to specific predators. The actual probability of attack has been measured for *Euplotes* spp. by offering the predator an alternative prey. Kolaczyk and Wiackowski (1997) demonstrated that the induced response was directly proportional to the actual feeding of the predator on the prey; presence of the predator was not sufficient. The complexity of these responses emphasizes the importance of knowing the probability of mortality caused by specific predators before benefits and costs can be meaningfully calculated.

In an elegant experiment Tollrian (1995) was able to measure directly the physiological costs of producing neck teeth in *Daphnia pulex* and separated these physiological costs from those associated with life-history shifts. Life-history shifts are induced in response to cue later in development than the neck teeth; by carefully timing the exposure to cue, he was able to induce either of the responses or both. He found that individuals exposed to predator cues took longer to develop but that this was a trade-off with larger body size and not a physiological cost of induced production of neck teeth.

Estimation of the benefits and costs of induced responses for aquatic invertebrates has progressed very quickly. Although the true importance of minimizing costs is still controversial for aquatic invertebrates, several generalizations have already emerged at this time (Black and Dodson 1990; Spitze 1992; Black 1993; Tollrian 1995; S. I. Dodson, personal communication). Where costs have been found, they have been primarily ecological and opportunity costs. They are often dependent on exposure to a particular predator and on other environmental conditions. The direct physiological or allocation costs that have dominated thinking among plant biologists have generally not been found associated with the induced responses of aquatic invertebrates. The picture that has emerged from studying the plethora of induced changes in aquatic invertebrates and their specificity should humble workers starting to study induced responses in plants and expecting a simple picture to emerge.

In summary, the hypothesis that induced defenses are favored over constitutive defenses because they allow plants to save costs is intuitively appealing and has support from both theoretical studies and some experiments. The experimental support for the hypothesis has not been universal, however, and has been based on very few studies. If allocation costs have not driven the evolution of induced defenses in all cases, how else can we account for them? Below we consider several evolutionary scenarios (rival Darwinian histories in Kitcher's terminology [1985]) to explain the maintenance of induced defenses.

5.4 OTHER EVOLUTIONARY EXPLANATIONS FOR INDUCED DEFENSES

Induced defenses may be favored if they provide benefits to plants that constitutive defenses cannot. One nearly universal attribute of induced defenses is their potential to change over time frames shorter than the generation of a plant. As such, they allow a plant to completely alter (or fine-tune) the defensive phenotype that it presents to herbivores and other parasites. This variability can be difficult for herbivores to deal with over ecological and evolutionary time (section 5.4.1.4) and potentially can lead to increased plant fitness.

5.4.1 Induced Defenses as Generators of Variability and Moving Targets

5.4.1.1 Evidence That Variability Can Offer Resistance

Several experiments manipulated plant variability explicitly and found that it can provide resistance to even some generalist herbivores. In an early paper two caterpillar species were reported to be more efficient at converting plant material into body tissue if they were reared on the same food continuously than if their food was switched (Schoonhoven and Meerman 1978). In a more rigorous set of experiments, Berenbaum and Zangerl (1993, 1995) found that a mixture of furanocoumarins is more difficult for caterpillars to metabolize than are the same amounts of only a single furanocoumarin. The more diverse mixture probably provides more effective resistance at least in part because the chemical mixture (as opposed to a single chemical) inhibits the caterpillar's ability to detoxify the furanocoumarins.

Even single chemicals can be more difficult for herbivores to metabolize if they vary in their availability. For example, southern armyworm caterpillars (*Spodoptera eridania*) that were reared on a constant diet of 0.5% KCN were able to habituate to cyanide and were not poisoned when transferred to diets containing 1.0% KCN (Brattsten et al. 1983). In contrast, caterpillars that were switched from diets of 0% KCN to 1.0% KCN were more likely to be poisoned. Exposure to constant low levels of cyanide presumably allowed the first group to habituate, and the 1.0% cyanide diet was less novel to them than to the unhabituated controls. Many of the enzymatic detoxification systems that herbivores employ to metabolize plant allelochemicals are inducible (Lindroth 1991); presumably these inducible detoxification systems are most efficient if the diet of the herbivore does not vary.

In a particularly elegant experiment Stockhoff (1993a) reared gypsy-moth larvae on one of three diet treatments: constant nitrogen (3% nitrogen continuously), low variance in nitrogen (diets were alternated daily between 2.25% and 3.75% nitrogen), and high variance in nitrogen (diets were alternated daily between 1.25% and 4.75% nitrogen). Larvae experiencing variation in diet suffered reduced pupal mass and extended development. This was caused by nonlinearity in the relationship between nitrogen and food utilization efficiency and by disruption of compensatory feeding responses. However, larvae that were given a choice between these same foods (instead of being assigned by the experimenter) shifted their preferences and grew more rapidly than those fed on a constant diet (Stockhoff 1993b).

Stockhoff (1993a) suggested that the negative consequences associated with diet variability could be defensive if they resulted in reduced foliage removal. The consequences of this induced resistance on plant fitness have not yet been considered empirically. Plant fitness, and not herbivore performance, is the parameter that is critical to considerations of the evolution of strategies of plant defense.

5.4.1.2 Induced Resistance Increases Variability and Benefits

Variability in nutrition poses a problem for herbivores any time that (1) plant resistance varies and (2) herbivore performance is a concave function of the level of resistance (Karban et al. 1997). Induced responses

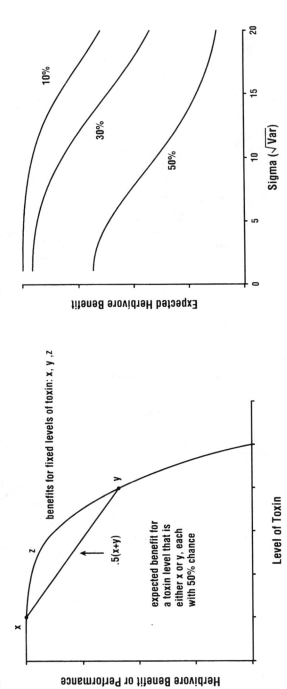

Fig. 5.4 *A*, Herbivore benefit as a function of the level of plant toxin (redrawn from Karban et al. 1997). We assume that the benefit an herbivore accrues from feeding on a plant is a decreasing function of the level of toxin produced. A plant that produces a variable level of toxin will lead to consistently lower levels of benefits for herbivores compared to a plant that produces the mean level. For example, a plant that changes as the result of induction produces toxin corresponding to *x* and *y*. The average benefit associated with this change, $0.5 (x + y)$, is always lower than *z*, the benefit associated with a constant level of toxin. If herbivore behavior reflects performance (Thompson 1988; Stockhoff 1993b), an herbivore will prefer to feed on a plant that produces a constant level of toxin. *B*, Plants are heterogeneous in nature and produce levels of toxins described by frequency distributions rather than single values. We assume that the level of toxin is normally distributed with mean equal to 10%, 30%, or 50% of the fatal level and variance σ^2. Under these conditions, benefits to herbivores decrease as the variance in toxin level increases.

increase the variability that an herbivore faces, by definition. For the sake of this argument, we assume that herbivore performance declines in response to the level of toxin produced by a plant, as in figure 5.4. We recognize that our use of the term "toxin" is an oversimplification and that in reality a chemical is toxic to an herbivore depending on the chemical and physical environments of both the plant and herbivore. Our assumption that the benefit to herbivores is a concave function of plant toxins is satisfied for many well-studied plant-herbivore systems, including hornworm larvae fed nicotine (Parr and Thurston 1972), four caterpillar species fed diterpene acids of sunflower (Elliger et al. 1976), cabbage looper larvae fed esters from crown vetch (Byers et al. 1977), and tobacco budworm caterpillars fed any of several allelochemicals from cotton (Jenkins et al. 1983).

A plant that provides a variable level of toxin will always provide lower levels of benefit to herbivores than a plant that produces the mean level. This result is true whenever the benefit function is concave (as in figure 5.4). When an herbivore attacks a plant or plant tissue or when it moves to another individual, the levels of toxin that it will encounter are uncertain, and it will suffer a reduction in benefit relative to an herbivore feeding on the mean level. As variance in the level of toxin increases, the benefit that an herbivore receives decreases, and this presumably translates into increased plant fitness (Karban et al. 1997). Herbivores may themselves respond to plant quality by choosing plants that are less variable. Given such herbivore responses, natural selection is expected to favor plants that vary spatially and temporally; induced responses, especially different responses in different plant tissues and tissues of different stages of development, can produce this variability. This model does not explain why plants are not always maximally defended; some form of cost or constraint presumably favors a submaximal level of defense. Given that a plant is going to deploy a certain level of toxin, this model suggests that variability increases the effectiveness of that deployment.

5.4.1.3 Directional and Nondirectional Induced Responses

The models that we have discussed previously, both those that rely on induced defenses' reducing costs to plants and those that rely on induced defenses' providing plants with additional benefits, assume that induced responses are directional. By this we mean that plants respond to attack by changing in a predictable direction, corresponding with increased defense. Conversely, plants respond to no attack by relaxing expensive defenses. Norm-of-reaction models are similar in that they assume predictable, directional changes (figure 5.1). In these models of phenotypic plasticity, plastic genotypes are assumed to take on particular

Table 5.6 Basic properties of the three models

Properties	Constitutive Defense	Optimal Inducible Defense	Moving Target
Individual phenotype	Fixed	Variable	Variable
Arrays of phenotypes	Fixed	Linear	Multidimensional
Response to attack	None	Defend	Change
Direction of response	None	Directional	Nondirectional
Response to no attack	None	Undefend	None
Cost of "defense"	Yes	Yes	No
Effectiveness of "defense"	Yes	Yes	On average

Source: Adapted from Adler and Karben 1994.

genotypes in particular environments, that is, to have fixed reaction norms (Via et al. 1995).

In contrast, a moving-target model assumes that plants respond to attack by changing and respond to no attack by not changing. In other words, plants that have been attacked may defend themselves by changing phenotypes, thereby presenting a moving target to their herbivores (table 5.6). The defense may simply be changing per se, rather than changing to a particularly well-defended phenotype (Karban and Myers 1989; Karban 1993a; Adler and Karban 1994).

Models of directional inducible responses assume that plants can be arrayed along a single axis of more-to-less defended. This axis of defense level may correspond to the concentration of a secondary chemical, although it might also be an emergent property of diverse responses. In either case phenotypic change is assumed to be directional: more defense when attacked and less when not attacked. In contrast the moving-target model assumes that phenotypic changes are nondirectional: the plant can escape by moving in any of a number of directions. The hypothetical process of induced defense is similar to dispersal by running away (changing phenotype) from bad environmental conditions (herbivores). The plant can be thought of as presenting an "n-dimensional phenotype" (sensu Mattson et al. 1982; Whitham et al. 1984); resistance and defense depend on the lack of overlap between the phenotypic space occupied by the plant and the phenotypic space that the herbivore can exploit most. Since change is nondirectional in the moving-target model, some plant responses make them more susceptible to herbivores. Just as a disperser might leave a bad situation only to arrive somewhere worse, change might modify an attacked individual to a phenotype that is even more vulnerable in a moving-target model. When a plant leaves a state that is worse than average, however, it will, on average, find itself better off. A genotype that consistently becomes more susceptible following damage will be selected against. However, a genotype that changes randomly may be better than one that stays the same even if, some percentage of the time, it becomes more susceptible by changing.

From the herbivore's point of view, nondirectional changes may give the impression of random shifts (rather than predictable shifts along an axis of better/worse defended or more/less of a particular chemical). This does not imply that the processes within the plant that produced the nondirectional responses are random. To the contrary, they may be highly regulated at a biochemical level. How could regulated responses produce a pattern that is unpredictable to the herbivore and appears random?

Many, if not all, induced responses involve a multiplicity of changes rather than movement along a single chemical axis. For example, induced responses in tomato foliage involve many classes of compounds in addition to the accumulation of PIs (Duffey and Felton 1989; Stout et al. 1994; Stout et al. 1996). The effects of these responses depend critically on subtle and numerous features of the abiotic and chemical environment within the plant tissues. In other words, the effects of a given induced response depend critically on the biochemical context. For example, the effects of induced PIs on herbivores can be reversed depending on the levels of plant protein available to the herbivore and on interactions between PIs and oxidative enzymes that can completely inactivate PIs (Duffey and Felton 1989; Workman et al. n.d.). Each of the chemical components of this system changes idiosyncratically following damage by different herbivores, on different-aged plant tissue, and so forth, so that the induced responses that herbivores experience are essentially unpredictable without knowledge of the many plant factors that regulate the effects of each induced chemical component (Stout et al. 1996; Stout and Duffey 1996). Although tomato may be an unusual example in the level of its complexity and synergisms, essentially every "simple" system that has been examined in detail has turned out to be far more complicated than anticipated. This complexity means that chemical changes that may be highly regulated at the molecular or biochemical level by the plant produce effects on herbivores that are context-dependent. In tissue of one age in one chemical environment, an induced response may have one effect. The same response in tissue of a different age or in a slightly different chemical environment will have a very different effect.

Induced responses that are context-dependent may produce considerable variability in even relatively simple systems. For instance, cyanogenesis in clover and trefoil has often been considered the example of a preformed induced response that is under simple genetic control and is predictable in its direction. However, detailed work on this system indicates that the genetic picture is more complicated and that each plant is heterogeneous in its response (De Wall 1942, cited in Till 1987; Jones and Ramnani 1985; Till 1987). Some leaves on an individual wild plant were found to be cyanogenic in response to damage and others were not; leaves that are cyanogenic now may not be in a few hours. This kind of variability has been ignored for years by many researchers who studied cyanogenesis (Till 1987), and it has been ignored by most biochemists studying other plants, who have considered variability to be uninformative noise. Cyanogenesis is a peculiar form of induced response (perhaps it is better considered a constitutive trait) because it is preformed (see section 3.3.2). However, the variability reported for this performed response may be quite widespread. For example, variation in foliar characteristics was found to be at least as great within an individual birch tree as it was between different trees (Suomela and Ayres 1994).

Induced responses may commonly involve far more secondary chemicals than originally suspected (Jones and Firn 1991; Berenbaum and Zangerl 1995). For example, tobacco leaves responded to damage by increasing concentrations of phenolics, PIs, pathogensis-related proteins, and sesquiterpenoid phytoalexins, as well as alkaloids such as nicotine (Baldwin 1994). Induced resistance to tobacco mosaic virus appears to involve the simultaneous coinduction of at least nine classes of genes (Ryals et al. 1992). The various chemicals within plant tissues interact synergistically, dramatically altering their effects on herbivores (e.g. Gonzalez-Coloma et al. 1990; Berenbaum and Zangerl 1993). These complications make plant responses much less predictable and much less directional than they are often portrayed.

The effects of plant responses depend critically not only on the chemical environment of the plant tissue but also on the chemical environment of the herbivore. For example, the action of many of the secondary chemicals induced when tomato foliage is damaged is quite sensitive to slight changes in the pH of the herbivore's gut (Felton et al. 1989, 1992). Some phenolics such as rutin and catechins of tomato can be considered nutrifying to insects in acidic conditions, but under basic conditions they are prooxidants and can be toxic, antinutritive, and antidigestive (Duffey and Stout 1996). Insect guts vary in pH both among individuals of a species and between species. Variation in gut pH typically spans the range over which induced enzymatic oxidation can be minimal or extensive. Similarly, tannins have been found to either reduce bioavailability of proteins or to enhance digestibility of proteins by proteases, depending on the details of the physiology of the herbivore in question (Martin and Martin 1984; Mole and Waterman 1985).

In summary, many factors make induced responses context-dependent. Different herbivores cause different kinds of induced responses. The same signal causes different induced responses in different plants of the same species and in different tissues of the same individual. The effects of an induced chemical change on an herbivore will be different depending on the precise chemical milieu of the particular plant tissue. Even though each step may be transcriptionally regulated by the plant, the net effect of the induced response on an herbivore may appear random without detailed knowledge of the position, age, history, and chemical environment of the plant tissue affected plus knowledge of the chemical environment of the herbivore's gut.

Previous models have generally assumed that induced responses are directional. For some of the better-studied systems, the pool of substances that have been shown to be defensive are always maintained or increased following herbivory. For example, the pool of nicotine in damaged tobacco plants never declines in response to damage. The nondirectional, moving-target model cannot apply to these systems. The moving-target model predicts a signature in which pool sizes of induced secondary chemicals should change in a variable manner. When plants are damaged, variability should increase without any predictable change in the mean value (figure 5.5). Indeed, for some plants, concentrations of defensive chemicals have been found to increase, decrease, or remain the same following damage. For example, concentrations of pyrrolizidine alkaloids in hound's tongue (*Cynoglossum officinale*) sometimes increased following artificial damage and sometimes decreased (van Dam and Vrieling 1994). In some plant families induced alkaloids consistently increased, while in others they consistently decreased.

5.4.1.4 Why Variability May Be Difficult for Herbivores

Earlier we considered empirical tests of the notion that variability in food-plant quality decreases performance of herbivores (section 5.4.1.1). Plant variability that is induced by herbivory could pose difficulties for herbivores for any of several different reasons. Change per se may allow a plant that currently presents a phenotype that is relatively preferred by herbivores to present a less preferred one (Karban 1993a). If attack persists or recurs, the plant may change again. Trypanosomes are thought to use this defense against the immunological attacks of their hosts (Clark 1986). When antibodies bind to the surface of the

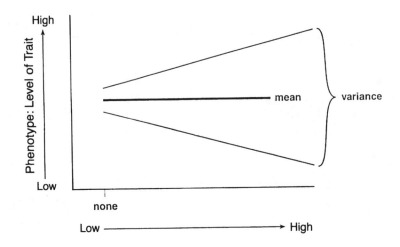

Environment: Level of Herbivory

Fig. 5.5 The moving-target model showing a plant's phenotypic response to a change in its environment. When plants are damaged, they respond by becoming more variable, without any predictable change in the mean phenotype. Compare this nondirectional model to the directional norm of reaction shown in figure 5.1.

trypanosome, they induce expression of new surface antigens. Chances are good that the host has not yet developed immunity against this new surface. This allows the trypanosome to escape for a short time until antibodies against its new surface begin to appear, at which time it changes again.

Several physiological mechanisms have been proposed to explain this suboptimal matching. Herbivores may be unable to compensate for changes in plant nutrition. This may occur if concentrations of nutrients found in plants are not correlated (Stockhoff 1993a). For example, an herbivore that suddenly faces a diet with only half the normal amount of nitrogen may double its consumption of food. However, its rate of assimilation may be altered, or it may now ingest suboptimal concentrations of other nutrients or superoptimal concentrations of secondary chemicals. For example, velvet-bean caterpillars (*Anticarsia gemmatalis*) that were faced with diets low in nutrients responded by eating more, even though increased consumption led to very detrimental doses of caffeine (Slansky and Wheeler 1992). Even if herbivores are able to compensate for changes in the nutritional content of their diets, they must do so rapidly; any lag time in their compensatory responses will reduce the value of dietary compensation (Stockhoff 1993a). The possibility that herbivore detoxification systems may function most efficiently when diet does not vary was discussed briefly in section 5.4.1.1.

Changing plant phenotypes may lessen the ability of herbivores to learn their habitats and therefore allow plants to suffer less herbivory. Jones and Ramnani (1985) argued that mammals (rabbits, voles, and lemmings) learn the microgeography of their habitats and forage selectively. If the leaves on a plant clone (or individual) shift in palatability over time, this could conceivably reduce the damage that they receive. Although Jones and Ramnani developed a group-selection argument to provide a benefit for a population of plants, a similar mechanism could apply to an individual plant if phenotypes of particular leaves shifted over time; this mechanism would require only individual-level selection.

If only a single kind of herbivore has the potential to reduce plant fitness, the constitutive or directional induced strategy providing the best defense against that one threat might be expected. However, if the plant can face many different potential herbivores and if the traits that provide defense against one type of herbivore make the plant more vulnerable to other herbivores, then both constitutive and directional induced strategies would be at a disadvantage when the "alternate" herbivore was present. There are many examples of constitutive and induced traits that make plants more resistant to some herbivores (types or species), and precisely the same traits make them more vulnerable to other types or species. Some of these examples were reviewed in sections 2.2.2 and 5.3.3. When directional changes (e.g. consistent increases in the concentration of some chemical) cause divergent effects on different herbivores, then selection may favor a moving-target strategy of defense (Adler and Karban 1994). The nondirectional moving-target strategy is favored because the plant does not suffer greater attack when it responds to the "wrong" herbivore. Having the "wrong" constitutive or directional inducible strategy might be considered a cost, in the broadest sense of that term. As such, a nondirectional, moving-target strategy that minimized the probability of responding "inappropriately" could be considered an indirect means of reducing the ecological costs of defense.

A nondirectional moving-target strategy was found to be capable of invading in a theoretical model (Adler and Karban 1994). Presumably, this strategy could outcompete constitutive and optimal directional inducible strategies when allocation costs of defense are not too high and plant phenotypes effective against one herbivore are ineffective against others. Other models that do not rely on costly defenses are also worthy of consideration. Of course, these models cannot provide information about whether this has ever actually occurred in nature.

All four of the mechanisms considered above provide benefits immediately, in ecological time, to the individual plant attacked. A plant population that changes idiosyncratically following herbivore attack might reduce the rate of counteradaptation by herbivores over evolutionary
time. Changing plant phenotypes present herbivores with selection pressures that are inconsistent, and this should slow herbivores' evolutionary
responses (Whitham 1983; Schultz 1988; Karban and Myers 1989; Harvell 1990b). Several different defenses employed as mixtures so that they
vary in space or time may also be more resistant to counteradaptation
by herbivores (Gershenzon and Croteau 1991). This topic is discussed
again in section 6.2.3.

The five mechanisms proposed above all assume implicitly that selection favors plants that exhibit inconsistent responses to herbivory (selection has increased variance). Alternatively, inconsistent, nondirectional
responses may have resulted from selection acting on several separate
properties that together gave rise to a variable response. For example,
plants may initiate two different processes following wounding, defense
and regrowth (van der Meijden et al. 1988). These two processes may
not be well integrated, so that regrowth may actually impede defense.
Two antagonistic processes, each with a distinct biochemical signal, such
as these, could conceivably produce an inconsistent pattern in plant responses that appeared to be nondirectional.

The distinction between directional and nondirectional induced responses seems to be an important one. We have many good examples
of repeatable, directional responses of plants by increasing synthesis or
transport of secondary chemicals (see section 3.3.3). Induced responses
in these plants are not context-dependent but are robust to a wide assort-
ment of conditions. For other plants, induced responses increase variability, although the relationship between increased variability and reduced herbivore performance is not universal. For some herbivores,
variable diets have been associated with increased, rather than decreased, performance (e.g. Bernays et al. 1994). The moving-target hypothesis probably does not apply to these plants or these herbivores.
We have only a few well-documented cases of nondirectional responses.
However, many workers who have studied induced responses in various
systems have been impressed with how variable their results seemed to
be. This variability may represent uninteresting experimental noise, as
is often assumed, or it may represent nondirectional variability. Careful
experimental work should allow differentiation of these two possibilities.
In addition, the biological consequences of variability should be evaluated theoretically and empirically. This variability may not be the direct

result of selection pressure by herbivores that favored variability and made the plant more difficult for herbivores to use. Determining why the variability arose is far more difficult than demonstrating that it currently provides resistance or defense; this former goal is beyond the scope of this argument.

In summary, temporal and spatial variability in plant quality can result in resistance and possibly defense, although variability need not cause resistance. Herbivores may experience this variability on a very small scale as they move from tissue to tissue on an individual plant or between plants of a species. To the extent that induced responses increase plant variability, they may also increase resistance and defense. Variability is further increased by responses that are specific to particular plant tissues or by responses that are expressed within a single tissue only at particular ontogenic phases.

5.4.2 Induced Resistance That Is Expressed Constitutively

When people think about induced defenses as traits that allow plants to save costs, they implicitly assume that induced defenses are derived traits. Constitutive defenses are assumed to be the forms that first arose; these have been subsequently modified because they were too expensive. It is also possible, however, that induced defenses arose as induced defenses rather than as special cases of constitutive defenses. If saving costs has not driven the evolution of induced defenses, then they may not be limited to conditions when herbivory is uncommon and unpredictable. It is conceivable that induced defenses could evolve in plants in response to predictable and relatively continuous herbivory. Under these conditions, plants may be induced much of the time. As an analogy, consider the induced callus on the soles of our feet. Inducibility probably did not evolve as a means of saving costs when callus was not beneficial. Instead callused soles probably arose in the mammalian line as an induced trait and have remained so ever since. Since the cues to stimulate callus are present virtually all the time (at least before the invention of shoes), the induced condition has always been expressed. We explain the maintenance of this trait because of its effectiveness and not because it saves costs. Similarly, selection may have favored induced defenses that are expressed permanently in plants because they are effective. Plant defenses may be perpetually induced by plant fungi that inhabit the plant surface and elicit some level of resistance (Heath 1995). Plants that were grown under conditions that removed their normal nonpathogenic microbes became more susceptible to attacks by damaging pathogens

(Gregersen and Smedegaard 1989; Sahashi et al. 1989). It is possible that the nonpathogenic microbes consistently induced plant defenses that provide protection against other challenges.

5.4.3 Induced Defenses as a Result of Plant Rearrangement: The Carbon/Nutrient Balance Model

The C/N balance hypothesis was developed as a mechanistic or physiological explanation of induced resistance (see section 3.3.3.1). However, it has not been developed into a new evolutionary hypothesis (see the arguments in section 5.2). It was conceived originally as an extension of the optimal allocation models and had the novel feature of including constraints imposed by differential availability of resources for plants (Janzen 1974; Bryant et al. 1983; Coley et al. 1985). Recent versions of the C/N balance hypothesis have emphasized that it makes different predictions than most of the optimal allocation models (Bryant et al. 1988; Bryant, Heitkonig, et al. 1991; Bryant et al. 1993; Tuomi et al. 1990; Herms and Mattson 1992).

We have emphasized repeatedly that the C/N balance model has not provided an evolutionary or ultimate explanation for the induced responses it describes. If it were to be extended to become an evolutionary scenario, it might go something like this: Evolution has maximized plant growth rate, not defense against herbivory. In productive habitats maximizing growth rate has meant that carbon-based resources are rarely stored by plants, and so those plants appear to have minimal defenses if only carbon-based defenses are considered. In unproductive habitats maximizing growth has meant that carbon-based resources are stored following defoliation, and this process makes plants less vulnerable and/or less attractive to herbivores. The key point is that the defenses that we see were shaped by selection to maximize growth and not by selection pertaining to herbivory. The emphasis here is on selective pressures that may have molded traits and not on the constraints that may have limited those processes. We are not sure if this adaptive scenario is precisely the same as that envisioned all along by the proponents of the C/N balance theory or if they would disagree with it.

The C/N balance theory assumes that there is little or no cost associated with the production of carbon-based secondary chemicals. This argument could be extended to include other types of defenses that are presumed to result from other damage-induced metabolic imbalances. Since the defense results from resources that the plant cannot use for its primary metabolism, the traits that provide defense would not be expected to accrue any additional allocation costs. Indeed, any induced response that serves functions other than defense may appear to be cost-free.

5.5 FUTURE DIRECTIONS: GENERATING AND TESTING EVOLUTIONARY HYPOTHESES ABOUT INDUCED DEFENSES

At this point we have considerable evidence that induced responses affect herbivore performance but slightly less understanding of the consequences of those effects on herbivore population dynamics (chapter 4) and much less understanding of the consequences of those effects on plant fitness and defense (this chapter). The strong and widespread effects of induced resistance that have been observed have the potential to translate into important effects on herbivore population dynamics, depending on behavioral responses of herbivores and the relative importance of other factors that also affect those populations. Similarly, herbivory and induced resistance to herbivores can be important forces molding plant fitness, depending on the effectiveness of induced responses and the tolerance of plants to herbivory. These uncharted areas of inquiry are interesting in their own right, and a much fuller exploration of these poorly known effects is essential to understand the relationship between induced responses and plant fitness and the selective forces that shape plant defenses.

Our current knowledge of the effects of induced responses on plant fitness is woefully inadequate. This is surprising, considering that keeping track of plant fitness is not a very difficult empirical task. Benefits can be estimated simply by inducing responses in environments with herbivores and measuring the relative fitness of induced and uninduced plants. Costs can be estimated by inducing responses in environments without herbivores and measuring the relative fitness of induced and uninduced plants. Recent developments in our understanding of plant signals and elicitors of induction allow us to induce plant responses and measure the effects of induction without actual damage. These new techniques will allow us to uncouple wounding from elicitation and will facilitate more accurate measurements of the actual costs and benefits of induced responses. Interpretation of differences in plant fitness is always more straightforward for annuals than for longer-lived perennials. Real measures of plant fitness (survival and reproduction) are far superior to very indirect measures such as leaf area removed. More studies should consider whether induced responses are variable and whether this variability is heritable. The present paucity of this information is understandable, since large experiments must be designed.

The information from experiments about variability, heritability, and effects on relative fitness will establish whether induced responses affect plant fitness and can be considered defenses. Assuming that induced responses do defend plants begs the question, why are these traits induced rather than constitutively expressed? The conventional explanation has been based on saving allocation costs. Thus far, this hypothesis has met with mixed success, although it may prove to be correct. We have tried to develop and present several alternative explanations. At this point we have cannot tell if the evolutionary models that do not rely on minimizing allocation or physiological costs have any grounding in truth. At the very least we hope that presenting them in some detail will stimulate empiricists to test them and will loosen the stranglehold that optimal allocation models have had on our attempts to understand the evolutionary factors responsible for induced defenses.

6 Using Induced Resistance in Agriculture

6.1 INDUCED RESISTANCE FOR DISEASE CONTROL IN MEDICINE AND PLANT PATHOLOGY

The introduction of mild, avirulent strains of microorganisms is the most common form of Western preventive medicine practiced for disease control for vertebrates, including humans. Vaccinations induce resistance in the host, so that symptoms caused by the disease are much reduced. Contrary to common perception, most successful vaccinations do not completely exclude the disease organisms; vaccinations cause the host to contain them, slow them down, and reduce the expression of their symptoms.

Plant pathologists have been aware of induced resistance for much of this century (Beauverie 1901; Ray 1901; Chester 1933). The similarities and differences between induced resistance for disease control in plants and immunity in animals have been discussed extensively in that literature (Chester 1933; Kuc and Caruso 1977; Kuc 1987). Plant pathologists have achieved several impressive successes with releases of antagonists to protect plants against viruses, bacteria, and fungi.

6.1.1 Successes against Viruses

Inoculations of less virulent strains of viruses to prevent the deleterious effects of more damaging strains (called cross-protection) have been particularly noteworthy. Strains of mosaic viruses have been used to protect tomato seedlings against tomato mosaic virus throughout much of the world, particularly for tomatoes grown in greenhouses (Rast 1972; Fletcher and Rowe 1975; Broadbent 1976; White and Antoniw 1989). The first inocula used were natural strains that caused few symptoms but these were replaced by avirulent strains produced by nitrous acid mutagenesis. The technique was economically attractive because tomato mosaic virus had been very difficult to control by other methods. When constitutively resistant varieties became available, most growers switched to these and discontinued using inoculations. Cross resistance is still used for cherry tomatoes because resistant varieties are not available (Fulton 1986; J. Whipps personal communication).

Inoculations have been used, as well, to protect citrus trees against citrus tristeza, a damaging virus vectored by aphids (Costa and Muller 1980; Fulton 1986). Widespread releases of mild, naturally occurring strains have been successful in most growing regions of the world.

More recently, workers in China have developed a method to inoculate peppers against cucumber mosaic virus by including some satellite RNA with the virus (Tien et al. 1987). Satellite RNA cannot replicate itself and is dependent on the mosaic virus for replication. Tien and coworkers constructed mild viruses by adding satellite RNA to existing viral genomes. When pepper plants were then inoculated with these mild viruses, disease symptoms were reduced and yields increased at most locations because the inoculated plants were resistant to more damaging strains. Cucumber mosaic virus had not been controlled effectively using other techniques which made inoculations particularly attractive. Yield increases of 10–55% have been reported with the use of this technique on approximately three thousand hectares of peppers annually (Tien and Wu 1991). Inoculations of attenuated cucumber mosaic virus have provided protection against damaging strains and increased yields in field tests for tomato, tobacco, eggplant, cabbage, and cucurbits. These inoculations have also been found to induce resistance against downy mildew and other fungi so that fungicide sprays have been discontinued at some locations where inoculations have been used (Tien and Wu 1991; Qin et al. 1992). Expanded use of this technique is planned for the future in China. Risk of mutation to a more damaging strain (a single nucleotide may be sufficient), limited natural occurrence of satellite RNA, and the possibility of changes in helper-virus relationships have limited the widespread use of similar techniques involving satellite RNA (Wilson 1993).

6.1.2 Successes against Bacteria

Inoculations of avirulent strains of bacteria have been used successfully to control two bacterial diseases (Campbell 1989). Crown gall disease caused by *Agrobacterium tumefaciens* has a worldwide distribution and a very broad host range. Control has been achieved by dipping seeds, seedlings, or rootstocks in a suspension of a particular strain of *Agrobacterium radiobacter*. Control is believed to result from competition for establishment sites and possibly from the release of inhibitory substances (Moore and Warren 1979; Clare et al. 1987). The technique is inexpensive and has been used widely in commercial operations (Rishbeth 1988).

A second commercial success that has not been employed quite as widely is control of fire blight (*Erwinia amylovora*), a disease of fruit trees. Fire blight has been controlled by releasing either *E. herbicola* when control is needed only during flowering, or nonvirulent strains of *Pseudomonas syringae* when longer control is needed (Beer et al. 1984; Lindow 1985). The mechanism is not clear in this case, although direct competition, rather than induced responses of the host plant, is believed to be involved.

Legumes that become infected with rhizobium bacteria become resistant to infection by other bacteria (Bauer 1981; Verma and Nadler 1984). Within two to four hours after encountering rhizobium bacteria, the infections that give rise to nitrogen-fixing nodules block infections of other strains at adjacent sites on the root. Legume seeds are routinely inoculated commercially with appropriate strains of root-nodule bacteria to improve plant growth and yield. The precise mechanisms responsible for control are unclear in this case.

6.1.3 Successes against Fungi

Fungal diseases have been controlled by inoculations of fungi and bacteria. The first such successes involved inoculating cut stems of conifers with a fungus, *Peniophora gigantea,* which outcompetes *Heterobasision annosum,* the organism responsible for butt rot (Rishbeth 1975, 1988). This technique has been in use commercially for approximately thirty years (Campbell 1989). Another group of fungi, *Trichoderma* spp., has been employed as inocula on seeds, stems, and in soil to provide protection against a wide variety of fungal diseases (Papavizas 1985; Chet 1987), although any one strain of *Trichoderma* generally controls only one plant disease. Products containing *T. viride* and *T. polysporum* have been produced by Bio-Innovation in Sweden and registered in Europe for commercial use on orchard trees and in mushroom culture (Ricard 1981). *Trichoderma* spp. are thought to act as both competitors and as parasites of the pathogenic fungi.

6.1.4 Necessary Conditions for Commercial Use of Cross-Resistance: Lessons from Plant Pathology

Despite considerable research on cross-resistance and several impressive successes, introduced inocula are used commercially to control only a few plant diseases. Numerous authors have considered the reasons why these techniques have worked in a few systems but have not been used more extensively (Fulton 1986; Buck 1988; Deacon 1988; Rishbeth 1988; Campbell 1989; Whipps 1996). Several conditions seem necessary for the use of biological agents as inducers of resistance against pests. These necessary conditions also probably apply to most, if not all, forms of biological control.

1. The pest must be sufficiently destructive to warrant the cost of using inducers. It also must be difficult to control by other, more familiar, methods.

2. The inducer must not be a pest itself, that is, it must not cause significant losses in yield. These losses can take many forms: actual reductions in yield resulting from the inducer's own use of the host, reductions in the use of other crops in the ecosystem or other uses of the target crop, reductions in desirable plant characteristics associated with effective levels of resistance, and so forth. In addition, use of the inducer should not pose environmental or safety hazards.

3. It must be possible to introduce the inducer into the system. In other words, the inducing organism must be a reasonably good colonist and competitor. For this reason, many potential antagonists that seemed promising in greenhouse trials have failed in subsequent field tests (Rishbeth 1988). Campbell (1989) considered this condition to be the primary stumbling block of attempts to use vaccinations to protect more plants against pathogens. It is probably no coincidence that the two most widely used inducers against soil-borne diseases are the two that are most successful at gaining occupancy of the infection court and reproducing (Deacon 1988).

4. Use of the inducer must be cost-effective. This condition is more likely to be satisfied if the pest is costly and the inducer is minimally destructive and easy to apply (conditions 1–3 above). In addition, the more effective the inducer is at protecting the plant from the pest, the more likely it will be cost-effective. For many systems there is a narrow window of time when the inducer can be applied to provide adequate protection (Rishbeth 1988). Numerous biocontrol agents have been developed as seed treatments that give excellent control of seed rots and seedling diseases (Kommedahl and Windels 1981). These have not been used commercially because they have been more expensive than conventional chemical treatments (Al-Hamdani et al. 1983; Deacon 1988). The new technique must have a facilitator who can make a profit. Kuc (1995) observes that, although many of the compounds that his research group has developed are safe, inexpensive, and efficient elicitors of disease control, they have not been marketed commercially. He concludes that

commercialization has been driven by the ease of obtaining a patent monopoly for a highly profitable product. Any product that is readily available may not be attractive to major agrochemical companies.

The four conditions listed above are probably essential: three more are desirable for inoculations of inducers to be employed commercially.

5. An inducer that is self-replicating (sustainable) is much preferred to one that must be applied repeatedly. The widespread use of toxins from *Bacillus thuringiensis* and chemical pesticides attest to the fact that this attribute is not a necessity. On the other hand, poor long-term survival of avirulent isolates of *Rhizoctonia solani* has limited their use to protect against damping off in sugar beets (Buck 1988).

6. Caution about the spread of infectious agents has limited the use of avirulent inducers. The inducer should ideally be host-specific so that it does not spread to other hosts and other environments where it may cause damage. The safest uses of these techniques probably involve situations where the inducer is already present in the agroecosystem; conditions for the inducer can be improved and made less favorable for the pest.

7. The plant should have a pretty high tolerance for the pest (and for the inducer). Induced resistance rarely provides complete elimination of the pest, unlike some chemical applications. Damage tolerances are often higher on plant tissues that are not economically prized. For example, induced resistance is more likely to be successful for inducers and pests of the leaves of fruit trees rather than the fruits themselves. Induced resistance is more likely to be practical if an individual inducer or pest herbivore is small relative to the plant. For example, a tomato grower can tolerate a few mites per plant better than a few hornworms.

These conditions for the effective commercial use of inducing agents have been developed after considering the efforts of plant pathologists. However, many of the considerations that became apparent for attempts to use induced resistance to control diseases also will apply to similar attempts to control herbivores in agriculture.

6.2 Strategies Using Induced Resistance for Control of Herbivore Pests

Introductions of avirulent organisms that will induce resistance against subsequent attacks by more damaging parasites is the most straightforward approach to using induced resistance in agriculture. This technique probably received the most attention in the past. However, we can envision other strategies that could also be used as management tools. Cultural manipulations that simulate the effects of biological inducers could be applied to the crop to induce resistance against pests. These manipulations might involve the application of chemical inducers or elicitors. Crop plants could be bred for more induced resistance. Genetic engineering could add genes which provide induced resistance into crop genomes where they have not occurred in nature. This technique is likely to receive the most attention in the future. Below we will discuss each of these strategies, describing current applications in use in commercial agriculture. Earlier discussions of alternate strategies for pest control using induction were presented by Kogan and Paxton (1983), Fischer, Kogan, and Greany (1990) and Karban (1991b). Similar techniques may also be useful for management of forest pests (Wagner 1988).

6.2.1 Vaccinating Plants by Introducing Organisms That Induce Resistance

Agricultural crops are hosts for at least several and often very many different herbivore species. Some of these may be serious economic pests and others may cause only minimal economic losses. Species can cause minimal losses because their populations remain small or because they feed on tissues that the plant can replace or tissues of little economic value. In such cases, the activities of one herbivore species may induce resistance against a far more damaging one. This may occur naturally or may be assisted by agronomic activities.

European red mites (*Panonychus ulmi*), a serious economic pest of apple trees, can be outcompeted by smaller, less damaging apple rust mites (*Aculus schlechtendali*) (Croft and Hoying 1977). It is not clear whether the negative interaction between these two species is caused primarily by direct competition, induced changes in the host plant, or changes in the numbers of predators. Growers in Michigan have been encouraged to ignore high populations of rust mites, a practice which has reduced the need to treat for European red mites (B. A. Croft, personal communication).

Strawberry crops in New Zealand are grown as perennials and experience damage caused by two-spotted spider mites (*Tetranychus urticae*) (Butcher et al. 1987). Traditionally, growers have applied miticides repeatedly during the period of production. However, plants that were allowed to host mites during their first season tended to have few mites during subsequent seasons. As was true for apples discussed above, the mechanisms responsible for this pattern are not well known although induced resistance has been suggested (Butcher et al. 1987). Many growers have elected to forego applications of miticides during the first year of growth and have found that mite populations have remained low thereafter, obviating the use of chemical controls (M. R. Butcher, personal communication).

Wine grapes in California's central valley are attacked by two species of herbivorous spider mites: economically damaging Pacific mites (*Tetranychus pacificus*) and economically less important Willamette mites (*Eotetranychus willamettei*), a native species. Often, the two species are negatively associated; vineyards with Willamette mites early in the season tend not to have problems with Pacific mites whereas those without Willamette mites are more likely to suffer losses from Pacific mites (Flaherty and Huffaker 1970; English-Loeb and Karban 1988). Almost all vineyards that are not sprayed with miticides support small populations of Willamette mites. Some zinfandel vineyards have had chronic problems with high populations of Pacific mites that have defied control attempts with miticides or predatory mites. Introductions of Willamette mites into these problem vineyards have reduced populations of Pacific mites and increased yield characteristics (English-Loeb and Karban 1988; Karban and English-Loeb 1990; English-Loeb et al. 1993). These "vaccinations" were only successful if Willamette mites were present early in the season as the new shoots were expanding; introductions two weeks later were much less effective (Hougen-Eitzman and Karban 1995).

An introduction of Willamette mites that successfully reduces populations of Pacific mites can operate by any of several possible mechanisms. The two species may compete directly for resources or may physically interfere with one another. The two species may compete by inducing changes in the plant tissue that they both use. The two species may experience "apparent competition" if the presence of one species increases the rate of predation on the other species. Each of these three mechanisms probably occurs in this system, at least occasionally. Induced resistance was found to be the strongest and most consistent of the three mechanisms. The negative effect of Willamette mites on populations of Pacific mites was as strong when the two species were separated spatially or temporally as when the two co-occurred (Hougen-Eitzman and Karban 1995). These results were not consistent with the hypothesis that direct competition was responsible. Predatory mites were found to be more effective at controlling Pacific mites when Willamette mites were also present (Karban et al. 1994). However, this result of increased predation was found only in one of four experiments, suggesting that it is not generally an important mechanism.

Introductions of Willamette mites are sustainable in the sense that they need not be applied each season to provide good control of Pacific mites (Karban, English-Loeb, and Hougen-Eitzman 1997). Willamette mites, but generally not Pacific mites, overwinter on the grapevines; most vineyards and wild grapes in northern California maintain small, endemic populations of Willamette mites. This system is easily upset by the use of chemical miticides. One of the two most widely used miticides on grapes caused dramatic decreases in populations of both Willamette and Pacific mites. However, this was often followed by a resurgence of mite numbers that far exceeded levels experienced without miticide applications (Karban, English-Loeb, and Hougen-Eitzman 1997). Effects such as these have helped to habituate growers to the repeated use of chemical miticides.

Vaccinations have proven effective in large-scale commercial operations. Introductions of Willamette mites meet all seven conditions outlined above (section 6.1.4). Willamette mites have the potential to become economic pests (Welter et al. 1989), although this is rarely the case in California's central valley (Flaherty and Huffaker 1970; Karban and English-Loeb, personal observation). However, meeting the criteria outlined above has not guaranteed that inoculations of Willamette mites will become the primary tactic for control of Pacific spider mites on grapevines. For one thing, this technique is only applicable in those vineyards with chronic problems; most growers need not concern themselves with vaccinations. However, even though vaccinations are cheaper and more effective than either chemical control or introductions of predatory mites, widespread use of vaccinations faces several major hurdles. First, there is no commercial supplier of Willamette mites. Second, many growers cannot differentiate the two species. Third, and most important, the system lacks an effective facilitator. Chemical pesticides can be pushed by the companies that produce and apply them for a profit. Companies cannot make money supplying readily available biological agents or solutions that have only limited markets. Almost all growers in California hire pest-control advisers to assist them in managing their crops. These state-licensed pest-control advisers often work directly for the chemical companies, or if they are independents, they receive a commission on the products they sell to growers. They have little incentive to recommend approaches that fail to provide these monetary kickbacks. University extension personnel and researchers are often less open to new approaches than innovative growers, as they fear losing the paradigms that they have built careers testing and advocating. An excellent account of frustrated attempts to introduce change into the research and extension communities has been given by van den Bosch (1978). The success of any new approach is determined far more by the politics of who gets behind it and the economics of who can benefit by pushing it than by the considerations of its effectiveness discussed above.

The examples described above represent the cases that we are aware of in which less damaging organisms have been used to induce resistance against pests in commercial agriculture. Research into these systems was stimulated by the observations that two species were negatively associated. There are many other crop systems in which damage by herbivores or other agents was associated with a decrease in the performance or population size of pest insects and mites (table 6.1). In these examples the negative associations could have been caused by induced plant responses or by other mechanisms; rarely is the mechanism known. Most, though not all, of the examples come from lab studies and have not been confirmed in the field. In very few of the examples did the authors consider the effects to the plant of adding inducers; there may be no benefits in terms of yield or increased profits. Therefore, the table is not intended to provide examples of crop systems that are sufficiently researched and ready for commercial implementation. Rather, the list includes all the systems we could find that have shown any signs of promise that could repay future study. Very few plants and herbivores have been examined with respect to induced resistance; thus many systems not included in the table may ultimately prove tractable by this general approach.

6.2.2 Inducing Plants Using Cultural Manipulations

Many agronomic practices such as fertilization, watering, and so forth can have large effects on plant resistance to pests (see reviews in Heinrichs 1988). Aside from their direct effects on herbivores, physical and biological factors influence the quality of host plants and affect herbivores indirectly. Entomologists have had some success using cultural techniques to induce resistance. Direct insecticidal effects will not be considered here, although several different mechanisms that are mediated by changes in the host plant will be discussed.

The most common and successful method of cultural manipulation involves changing the plant's phenology so that it becomes less suitable for damaging herbivores. This has been accomplished by applying natural plant hormones or synthetic plant-growth regulators (Hedin 1990). For example, applications of gibberellic acid, a natural plant-growth regulator, to green grapefruits has been used to maintain their natural resistance against Caribbean and Mediterranean fruit flies (Greany and Shapiro 1993). Gibberellic acid slows the rate at which the fruit ripens, making fruit less attractive to ovipositing flies and exposing fly larvae to peel constituents, thus reducing fly survival (Greany et al. 1987). Ripening of the internal parts of the fruits was not affected by the applications of gibberellic acid (McDonald et al. 1987).

Synthetic growth regulators have also been used to alter plant phe-

Table 6.1 Examples of competition between individuals of the same and different species that use agricultural crops as hosts

Crops	Inducer	Pest Species	Effect	Lab/Field	Reference
Field crops					
Beans	Spider mites	Spider mites	Increase in mortality and development time	Lab	Wrensch and Young 1975
Beets	Beet flies	Beet flies	Increase in mortality	Field	Rottger and Klinghauf 1976
Brassica	Nontoxic chemicals	Aphids	Various negative effects	Both	Van Emden 1969
	Flea beetles or mechanical damage	Flea beetles	Reduced damage	Lab	Palaniswamy and Lamb 1993
Celery	*Fusarium* fungus	Caterpillars	Reduced growth?	Lab	Heath-Pagliuso et al. 1992; Zangerl 1990
Chrysanthemum	Spider mites	*Liriomyza* flies	Reduced attractiveness	Field	Price 1981
	Spider mites	*Liriomyza* flies	Reduced attractiveness	Lab	van de Vrie et al. 1988
Corn	Mechanical simulation of caterpillars	Aphids	Reduced growth, reduced survival	Lab	Morse et al. 1991
Cotton	Spider mites	Spider mites	Reduced numbers	Lab	Karban and Carey 1984
	Spider mites	Spider mites	Reduced numbers	Field	Karban 1986a
	Mechanical damage	Spider mites	Reduced numbers	Lab	Karban 1985
	Verticillium fungus	Spider mites	Reduced numbers	Lab	Karban et al. 1987
	Spider mites	Armyworms	Increased mortality	Lab	Karban 1988
	Mechanical damage	Armyworms	Reduced attractiveness	Lab	Croxford et al. 1989
	Bacteria	Boll weevils	Increased mortality, reduced damage	Both	Benedict and Chang 1991
	Aphids	Aphids	Reduced numbers	Lab	Wool and Hales 1996

Plant	Inducer	Target	Effect	Type	Reference
Cucumber	Spider mites	Spider mites	Reduced attractiveness, numbers	Lab	Tomczyk 1989; A. A. Agrawal, personal communication
Lima beans	Spider mites	Spider mites	Reduced attractiveness	Lab	Dicke 1986
Potato	Eriophyid mites	Eriophyid mites	Reduced survival	Lab	Dreger-Jauffret et al. 1990
	Potato leafhoppers	Colorado potato beetles	Reduced survival and growth	Lab	Tomlin and Sears 1992
	Colorado potato beetles	Potato leafhoppers	Reduced attractiveness	Lab	Tomlin and Sears 1992
Rice	Gall midges	Gall midges	Increase in mortality	Lab	Bentur and Kalode 1996
Soybeans	UV-induced phytoalexins	Bean beetles	Reduced attractiveness	Lab	Hart et al. 1983
	Soybean looper	Bean beetles	Reduced attractiveness	Lab	Lin et al. 1990
	Mechanical damage	Soybean looper larvae	Reduced growth	Both	Reynolds and Smith 1985
	Spider mites	Caterpillars	Reduced growth	Lab	Wheeler and Slansky 1991
	Velvet-bean caterpillars	Caterpillars	Reduced growth	Lab	Wheeler and Slansky 1991
	Spider mites	Spider mites	Reduced numbers	Lab	Hildebrand, Rodriguez, Brown, and Volden 1986
	Mechanical damage	Armyworms	Reduced attractiveness	Lab	Croxford et al. 1989
Squash	Mechanical damage	Beetles	Reduced attractiveness	Field	Carroll and Hoffman 1980
	Mechanical damage	Beetles	Reduced growth, survival, fecundity	Lab	Tallamy and McCloud 1991
Strawberries	Spider mites	Spider mites	Reduced numbers	Field	Butcher et al. 1987
	Spider mites	Spider mites	Reduced numbers	Lab	Shanks and Doss 1989
Sunflowers	Various	Sunflower beetles	Reduced attractiveness	Lab	Olson and Roseland 1991
Tobacco	Tobacco mosaic virus	Aphids	Reduced reproduction	Lab	McIntyre et al. 1981
	Tobacco mosaic virus	Hornworm caterpillars	Reduced consumption, growth	Lab	Hare 1983

(continues)

Table 6.1 continued

Crops	Inducer	Pest Species	Effect	Lab/Field	Reference
Tomato	Mechanical damage	Armyworms	Reduced attractiveness	Lab	Edwards et al. 1985
	Armyworms	Armyworms	Reduced growth	Lab	Broadway et al. 1986
	Mechanical damage	Hornworm caterpillars	Reduced growth	Lab	Wolfson and Murdock 1990
Orchard crops					
Apples	European red mites	Brown mites	Reduced numbers	Lab	Kuenen and Post 1958
	Brown mites	European red mites	Reduced numbers	Lab	Kuenen and Post 1958
	European red mites	Spider mites	Reduced numbers	Field	Lienk and Chapman 1951
	Rust mites	European red mites	Reduced numbers	Both	Croft and Hoying 1977
Avocado	Avocado brown mites	Avocado brown mites	Reduced numbers	Lab	McMurtry 1970
Citrus	Citrus red mites	Citrus red mites	Reduced numbers	Lab	Henderson and Holloway 1942
Grapes	Willamette mites	Pacific mites	Reduced numbers	Both	English-Loeb and Karban 1988; Karban and English-Loeb 1990; English-Loeb et al. 1993
Peach	Silverleaf fungus	European red mites	Reduced numbers	Field	Kuenen 1948
	Spider mites	European red mites	Reduced numbers	Lab	Foott 1962
Walnut	Walnut aphids	Walnut aphids	Increased mortality	Lab	Davis 1957
	Walnut aphids	Walnut aphids	Reduced numbers	Field	Sluss 1967

nology and thereby make plants more resistant. In apples and pears growth regulators have been used to slow vegetative growth. This has resulted in reduced populations of aphids on apple trees (Hall 1972) and on pears, slowed vegetative growth was associated with reduced psylla populations and reduced numbers of damaged fruits (Westigard et al. 1980). Applications of bioregulators to cotton plants late in the season terminate growth of late, immature bolls (Kittock et al. 1980). Removal of immature bolls denied late-season bollworm caterpillars sufficient food to complete development and reduced their overwintering populations.

Applications of plant-growth regulators can also induce chemical changes in plants so that they provide less nutrition or higher concentrations of secondary chemicals to herbivores. Chlormequat chloride (CCC) has been shown to induce resistance in a variety of crops to aphids (Campbell et al. 1984; Fischer, Kogan, and Greany 1990). Reductions in growth, development, survival, and fecundity of aphids on treated plants have often been attributed to reductions in concentrations of free amino acids, although this view of the mechanism responsible for the effects has not been well substantiated (Campbell et al. 1984). Workers have measured concentrations of amino acids in the total plant and not from phloem samples, the plant part that aphids use. For many of these crops, plant-growth regulators have been found to have no direct insecticidal effect, making the possibility of an induced plant response likely. Application of CCC to sorghum seedlings increased concentrations of intercellular pectin, an important barrier to aphid stylet penetration, suggesting a possible mode of action (Campbell et al. 1982, 1984). Other growth regulators that induced resistance to insects induced corresponding accumulations of various allelochemicals such as gossypol, tannins, flavonoids, and terpenoids (Hedin 1990). Two commercially available preparations of cytokinins, Kinetin and Burst, increased cotton-leaf flavonoids, plant resistance to *Heliothis* caterpillars, and plant yields (Hedin et al. 1988). However, the causal links between increased concentrations of flavonoids, resistance, and yield are still speculative.

Many workers in both academia and industry have attempted to find or develop chemicals that will act as potent inducers of plant resistance. Most of these attempts have focused on induced defense against diseases rather than herbivores. Application of such inducers would allow growers to manipulate plants to constitutively express resistance traits that are naturally expressed only following damage. However "successful" attempts to have inducible traits be expressed constitutively sometimes produce "unsuccessful" end results. Kuc (1987) reported that he could elicit production of phytoalexins in unattacked green bean and soybean plants. A group at Ciba-Geigy led by John Ryals (1992) found that salicylic acid induced expression of *Bacillus thuringiensis* genes that had been engineered into tobacco plants. In both cases exogenous applications of these inducers led to unacceptably high levels of phytotoxicity.

The search continues for inducers that protect plants but do not cause autotoxicity. An interesting example of an inducer that has been developed commercially for control of fungal diseases is chitosan, a biodegradation product of chitin. Chitosan, applied exogenously, induced phytoalexin production in pea pods and inhibited germination of *Fusarium* conidia (Hadwiger and Beckman 1980). Chitosan from crab shells was found to inhibit the growth of many different fungi, making it very attractive for commercial uses. Chitosan from crustacean shells elicited responses similar to chitosan produced by fungi; both induced the formation of PAL and affected the synthesis of at least twenty other proteins in pea plants (Kendra et al. 1989). Based on the work of Lee Hadwiger's group, Bentech Laboratory, Inc., developed a product marketed as YEA! for agricultural use in seed treatments (L. A. Hadwiger, personal communication). Recently Vanson, a subsidiary of DuPont, has bought the rights to market this material. Chitosan has been used primarily on wheat seeds to protect them against straw breaker fungus (*Pseudocercosporella herptotrichoides*) and other root rots. Application of chitosan is less expensive than other conventional materials; it generally caused reduced disease symptoms but did not protect seedlings as completely as competing chemical techniques. Patents are pending for additional uses of this elicitor in foliar sprays and with drench irrigation.

Other research teams in academia and private industry recently have been conducting experiments with chemical elicitors of resistance against plant pathogens (Kessman et al 1994). Several of these, such as 2,6-dichloroisonicotinic acid (INA), have been found to induce systemic resistance in many plant species under greenhouse and field conditions. Application of this compound induced the same set of nine gene families systemically as did infection with pathogens. Thus far, INA has not been marketed commercially and probably never will be because of unacceptably high levels of phytotoxicity (J. Ryals, personal communication). A new elicitor, benzo (1,2,3) thiadiazole-7-carbothioic acid *S*-methyl ester (BTH) has been registered by Ciba under the name Bion for use in wheat against powdery mildew and in tobacco against blue mold (J. Ryals, personal communication). Initially it will be marketed in Germany, Switzerland, and New Zealand. It activates the same biochemical and molecular markers as seen when actual pathogens are used as inducers (Gorlach et al. 1996). Induced resistance is involved because the chemical has no direct antimicrobial activity, although treating plants before exposure to pathogens is necessary for effective protection (Gorlach et al. 1996; K. Lawton, personal communication). Other chemicals that artificially activate systemic acquired resistance to diverse pathogens in a wide spectrum of crops (tobacco, grapes, cucumber, wheat, rice, etc.) are also being tested by Ciba for commercial registration (J. Salmeron, personal communication).

Cultural manipulations of plants may alter the plants such that resistance to attacking organisms is more rapid or more effective. Agents that do not directly induce production of chemicals involved with resistance but increase the plant's capacity to respond are called sensitizers. For example, chemicals were found that did not directly stimulate phytoalexin production but which increased the phytoalexin response of rice following infection (Cartwright et al. 1977). Many such sensitizers have been identified by A. K. Sinha and coworkers in West Bengal, India (Sinha and Hait 1982; Sinha 1984; Hait and Sinha 1986, 1987). Seed and foliar treatments with these chemicals sensitized plants so that their responses to fungal pathogens were more intense. Varieties that were otherwise susceptible resembled resistant varieties after such treatments. Despite these dramatic results, sensitizers have not yet been incorporated into commercial practices. Several authors have questioned their potential because they provide only short term protection so that they may have to be applied repeatedly in some instances (Hedin 1990). The sensitizers may also be incompatible with other agronomic practices (Hedin 1990). Deacon (1988) suggested that another possible reason for the lack of commercial use of these materials is that adequate, cheap, and environmentally safe control of the particular fungi that sensitizers affect has been available already.

Cultural manipulations of plants may also make them more tolerant (as opposed to resistant) of the attacks that they receive (Kogan and Paxton 1983; Fischer, Kogan, and Greany 1990). Although this mechanism is very attractive, it has received little commercial attention thus far.

6.2.3 Genetic Improvement of Crops to Include Induced Defenses

Breeding programs for crops could select plants that exhibited high levels of induced resistance against herbivores. Although this idea has been in the literature for some years (Kogan and Paxton 1983), we are not aware of breeding programs that have actively evaluated and incorporated induced resistance.

Grasses with fungal endophytes provide one line of research where breeders have attempted to incorporate an unusual form of resistance against insect pests (Clay 1989; Siegel et al. 1989). Some fungal endophytes produce systemic infections that occur within the cells of many healthy grass species for much of their life cycles. The fungus-grass associations have been well known for many years because some are poisonous to livestock. Attempts to remove the endophytes from grasses in order to improve forage quality led to observations that fungus-free grasses were more susceptible to attack by insect herbivores. It has recently been shown that grasses with endophytes are repellent and/or antibiotic to many insect herbivores (Popay and Rowan 1994). This has not been considered as induced resistance per se because the endophytes produce compounds that are directly active rather than inducing changes in their host plants. However, recent evidence suggests that the negative effects of endophytes on herbivores may be induced, at least in some systems (Bultman and Ganey 1995). Ryegrass that was infected with endophytes but was not damaged produced smaller deleterious effects on fall armyworm (*Spodoptera frugiperda*) pupal mass and development times compared to infected plants that had been artificially clipped. This interaction between prior damage and endophyte infection suggests that induced plant responses accentuate the effects of endophytes.

There are specific situations, such as roadsides, lawns, and golf courses where endophyte-infested turf grasses would not be consumed by mammalian herbivores and resistance against insects is desirable. In addition to these limited uses, breeders in New Zealand have been attempting to develop strains of grasses and associated endophytes that are resistant to insect herbivores but not to livestock (Breen 1994). One strain of endophyte that has shown considerable promise in perennial ryegrass has been distributed in New Zealand under the label of Endosafe (R. Prestidge, personal communication). This endophyte strain produces an unusually narrow spectrum of alkaloids that have quite severe negative effects against Argentine stem weevils (*Listronotus bonariensis*) but smaller deleterious effects on livestock. Several turf grass varieties are now commercially marketed because of their high levels of endophyte infection. In essence, these grasses have been selected from among naturally occurring varieties because they already had infections. In addition, breeding programs are underway to incorporate useful endophytes into leading cultivars and germ plasms although these selections have not been released commercially as of yet (Saha et al. 1987; Popay and Rowan 1994). These breeding programs hold considerable promise because many endophytes are maternally inherited from the parental grass plant to its seed progeny. Repeated introductions will be unnecessary. Endophytes can be introduced into uninfested grasses that have other desirable characteristics, and many endophytes have no mechanism for infectious spread to new host species (Clay 1989).

The range of defensive traits that can be incorporated using classical breeding programs is limited to the existing variation present in the species (or sometimes closely related species). The possibilities for crop improvement using genetic engineering are exciting because genes can be transferred among plants of different species, families, and even phyla. Consequently, considerably more effort is being placed in genetic improvement of crop plants using molecular techniques rather than classical breeding. Once the mechanisms of induced resistance are known, it may be possible to add those genes to plants that do not naturally possess them. Vectors are now available to incorporate foreign genes into the genomes of virtually any dicotyledonous plant and some monocotyledonous species. It is also possible to make a trait that is constitutive inducible by fusing it with an inducible promoter. The trait would be expressed only after the inducible promoter responded. The converse is also possible; traits that are inducible could be made constitutive.

The earliest and simplest application of genetic engineering techniques has been to move foreign genes that code for new mechanisms of resistance into plants that previously lacked them. Often this has meant that the resistance was expressed constitutively, even when it was expressed naturally only as a damage-induced trait. For example, plants have been engineered that produce the coat protein of tobacco mosaic virus (Powell Abel et al. 1986; Nelson et al. 1987; Beachy et al. 1987). These transgenic plants expressed only a fraction of the virus genome (the coat only) which was sufficient to induce resistance but not enough to allow virus replication or to cause disease symptoms. Indeed, when these engineered plants were challenged with intact virus, virus replication and spread were reduced; in addition, transgenic plants took much longer to develop symptoms than untransformed controls. Similarly, the gene encoding cowpea trypsin inhibitor (a PI) was inserted into tobacco plants (Hilder et al. 1987). The PI was expressed constitutively in the transformed tobacco plants, and the plants became more resistant against tobacco budworms (*Helicoverpa virescens)* and tobacco hornworms (*Manduca sexta*) (Hilder et al. 1987; Johnson, Narvaez, et al. 1989).

Plants may be at a disadvantage in terms of the evolution of resistance compared to their herbivores which have much shorter generation times. The molecular biologist can intervene to shift this inequity to favor the plant. For example, Jongsma et al. (1995b) selectively altered the genetic control of PI production by simulating the process of evolution (see 3.3.3.2). With phage display techniques Jongsma and colleagues improved dramatically the ability of plants to inhibit novel herbivore proteases by screening a library of PI variants, selecting those variants that were particularly effective against the targeted proteases and then transforming tomato plants with those variants.

The prospects for using genetically engineered crops are exciting because the resistance traits can be fused to promoters so that they can be made to express only following attack or following application of a specific elicitor. Potentially, this could allow for tremendous specificity in terms of the stimuli that cause the resistance traits to be expressed (Ryan 1989). Toward this end, scientists at Ciba fused the gene that produces the *Bacillus thuringiensis* (*Bt*) endotoxin to a promoter that activates it only after the plant had been damaged (Ryals et al. 1992; Williams et al. 1992). The *Bt* endotoxin has been found to be antibiotic against a variety of insect herbivores, especially caterpillars, and has been used as a "biorational insecticide" for over a decade. Production of endotoxin could also be induced or terminated in tobacco plants by application of specific chemicals. This allowed artificial control and regulation of the production of the putative defense. Unfortunately, the *Bt* endotoxin has been used so extensively as a noninduced source of mortality that many herbivores are well on their way to evolving resistance to it (Tabashnik et al. 1990), greatly diminishing its potential as an inducible source of resistance.

Additional specificity is possible by careful selection of the inducible resistance traits. For example, different PIs have different specificities against insects, mammals, and pathogens (Ryan 1989). It may be possible in the future to identify or alter PIs that have activity against only specific target pests (Ryan 1989). Because PI genes are wound-inducible, several research groups have transformed plants to express the PI proteins only after the plants had been damaged (Sánchez-Serrano et al. 1987; Ryan 1989).

The results described above have convinced many workers that genetic engineering for more and different forms of resistance will provide the pest control of the future. However, genetically engineered solutions are not without their limitations. To date, plants have only been transformed with a few gene products for potential defense against herbivores: for example, systemin, PIs, and *Bt* endotoxins. This list is likely to increase in the future. However, numerous hurdles must be overcome before induced resistance in genetically engineered crops can become a mainstay of agricultural pest control.

One of the complexities of transforming plants with novel traits is the necessity to consider the chemical milieu in which the trait will be expressed. For example, increased concentrations of nicotine were found to reduce the effectiveness of *Bt* toxin in artificial diets against hornworm caterpillars (*Manduca sexta;* Krischik et al. 1988). High levels of phenolic compounds reduced mortality of gypsy-moth larvae (*Lymantria dispar*) to nuclear polyhedrosis virus (Schultz and Keating 1991). The effectiveness of PIs has also been found to depend on the chemical environment of the plant. Many other wound-activated chemicals are capable of inactivating PIs (Duffey and Felton 1989; Workman et al. n.d.). PIs can also be rendered ineffective if plant nutrient levels are high (Broadway and Duffey 1988). Because defenses interact, adding new plant defensive traits does not necessarily add more defense. The lesson here may be analogous to the one agroecologists have learned about diversity and stability. More stable ecosystems are generally more diverse but haphazardly adding species to agroecosystems has been no guarantee that they will become more stable.

As with any resistance trait under simple genetic control, a second potential hurdle for genetically engineered crops is that they are subject to counteradaptation by herbivores. Let's assume that 99% of the population of herbivores is killed by a new plant trait for resistance. The 1% of the population that is not killed will parent the next generation of herbivores. If the ability to use the novel host plant is heritable, then much more of the next herbivore generation will be unaffected, and so on. Counteradaptation of herbivores to synthetic pesticides is widespread and is considered to be one of the primary reasons that these materials have lost their effectiveness, and new ones are not being developed by chemical companies (National Academy of Sciences 1986). This counteradaptation is called pesticide resistance, an unfortunate and confusing use of the term "resistance."

In chapter 5 we considered the evidence that resistance traits were costly for plants; this issue has more than academic interest. If induced resistance is beneficial for plants (see section 5.1.3), determining whether that induced resistance is costly and the nature of those costs (whether they are physiological, ecological, etc.) has important applied consequences. If resistance is costly, some of those costs can be overcome by agronomic practices such as fertilization. When resistance mechanisms are costly, wild plants may never employ those mechanisms because of their costs. In agricultural situations, however, the defenses may be valuable since natural selection on plants is relaxed due to agronomic practices. According to this scenario, adding novel resistance mechanisms to cultivated plants may greatly reduce pest problems. On the other hand, if resistance is not costly and the variability that results from inducibility is beneficial against herbivores in itself (see section 5.4.1.4), then plant breeders and molecular biologists may wish to consider switching the emphasis from developing constitutive resistance traits that can be expressed inducibly to an emphasis on traits that make plants more variable. At a molecular level such traits that increase variability are already being discovered (G-boxes, jasmonate and sucrose-responsive promoters, etc.). The potential commercial value of these traits as generators of variability has not yet been considered.

Genetic engineering has been hailed as a major breakthrough by some because engineered plants are likely to pose fewer hazards to farmworkers, farming communities, wildlife, other users of the "down-stream environment," and consumers. Others have warned that genetically engineered organisms can "escape" into the environment and disrupt the ecological and genetic workings of our planet. We have considered only the scientific feasibility of using genetic engineering of induced defenses to control agricultural pests. These other considerations are very important but lie beyond the scope of this book.

6.3 PROSPECTUS FOR AGRICULTURAL USES

We have attempted to describe the current commercial uses of induced resistance in agriculture. We have also done a bit of squinting and tried to imagine methods and situations where induced resistance could be used in the future. We have attempted to temper our optimism about induced resistance with a consideration of the limitations of the techniques we have considered. Despite these limitations we believe that employing competitors of damaging herbivores and inducing plant defenses with elicitors and with transformed plants that include novel genes for induced resistance will provide important and effective means of pest control in the future.

6.4 CONCLUSION

The study of induced plant responses is a young field but no longer a nascent one. This book is a first attempt to evaluate what we have learned so far and to suggest those questions that we believe will lead to the most important progress in the future. We have summarized current knowledge about biochemical mechanisms of induced responses, the effects of those responses on performance and population dynamics of herbivores, and the evolutionary factors that may favor induced rather than constitutive forms of resistance. Finally we have considered the past successes of agricultural applications of induced resistance against pathogens and potential uses against herbivores.

The study of induced responses of plants highlights the practical and philosophical problems associated with studying phenomena that have relevance at different levels of analysis, from molecular to community levels. Implicit in our desire to merge the traditions of Ryan and Haukioja is an attempt to have researchers working at these different levels in the hierarchy respect the complexities of scaling a question either up or down. For example, we hope to convince researchers oriented toward mechanisms that just because a particular signal might be important in controlling the expression of particular genes coding for a portion of the resistant trait, this importance may not translate up the biological hierarchy to the level of whole plants. Even if the signal is found to increase whole-plant resistance in the laboratory, it may not function to increase plant fitness in the field. Moreover, if the signal does increase plant fitness in the field it may not be fixed by selection due to complexities of the population structure or the nature of genotype-by-environment linkages. Inducible responses are, by definition, conditional and the ecological and evolutionary milieu in which they function can influence their importance as much as their biochemical milieu.

It seems important to us that ecologists begin to appreciate the advances that biochemists have made and that molecular biologists begin to think a bit more like ecologists and evolutionary biologists. Why do we feel that this cross-fertilization of disciplines is so important to the advancement of this field?

Molecular biology is providing techniques to manipulate individual traits and to examine their function with unprecedented experimental rigor. Ecologists who fail to recognize the usefulness of these techniques deny themselves of an extremely powerful and elegant experimental tool. However, this experimental rigor is only as meaningful as the ecological settings in which the experiments are conducted. Traits function in ecological contexts, and as we have seen, the most potent resistance traits can be rendered dysfunctional by the ecological context. For example, every resistance trait that slows the growth of an herbivore feeding on its host plant can be a liability for that host if the herbivore is invulnerable to other mortality agents. In that case the herbivores may consume more plant tissue instead of less; the plant trait may only function as a defense in the ecological context of natural enemies of herbivores. A more complete understanding of the costs and ecological consequences of induced responses will allow molecular biologists to design more effective and more stable forms of novel resistance. Because traits generally produce different effects on different herbivores, pathogens, and plant mutualists, this more complete understanding is essential. There is no point in designing a resistant plant that suffers from poor agronomic qualities or that is susceptible and highly attractive to other "secondary" herbivores. We were struck by the poor appreciation that the field has for the effects of induced responses on plants. Physiological and evolutionary studies considering the detailed consequences of induction from the plant's point of view are sorely needed.

The power of the approach that can bring novel resistance traits to crop plants is being realized. The more successful molecular biology is at identifying and manipulating potent resistance traits, the more dependent those successes will be on an understanding of the evolutionary process (Gould 1988). The most potent resistance will exert the strongest selection on herbivores for counteradaptation. As the current debate about resistance management so graphically illustrates, short-term successes of molecular biology can quickly be negated without a sophisticated grasp of the evolutionary process. A very similar process is occurring now as resistant human diseases evolve to diminish the effectiveness of antibiotics.

It seems useful to ask which kinds of plants and which kinds of ecological situations are most common for examples of induced resistance and defense in nature. Induced resistance has been more often observed for perennial plants than for annuals. Induced resistance may occur quickly or over several years; induced susceptibility is more likely to be delayed. Induced resistance is most often associated with rapid plant growth and young plant tissues. Induced resistance is expected on theoretical grounds for plants that experience intermittent risk of herbivory and for plants where early damage is a reliable cue of increased probability of later risk. These generalizations about the natural occurrence of induced responses may provide guidelines for the effective use of induced resistance in agriculture.

The power of the comparative approach has allowed ecologists to distill generalizations from the diversity of plant-herbivore interactions, the most common trophic interactions on the planet. While surveying the diversity of natural systems is surely a worthy endeavor, progress would probably come more quickly if ecologists and evolutionary biologists could agree on a few model systems of genetically tractable plant-herbivore interactions, which they could understand in much greater detail by working collectively. This effort would surely be facilitated if these model natural systems were ones for which an understanding of mechanisms was developing. Evolutionary questions would become more readily addressed if the model systems were annuals, rather than long-lived trees, and if workers paid more attention to the fitness consequences of responses from the plant's point of view.

Several examples that were developed in earlier chapters illustrate how an understanding of mechanisms can facilitate progress on purely ecological and evolutionary questions. Understanding mechanisms provides direction about where and when population-level effects are expected. For example, confusion about inconsistent results of damage on induced resistance involving cottonwood trees can be resolved by a knowledge of the plant's vascular connections (see figure 5.2). More detailed understanding of mechanisms will allow rejection of evolutionary scenarios that fail to match empirical observations and should improve our efficiency at generating models that are realistic. For example, there may be little point in pursuing evolutionary models based on the carbon/nutrient balance model (section 5.4.3) if the model fails as a mechanistic explanation (section 3.3.3.1). We hope that researchers interested in the ecological consequences and evolution of induced responses will appreciate how much more precisely their hypotheses can be framed when the details of the mechanisms are known.

Attempts to use induced resistance in agriculture are likely to benefit from a better understanding of both mechanisms and ecological and evolutionary considerations. Plant pathology lacks an evolutionary tradition and many developments in that field beg evolutionary explanations. For example, field studies of an inbreeding population of common groundsel (*Senecio vulgaris*) contained a large and diverse collection of resistance types to mildew (Bevan et al. 1993). What generated and maintains this extraordinary level of variability for specific interactions between these plants and their pathogens?

Agriculture tips the balance of the coevolutionary race in favor of plant parasites. Planting large monocultures of genetically homogenous crops provides consistent, directional selection that accelerates the rate at which herbivores develop adaptations to circumvent plant defenses. If we hope to develop crops that can survive their parasites, either by using cultural techniques, breeding for resistance, or employing genetic engineering, it behooves us to first understand the processes we are up against.

References

Adler, F. R., and R. Karban. 1994. Defended fortresses or moving targets? Another model of inducible defenses inspired by military metaphors. *American Naturalist* 144:813–832.

Aerts, R. J., D. Gisi, E. De Carolis, V. DeLuca, and T. W. Bauman. 1994. Methyl jasmonate vapor increases the developmentally controlled synthesis of alkaloids in *Catharanthus* and *Cinchoma* seedlings. *Plant Journal* 5:635–645.

Agrawal, A. A. n.d. Leaf damage and associated signals induce aggressive ant recruitment in a neotropical ant plant. Manuscript.

Agrawal, A. A., S. Y. Strauss, and M. J. Stout. n.d. Male and female fitness costs of induced resistance and tolerance to herbivory in *Raphanus raphanistrum*. Manuscript.

Ajlan, A. M., and D. A. Potter. 1991. Does immunization of cucumber against anthracnose by *Colletotrichum lagenarium* affect host suitability for arthropods? *Entomologia Experimentalis et Applicata* 58:83–91.

Akazawa, T., I. Uritani, and H. Kubota. 1960. Isolation of ipomearamone and two coumarin derivatives from sweet potato roots injured by the weevil *Cylas formicarius elgantulus*. *Archives of Biochemistry and Biophysics* 88:150–156.

Akcakaya, H. R. 1992. Population cycles of mammals: Evidence for a ratio-dependent predation hypothesis. *Ecological Monographs* 62:119–142.

Albrecht, T., A. Kehlen, K. Stahl, H.-D. Knofel, G. Sembdner, and E. W. Weiler. 1993. Quantification of rapid, transient increases in jasmonic acid in wounded plants using a monoclonal antibody. *Planta* 191:86–94.

Albrechtova, J. T. P., and J. Ullmann. 1994. Methyl jasmonate inhibits growth and flowering in *Chenopodium rubrum*. *Biologia Plantarum* 36:317–319.

Anderson, R. M., and R. M. May. 1979. Population biology of infectious diseases. *Nature* 280:361–367.

Anderson, S. S., K. D. McCrea, W. G. Abrahamson, and L. M. Hartzel. 1989. Host genotype choice by the gallmaker *Eurosta solidaginis* (Diptera: Tephritidae). *Ecology* 70:1048–1054.

Apostol, I., P. S. Low, and P. Heinstein. 1989. Effect of age of cell suspension cultures on susceptibility to a fungal elicitor. *Plant Cell Reports* 7:692–695.

Apriyanto, D., and D. A. Potter. 1990. Pathogen-activated induced resistance of cucumber: Response of arthropod herbivores to systemically protected leaves. *Oecologia* 85:25–31.

Astrom, M., and P. Lundberg. 1994. Plant defence and stochastic risk of herbivory. *Evolutionary Ecology* 8:288–298.

Auer, C., A. Roques, F. Goussard, and P. J. Charles. 1981. Effects de l'accroisse-ment provoque du niveau de population de la tordeuse du mélèze *Zeiraphera diniana* Guenee (Lep., Tortricidae) au cours de la phase de régression dans un massif forestier du Brianconnais. *Zeitschrift für Angewandte Entomologie* 92:286–302.

Auerbach, M., and D. Simberloff. 1984. Responses of leaf miners to atypical leaf production patterns. *Ecological Entomology* 9:361–367.

————. 1989. Oviposition site preference and larval mortality in a leaf-mining moth. *Ecological Entomology* 14:131–140.

Ausubel, F. M., F. Katagiri, M. Mindrinos, and J. Glazebrook. 1995. Use of *Arabi-dopsis thaliana* defense related mutants to dissect the plant response to patho-gens. *Proceedings of the National Academy of Sciences of the USA* 92:4189–4196.

Ayers, A. R., J. J. Goodell, and P. L. DeAngelis. 1985. Plant detection of patho-gens. *Recent Advances in Phytochemistry* 19:1–20.

Baker, H. G., and I. Baker. 1975. Studies of nectar-constitution and pollinator-plant coevolution. In *Coevolution of animals and plants,* ed. L. E. Gilbert and P. H. Raven. Austin: University of Texas Press, 100–140.

Baldwin, I. T. 1988a. The alkaloidal responses of wild tobacco to real and simu-lated herbivory. *Oecologia* 77:378–381.

————. 1988b. Damage-induced alkaloids in tobacco: Pot-bound plants are not inducible. *Journal of Chemical Ecology* 14:1113–1120.

————. 1988c. Short-term damage-induced increases in tobacco alkaloids pro-tect plants. *Oecologia* 75:367–370.

————. 1989. The mechanism of damaged-induced alkaloids in wild tobacco. *Journal of Chemical Ecology* 15:1661–1680.

————. 1991. Damage-induced alkaloids in wild tobacco. In *Phytochemical induc-tion by herbivores,* ed. D. W. Tallamy and M. J. Raupp. New York: John Wiley, 47–69.

————. 1994. Chemical changes rapidly induced by folivory. In *Insect-plant inter-actions,* ed. E. A. Bernays. Boca Raton, FL: CRC Press, 5:1–23.

Baldwin, I. T., and P. Callahan. 1993. Autotoxicity and chemical defense: Nico-tine accumulation and carbon gain in solanaceous plants. *Oecologia* 94:534–541.

Baldwin, I. T., and W. Hamilton III. n.d. Competition increases the fitness costs of an induced defense. Manuscript.

Baldwin, I. T., and M. J. Karb. 1995. Plasticity in the allocation of nicotine to reproductive parts in *Nicotiana attenuata. Journal of Chemical Ecology* 21:897–909.

Baldwin, I. T., M. J. Karb, and T. E. Ohnmeiss. 1994. Allocation of ^{15}N from ni-trate to nicotine after leaf damage: Production and turnover of a damage-induced mobile defense. *Ecology* 75:1703–1713.

Baldwin, I. T., R. C. Oesch, P. M. Merhige, and K. Hayes. 1993. Damage-induced root nitrogen metabolism in *Nicotiana sylvestris:* Testing C/N predictions for alkaloid production. *Journal of Chemical Ecology* 19:3029–3043.

Baldwin, I. T., and T. E. Ohnmeiss. 1993. Alkaloidal responses to damage in *Nico-tiana* native to North America. *Journal of Chemical Ecology* 19:1143–1153.

————. 1994. Coordination of photosynthetic and alkaloidal responses to leaf damage in uninducible and inducible *Nicotiana sylvestris*. *Ecology* 75:1003–1014.

Baldwin, I. T., and E. A. Schmelz. 1994. Constraints on an induced defense: The role of leaf area. *Oecologia* 97:424–430.

————. 1996. Immunological memory in the induced accumulation of nicotine in wild tobacco. *Ecology* 77:236–246.

Baldwin, I. T., E. A. Schmelz, and T. E. Ohnmeiss. 1994. Wound-induced changes in root and shoot jasmonic acid pools correlate with induced nicotine synthesis in *Nicotiana sylvestris* Spegazzini and Comes. *Journal of Chemical Ecology* 20:2139–2157.

Baldwin, I. T., E. A. Schmelz, and Z.-P. Zhang. 1996. Effects of octadecanoid metabolites and inhibitors on induced nicotine accumulation in *Nicotiana sylvestris*. *Journal of Chemical Ecology* 22:61–74.

Baldwin, I. T., and J. C. Schultz. 1983. Rapid changes in tree leaf chemistry induced by damage: Evidence for communication between plants. *Science* 221:277–279.

————. 1984. *Damage and communication induced changes in yellow birch leaf phenolics.* Proceedings of the 8th Annual Forest Biology Workshop. Logan: Utah State University Press, 25–33.

Baldwin, I. T., C. L. Sims, and S. E. Kean. 1990. The reproductive consequences associated with inducible alkaloidal responses in wild tobacco. *Ecology* 71:252–262.

Baldwin, I. T., Z.-P. Zhang, N. Diab, T. E. Ohnmeiss, E. S. McCloud, G. Y. Lynds, and E. A. Schmelz. 1997. Quantification, correlations, and manipulations of wound-induced changes in jasmonic acid and nicotine in *Nicotiana sylvestris*. *Planta*. In press.

Baltensweiler, W. 1985. On the extent and the mechanisms of the outbreaks of the larch budmoth (*Zeiraphera diniana* Gn., Lepidoptera, Tortricidae) and its impact on the subalpine larch–cembran pine ecosystem. In *Establishment and tending of subalpine forest: Research and management,* ed. H. Turner and W. Tranquillini. Birmensdorf, Switzerland: Swiss Federal Institute of Forestry Research, 215–219.

Baltensweiler, W., G. Benz, P. Bovey, and V. Delucchi. 1977. Dynamics of larch budmoth populations. *Annual Review of Entomology* 22:79–100.

Baltensweiler, W., and A. Fischlin. 1988. The larch budmoth in the Alps. In *Dynamics of forest insect populations: Patterns, causes, implications,* ed. A. A. Berryman. New York: Plenum, 331–351.

Barbour, M. G., R. B. Craig, F. R. Drysdale, and M. T. Ghiselin. 1973. *Coastal ecology: Bodega Head.* Berkeley: University of California Press.

Bardner, R., and K. E. Fletcher. 1974. Insect infestations and their effects on the growth and yield of field crops: A review. *Bulletin of Entomological Research* 64:141–160.

Barlow, B. A., and D. Wiens. 1977. Host-parasite resemblance in Australian mistletoes: The case for cryptic mimicry. *Evolution* 31:69–84.

Bassman, J. H., and D. I. Dickmann. 1982. Effect of defoliation in the developing leaf zone on young *Populus X euramericana* plants: 1, Photosynthetic physiology, growth, and dry weight partitioning. *Forest Science* 28:599–612.

Bauer, W. D. 1981. Infection of legumes by rhizobia. *Annual Review of Plant Physiology* 32:407–449.

Baur, R., S. Binder, and G. Benz. 1991. Nonglandular leaf trichomes as short-term inducible defense of the grey alder, *Alnus incana* (L.), against the chrysomelid beetle, *Agelastica alni* L. *Oecologia* 87:219–226.

Baur, R., V. Kostal, and E. Stadler. 1996. Root damage by conspecific larvae induces preference for oviposition in cabbage root flies. *Entomologia Experimentalis et Applicata* 80:224–227.

Baydoun, E. A.-H., and S. C. Fry. 1985. The immobility of pectic substances in injured tomato leaves and its bearing on the identity of the wound hormone. *Planta* 165:269–276.

Beachy, R. N., D. M. Stark, C. M. Doem, M. J. Oliver, and R. T. Fraley. 1987. Expression of sequences of tobacco mosaic virus in transgenic plants and their role in disease resistance. In *Tailoring genes for crop improvement*, ed. G. Bruening, J. Harada, and T. Kosuge. New York: Plenum, 169–180.

Beauverie, J. 1901. Essais d'immunization des végétaux contre les maladies cryptogamiques. *Comptes Rendus Hebdomadaires des Séances de l'Acadamie des Sciences, Paris* 133:107–110.

Becerra, J. X. 1994. Squirt-gun defense in *Bursera* and the chrysomelid counterploy. *Ecology* 75:1991–1996.

Beer, S. V., J. R. Rundle, and J. L. Norelli. 1984. Recent progress in the development of biological control of fire blight: A review. *Acta Horticulturae* 151: 195–201.

Bell, E., R. A. Creelman, and J. E. Mullet. 1995. A chloroplast lipoxygenase is required for wound-induced jasmonic acid accumulation in *Arabidopsis*. *Proceedings of the National Academy of Sciences of the USA* 92:8675–8679.

Bell, E., and J. E. Mullet. 1991. Lipoxygenase gene expression is modulated in plants by water deficit, wounding, and methyl jasmonate. *Molecular and General Genetics* 230:456–462.

Benedict, J. H., and J. F. Chang. 1991. Bacterially induced changes in the cotton plant–boll weevil paradigm. In *Phytochemical induction by herbivores*, ed. D. W. Tallamy and M. J. Raupp. New York: John Wiley, 379–401.

Bennett, R. N., and R. M. Wallsgrove. 1994. Tansley review no. 72: Secondary metabolites in plant defence mechanisms. *New Phytologist* 127:617–633.

Bentur, J. R., and M. B. Kalode. 1996. Hypersensitive reaction and induced resistance in rice against the Asian rice gall midge *Orseolia oryzae*. *Entomologia Experimentalis et Applicata* 78:77–81.

Benz, G. 1974. Negative feedback by competition for food and space, and by cyclic induced changes in the nutritional base as regulatory principles in the population dynamics of the larch budmoth, *Zeiraphera diniana* (Guenee) (Lep., Tortricidae). *Zeitschrift für Angewandte Entomologie* 76:196–228.

———. 1977. Insect induced resistance as a means of self defense of plants. *Bulletin Section Régionale Ouest Paléarctique* 3:155–159.

Benz, G., and C. Abivardi. 1991. Preliminary studies on wound- and PIIF-induced resistance in some solanaceous plants against *Spodoptera littoralis* (Boisd.) (Lep., Noctuidae). *Journal of Applied Entomology* 111:349–357.

Berenbaum, M. R. 1995. The chemistry of defense: Theory and practice. *Proceedings of the National Academy of Sciences of the USA* 92:2–8.

Berenbaum, M. R., and A. R. Zangerl. 1993. Furanocoumarin metabolism in *Papilio polyxenes:* Biochemistry, genetic variability, and ecological significance. *Oecologia* 95:370–375.

———. 1994. Costs of inducible defense: Protein limitation, growth, and detoxification in parsnip webworms. *Ecology* 75:2311–2317.

———. 1995. Phytochemical diversity: Adaptation or random variation? *Recent Advances in Phytochemistry* 29.

Bergelson, J., S. Fowler, and S. Hartley. 1986. The effects of foliage damage on casebearing moth larvae, *Coleophora serratella*, feeding on birch. *Ecological Entomology* 11:241–250.

Bernays, E. A., K. L. Bright, N. Gonzalez, and J. Angel. 1994. Dietary mixing in a generalist herbivore: Tests of two hypotheses. *Ecology* 75:1997–2006.

Bernays, E. A., and R. F. Chapman. 1994. *Host-plant selection by phytophagous insects.* New York: Chapman and Hall.

Bernstein, C. 1984. Prey and predator emigration responses in the acarine system *Tetranychus urticae–Phytoseilus persimilis. Oecologia* 61:134–142.

Berryman, A. A. 1988. Towards a unified theory of plant defense. In *Mechanisms of woody plant defenses against insects*, ed. W. J. Mattson, J. Levieux, and C. Bernard-Dagan. New York: Springer-Verlag, 39–55.

Berryman, A. A., N. C. Stenseth, and A. S. Isaev. 1987. Natural regulation of herbivorous forest insect populations. *Oecologia* 71:174–184.

Bevan, J. R., D. D. Clarke, and I. R. Crute. 1993. Resistance to *Erysiphe fisheri* in two populations of *Senecio vulgaris. Plant Pathology* 42:636–646.

Bi, J. L., and G. W. Felton. 1995. Foliar oxidative stress and insect herbivory: Primary compounds, secondary metabolites, and reactive oxygen species as components of induced resistance. *Journal of Chemical Ecology* 21:1511–1530.

Bi, J. L., G. W. Felton, and A. J. Mueller. 1994. Induced resistance in soybean to *Helicoverpa zea:* Role of plant protein quality. *Journal of Chemical Ecology* 20:183–198.

Black, A. R. 1993. Predator-induced phenotypic plasticity in *Daphnia pulex:* Life history and morphological responses to *Notonecta* and *Chaoborus. Limnology and Oceanography* 38:986–996.

Black, A. R., and S. I. Dodson. 1990. Demographic costs of *Chaoborus*-induced phenotypic plasticity in *Daphnia pulex. Oecologia* 83:117–122.

Bloom, A., F. S. Chapin III, and H. A. Mooney. 1985. Resource limitation in plants: An economic analogy. *Annual Review of Ecology and Systematics* 16:363–392.

Bodnaryk, R. P. 1992. Effects of wounding on glucosinolates in the cotyledons of oilseed rape and mustard. *Phytochemistry* 31:2671–2677.

———. 1994. Potent effect of jasmonates on indole glucosinolates in oilseed rape and mustard. *Phytochemistry* 35:301–305.

Bohlmann, H., and A. K. Apel. 1991. Thionins. *Annual Review of Plant Physiology and Plant Molecular Biology* 42:227–240.

Bokhari, U. G. 1977. Regrowth of western wheatgrass utilizing ^{14}C-labeled assimilates stored in belowground parts. *Plant and Soil* 48:115–127.

Boland, W., Z. Feng, J. Donath, and A. Gabler. 1992. Are acyclic C_{11} and C_{16} homoterpenes plant volatiles indicating herbivory? *Naturwissenschaften* 79:368–371.

Boland, W., J. Hopke, J. Donath, J. Nuske, and F. Bublitz. 1995. Jasmonic acid and coronatine induce odor production in plants. *Angewandte Chemie International Edition in English* 34:1600–1602.

Boller, T. 1991. Ethylene in pathogenesis and disease resistance. In *The plant hormone ethylene*, ed. A. K. Mattoo and J. C. Suttle. Boca Raton, FL: CRC Press, 293–314.

———. 1995. Chemoperception of microbial signals in plant cells. *Annual Review of Plant Physiology and Plant Molecular Biology* 46:189–214.

Bolter, C. J. 1993. Methyl jasmonate induces papain inhibitor(s) in tomato leaves. *Plant Physiology* 103:1347–1353.

Bolter, C. J., and M. A. Jongsma. 1995. Colorado potato beetles (*Leptinotarsa decemlineata*) adapt to proteinase inhibitors induced in potato leaves by methyl jasmonate. *Journal of Insect Physiology* 41:1071–1078.

Boubals, D. 1966. Etude de la distribution et des causes de la résistance au phylloxéra radicicole chez les Vitacees. *Annales de l'Amélioration des Plantes* 16:145–185.

Bowers, M. D., and N. E. Stamp. 1993. Effects of plant age, genotype, and herbivory on *Plantago* performance and chemistry. *Ecology* 74:1778–1791.

Bowles, D. J. 1990. Defense-related proteins in higher plants. *Annual Review of Biochemistry* 59:873–907.

Bradley, R. T., J. M. Christie, and D. R. Johnston. 1966. *Forest management tables.* London: Her Majesty's Stationary Office.

Bradshaw, A. D., and K. Hardwick. 1989. Evolution and stress: Genotypic and phenotypic components. *Biological Journal of the Linnean Society* 37:137–155.

Braga, M. R., M. C. M. Young, J. V. A. Ponte, S. M. C. Dietrich, V. P. Emerenciano, and O. R. Gottliev. 1986. Phytoalexin induction in plants in tropical environments. *Biochemical Systematics and Ecology* 14:507–514.

Brattsten, L. B., J. H. Samuelian, K. Y. Long, S. A. Kincaid, and C. K. Evans. 1983. Cyanide as a feeding stimulant for the southern armyworm, *Spodoptera eridania. Ecological Entomology* 8:125–132.

Breen, J. P. 1994. Acremonium endophyte interactions with enhanced plant resistance to insects. *Annual Review of Entomology* 39:401–423.

Broadbent, L. 1976. Epidemiology and control of tomato mosaic virus. *Annual Review of Phytopathology* 14:75–96.

Broadway, R. M. 1995. Are insects resistant to plant proteinase inhibitors? *Journal of Insect Physiology* 41:107–116.

———. 1996. Dietary proteinase inhibitors alter complement of midgut proteases. *Archives of Insect Biochemistry and Physiology* 32:39–53.

Broadway, R. M., and S. S. Duffey. 1986. Plant proteinase inhibitors: Mechanism of action and effect on the growth and digestive physiology of larval *Heliothis zea* and *Spodoptera exigua*. *Journal of Insect Physiology* 32:827–833.

———. 1988. The effect of plant protein quality on insect digestive physiology and the toxicity of plant proteinase inhibitors. *Journal of Insect Physiology* 34:1111–1117.

Broadway, R., M., S. S. Duffey, G. Pearce, and C. A. Ryan. 1986. Plant proteinase inhibitors: A defense against herbivorous insects. *Entomologia Experimentalis et Applicata* 41:33–38.

Brody, A. K. 1992a. Oviposition choices by a predispersal seed predator (*Hylemya* sp.): 1, Correspondence with hummingbird pollinators, and the role of plant size, density, and floral morphology. *Oecologia* 91:56–62.

———. 1992b. Oviposition choices by a predispersal seed predator (*Hylemya* sp.): 2, A positive association between female choice and fruit set. *Oecologia* 91:63–67.

Brody, A. K., and R. Karban. 1992. Lack of a tradeoff between constitutive and induced defenses among varieties of cotton. *Oikos* 65:301–306.

Bronner, R., E. Westphal, and F. Dreger. 1991. Pathogenesis-related proteins in *Solanum dulcamara* L. resistant to the gall mite *Aceria cladophthirus* (Nalepa) (syn *Eriophyes cladophthirus* Nal.). *Physiological and Molecular Plant Pathology* 38:93–104.

Brown, D. G. 1988. The cost of plant defense: An experimental analysis with inducible proteinase inhibitors in tomato. *Oecologia* 76:467–470.

Brown, G. C., F. Nurdin, J. G. Rodriguez, and D. F. Hildebrand. 1991. Inducible resistance of soybean (var "Williams") to two-spotted spider mite (*Tetranychus urticae* Koch). *Journal of the Kansas Entomological Society* 64:388–393.

Bruin, J., M. Dicke, and M. W. Sabelis. 1992. Plants are better protected against spider mites after exposure to volatiles from infested conspecifics. *Experientia* 48:525–529.

Bryant, J. P. 1981. Phytochemical deterrence of snowshoe hare browsing by adventitious shoots of four Alaskan trees. *Science* 213:889–890.

———. 1987. Feltleaf willow–snowshoe hare interactions: Plant carbon/nutrient balance and floodplain succession. *Ecology* 68:1319–1327.

Bryant, J. P., F. S. Chapin, and D. R. Klein. 1983. Carbon/nutrient balance of boreal plants in relation to vertebrate herbivory. *Oikos* 40:357–368.

Bryant, J. P., K. Danell, F. Provenza, P. B. Reichardt, T. A. Clausen, and R. A. Werner. 1991. Effects of mammal browsing on the chemistry of deciduous woody plants. In *Phytochemical induction by herbivores*, ed. D. W. Tallamy and M. J. Raupp. New York: John Wiley, 135–154.

Bryant, J. P., I. Heitkonig, P. Kuropat, and N. Owen-Smith. 1991. Effects of severe defoliation on the long-term resistance to insect attack and on leaf chemistry in six woody species of the southern African savanna. *American Naturalist* 137:50–63.

Bryant, J. P., P. B. Reichardt, T. P. Clausen, and R. A. Werner. 1993. Effects of mineral nutrition on delayed inducible resistance in Alaska paper birch. *Ecology* 74:2072–2084.

Bryant, J. P., J. Tuomi, and P. Niemelä. 1988. Environmental constraint of constitutive long-term inducible defenses in woody plants. In *Chemical mediation of coevolution*, ed. K. C. Spencer. San Diego: Academic Press, 367–389.

Buck, K. W. 1988. Control of plant pathogens with viruses and related agents. *Philosophical Transactions of the Royal Society of London B* 318:295–317.

Bultman, T. L., and D. T. Ganey. 1995. Induced resistance to fall armyworm (Lepidoptera: Noctuidae) mediated by a fungal endophyte. *Environmental Entomology* 24:1196–1200.

Buratti, L., J. P. Allais, and M. Barbier. 1988. The role of resin acids in the relationship between Scots pine and the sawfly, *Diprion pini* (Hymenoptera: Diprionidae): 1, Resin acids in the needles. In *Mechanisms of woody plant defenses against insects*, ed. W. J. Mattson, J. Levieux, and C. Bernard-Dagan. New York: Springer-Verlag, 171–187.

Burdon, J. J. 1987. *Diseases and plant population biology.* Cambridge: Cambridge University Press.

Burns, C. W. 1968. The relationship between body size of filter-feeding Cladocera and the maximum size of particle ingested. *Limnology and Oceanography* 13:675–678.

Burns, R. M., and B. H. Honkala, eds. 1990. *Silvics of North America.* Vol. 1, *Conifers.* Vol. 2, *Hardwoods.* Agriculture Handbook 654. Washington, DC: USDA Forest Service.

Butcher, M. R., D. R. Penman, and R. R. Scott. 1987. Population dynamics of two-spotted spider mites in multiple year strawberry crops in Canterbury. *New Zealand Journal of Zoology* 14:509–517.

Byers, R. A., D. L. Gustine, and B. G. Moyer. 1977. Toxicity of β-nitropropionic acid to *Trichoplusia ni.* *Environmental Entomology* 6:229–232.

Bylund, H., and O. Tenow. 1994. Long-term dynamics of leaf miners, *Eriocrania* spp., on mountain birch: Alternate year fluctuations and interaction with *Epirrita autumnata.* *Ecological Entomology* 19:310–318.

Caldwell, M. M., J. H. Richards, D. A. Johnson, R. S. Nowak, and R. S. Dzurec. 1981. Coping with herbivory: Photosynthetic capacity and resource allocation in two semiarid *Agropyron* bunchgrasses. *Oecologia* 50:14–24.

Caloin, M., B. Clement, and S. Herrmann. 1990. Regrowth kinetics of *Dactylis glomerata* following defoliation. *Annals of Botany* 66:397–405.

Campbell, B. C. 1986. Host-plant oligosaccharins in the honeydew of *Schizaphis graminum* (Rondani) (Insecta, Aphididae). *Experientia* 42:451–452.

Campbell, B. C., B. G. Chan, L. L. Creasy, D. L. Dreyer, L. B. Rabin, and A. C. Waiss. 1984. Bioregulation of host plant resistance to insects. In *Bioregulators: Chemistry and uses*, ed. R. L. Ory and F. R. Rittig. Washington, DC: American Chemical Society, 193–203.

Campbell, B. C., D. L. McLean, M. G. Kinsey, K. C. Jones, and D. L. Dreyer. 1982. Probing behavior of the greenbug (*Schizaphis graminum*, biotype C) on resistant and susceptible varieties of sorghum. *Entomologia Experimentalis et Applicata* 31:140–146.

Campbell, R. 1989. The use of microbial inoculants in the biological control of plant diseases. In *Microbial inoculation of crop plants*, ed. R. Campbell and R. M. Macdonald. Oxford: Oxford University Press, 67–77.

Campbell, R., and R. M. Macdonald, eds. 1989. *Microbial inoculation of crop plants.* Oxford: Oxford University Press.

Carne, P. B. 1965. Distribution of the eucalypt-defoliating sawfly *Perga affinis affinis* (Hymenoptera). *Australian Journal of Zoology* 13:593–612.

Carroll, C. R., and C. A. Hoffman. 1980. Chemical feeding deterrent mobilized in response to insect herbivory and counteradaptation by *Epilachna tredecimnotata. Science* 209:414–416.

Cartwright, D., P. Langcake, R. J. Pryce, D. P. Leworthy, and J. P. Ride. 1977. Chemical activation of host defence mechanisms as a basis for crop protection. *Nature* 267:511–513.

Chapin, F. S. 1980. Nutrient allocation and responses to defoliation in tundra plants. *Arctic and Alpine Research* 12:553–536.

———. 1991a. Environmental stresses on nutrient availability and use. In *Response of plants to multiple stresses,* ed. H. A. Mooney, W. E. Winner, E. J. Pell, and E. Chu. San Diego: Academic Press, 67–88.

———. 1991b. Integrated responses of plants to stress. *Bioscience* 41:29–36.

Chapin, F. S., E.-D. Schulze, and H. A. Mooney. 1990. The ecology and economics of storage in plants. *Annual Review of Ecology and Systematics* 21:423–447.

Chapin, F. S., and M. Slack. 1979. Effects of defoliation upon root growth, phosphate absorption, and respiration in nutrient-limited tundra graminoids. *Oecologia* 42:67–79.

Chappell, J., and H. K. Hahlbrock. 1984. Transcription of plant defense genes in response to UV light or fungal elicitors. *Nature* 311:76–78.

Chen, Z., J. Malamy, J. Henning, U. Conrath, P. Sanchez-Casas, H. Silva, J. Ricigliano, and D. F. Klessig. 1995. Induction, modification, and transduction of the salicylic acid signal in plant defense responses. *Proceedings of the National Academy of Sciences of the USA* 92:4134–4137.

Chessin, M., and A. E. Zipf. 1990. Alarm systems in higher plants. *Botanical Review* 56:193–235.

Chester, K. S. 1933. The problem of acquired physiological immunity in plants. *Quarterly Review of Biology* 8:129–154, 275–324.

Chet, I. 1987. Trichoderma: Application, modes of action, and potential as a biocontrol agent of soil-borne plant pathogenic fungi. In *Innovative approaches to plant disease control,* ed. I. Chet. New York: John Wiley, 137–160.

Chew, F. S. 1988. Biological effects of glucosinolates. In *Biologically active natural products: Potential use in agriculture,* ed. H. G. Cutler. Washington, DC: American Chemical Society, 155–181.

Chew, F. S., and S. P. Courtney. 1991. Plant apparency and evolutionary escape from insect herbivory. *American Naturalist* 138:729–750.

Chew, F. S., and J. E. Rodman. 1979. Plant resources for chemical defense. In *Herbivores: Their interaction with secondary plant metabolites,* ed. G. A. Rosenthal and D. H. Janzen. New York: Academic Press, 271–307.

Chiang, H., D. M. Norris, A. Ciepiela, P. Shapiro, and A. Oosterwyk. 1987. Inducible versus constitutive PI 227687 soybean resistance to Mexican bean beetle, *Epilachna varivestis. Journal of Chemical Ecology* 13:741–749.

Choi, D., R. M. Bostock, S. Avdiushko, and D. Hildebrand. 1994. Lipid-derived signals that discriminate wound- and pathogen-responsive isoprenoid pathways in plants: Methyl jasmonate and the fungal elicitor archidonic acid induce different 3-hydroxy-3-methylglutaryl-coenzyme A reductase genes and antimicrobial isoprenoids in *Solanum tuberosum* L. *Proceedings of the National Academy of Sciences of the USA* 91:2329–2333.

Chou, C. M., and C. H. Kao. 1992. Stimulation of 1-aminocyclopropane-1-carboxylic acid-dependent ethylene production in detached rice leaves by methyl jasmonate. *Plant Science* 83:137–141.

Clare, B. G., A. Petit, and J. Tempe. 1987. The biology of pathogenic plasmids of *Agrobacterium*. In *Genetics and plant pathogenesis*, ed. P. R. Day and G. J. Jellis. Oxford: Blackwell, 79–90.

Clark, C. W., and C. D. Harvell. 1992. Inducible defenses and the allocation of resources: A minimal model. *American Naturalist* 139:521–539.

Clark, W. R. 1986. *The experimental foundations of modern immunology*. New York: John Wiley.

Clausen, T. P., P. B. Reichardt, J. P. Bryant, and R. A. Werner. 1991. Long-term and short-term induction in quaking aspen: Related phenomena? In *Phytochemical induction by herbivores*, ed. D. W. Tallamy and M. J. Raupp. New York: John Wiley, 71–83.

Clausen, T. P., P. B. Reichardt, J. P. Bryant, R. A. Werner, K. Post, and K. Frisby. 1989. Chemical model for short-term induction in quaking aspen (*Populus tremuloides*) foliage against herbivores. *Journal of Chemical Ecology* 15:2335–2346.

Clay, K. 1989. Clavicipitaceous endophytes of grasses: Their potential as biocontrol agents. *Mycological Research* 92:1–12.

Cohen, M. B., M. R. Berenbaum, and M. A. Schuler. 1989. Induction of cytochrome P450-mediated detoxification of xanthotoxin in the black swallowtail. *Journal of Chemical Ecology* 15:2347–2355.

Cohen, M. B., M. A. Schuler, and M. R. Berenbaum. 1992. A host-inducible cytochrome P450 from a host-specific caterpillar: Molecular cloning and evolution. *Proceedings of the National Academy of Sciences of the USA* 89:10920–10924.

Coleman, J. S., and C. G. Jones. 1991. A phytocentric perspective of phytochemical induction by herbivores. In *Phytochemical induction by herbivores*, ed. D. W. Tallamy and M. J. Raupp. New York: John Wiley, 3–45.

Coley, P. D. 1987. Anti-herbivore properties: The role of habitat quality and rate of disturbance. *New Phytologist* 106 (suppl.): 251–263.

Coley, P. D., J. P. Bryant, and F. S. Chapin. 1985. Resource availability and plant antiherbivore defense. *Science* 230:895–899.

Conn, E. E. 1979. Cyanide and cyanogenic glycosides. In *Herbivores: Their interaction with secondary plant metabolites*, ed. G. A. Rosenthal and D. H. Janzen. New York: Academic Press, 387–412.

Constabel, C. P., D. R. Bergey, and C. A. Ryan. 1995. Systemin activates synthesis of wound-inducible tomato leaf polyphenol oxidase via the octadecanoid defense signaling pathway. *Proceedings of the National Academy of Sciences of the USA* 92:407–411.

Cook, S. P., and F. P. Hain. 1988. Wound response of loblolly and shortleaf pine attacked or reattacked by *Dendroctonus frontalis* Zimmermann (Coleoptera: Scolytidae) or its fungal associate, *Ceratocystis minor* (Hedgecock) Hunt. *Canadian Journal of Forest Research* 18:33–37.

Costa, B. A., and G. W. Muller. 1980. Tristeza control by cross protection: A U.S.-Brazil cooperative success. *Plant Disease* 64:538–541.

Craig, T. P., P. W. Price, and J. K. Itami. 1986. Resource regulation by a stem-galling sawfly on the arroyo willow. *Ecology* 67:419–425.

Crawford, R. M. M. 1989. *Studies in plant survival: Ecological case histories of plant adaptation to adversity*. Boston: Blackwell.

Crawley, M. J. 1983. *Herbivory: The dynamics of animal-plant interactions*. Berkeley: University of California Press.

Creasy, L. L. 1985. Biochemical responses of plants to fungal attack. *Recent Advances in Phytochemistry* 19:47–79.

Creelman, R. A., and J. E. Mullet. 1995. Jasmonic acid distribution and action in plants: Regulation during development and response to biotic and abiotic stress. *Proceedings of the National Academy of Sciences of the USA* 92:4114–4119.

Creelman, R. A., M. L. Tierney, and J. E. Mullet. 1992. Jasmonic acid/methyl jasmonate accumulate in wounded soybean hypocotyls and modulate wound gene expression. *Proceedings of the National Academy of Sciences of the USA* 89: 4938–4941.

Croft, B. A., and S. A. Hoying. 1977. Competitive displacement of *Panonychus ulmi* (Acarina: Eriophyidae) by *Aculus schlechtendali* (Acarina: Eriophyidae) in apple orchards. *Canadian Entomologist* 109:1025–1034.

Croft, K. P. C., F. Juttner, and A. J. Slusarenko. 1993. Volatile products of the lipoxygenase pathway evolved from *Phaseolus vulgaris* (L.) leaves inoculated with *Pseudomonas syringae* pv. *phaseolicola. Plant Physiology* 101:13–24.

Croxford, A. C., P. J. Edwards, and S. D. Wratten. 1989. Temporal and spatial variation in palatability of soybean and cotton leaves following wounding. *Oecologia* 79:520–525.

Cutright, C. R. 1963. The European red mite in Ohio. *Ohio Agricultural Experimental Station Research Bulletin* 953:1–32.

Danell, K., and K. Huss-Danell. 1985. Feeding by insects and hares on birches earlier affected by moose browsing. *Oikos* 44:75–81.

Danell, K., K. Huss-Danell, and R. Bergstrom. 1985. Interactions between browsing moose and two species of birch in Sweden. *Ecology* 66:1867–1878.

Davies, E., and A. Schuster. 1981. Intercellular communication in plants: Evidence for a rapidly generated, bidirectionally transmitted wound signal. *Proceedings of the National Academy of Sciences of the USA* 78:2422–2426.

Davis, C. S. 1957. The biology of the walnut aphid *Chromaphis juglandicola* (Kalt.). Ph.D. dissertation, Department of Entomology, University of California, Davis.

Davis, J. M., E. E. Egelkrout, G. D. Coleman, T. H. H. Chen, B. E. Haissig, D. E. Riemenschneider, and M. P. Gordon. 1993. A family of wound induced genes in *Populus* shares common features with gene encoding vegatative storage proteins. *Plant Molecular Biology* 23:135–143.

Davis, J. M., M. P. Gordon, and B. A. Smit. 1991. Assimilate movement dictates remote sites of wound-induced gene expression in poplar leaves. *Proceedings of the National Academy of Sciences of the USA* 88:2393–2396.

Dawson, R. F. 1941. The localization of the nicotine synthetic mechanism in the tobacco plant. *Science* 94:396–397.

————. 1942. Accumulation of nicotine in reciprocal grafts of tomato and tobacco. *American Journal of Botany* 29:66–71.

Deacon, J. W. 1988. Biocontrol of soil-borne plant pathogens with introduced inocula. *Philosophical Transactions of the Royal Society of London B* 318:249–264.

De Luca, V. 1993. Molecular characterization of secondary metabolic pathways. *AgBiotech News and Information* 5:225N–229N.

Dermastia, R., M. Ravnikar, B. Vilhar, and M. Kovac. 1994. Increased level of cytokinin ribosides in jasmonic acid–treated potato (*Solanum tuberosum*) stem node cultures. *Physiologia Plantarum* 92:241–246.

DeWald, D. B., A. Sadka, and J. E. Mullet. 1994. Sucrose modulation of soybean Vsp gene expression is inhibited by auxin. *Plant Physiology* 104:439–444.

Dicke, M. 1986. Volatile spider-mite pheromone and host-plant kairomone, involved in spaced-out gregariousness in the spider mite *Tetranychus urticae*. *Physiological Entomology* 11:251–262.

————. 1994. Local and systemic production of volatile herbivore-induced terpenoids: Their role in plant-carnivore mutualism. *Journal of Plant Physiology* 143:465–472.

Dicke, M., and H. Dijkman. 1992. Induced defence in detached uninfested plant leaves: Effects on behavior of herbivores and their predators. *Oecologia* 91: 554–560.

Dicke, M., and M. W. Sabelis. 1989. Does it pay plants to advertize for bodyguards? Towards a cost-benefit analysis of induced synomone production. In *Causes and consequences of variation in growth rate and productivity of higher plants*, ed. H. Lambers et al. The Hague, The Netherlands: SPB Academic Publishing, 341–358.

Dicke, M., P. van Baarlen, R. Wessels, and H. Dijkman. 1993. Herbivory induces systemic production of plant volatiles that attract herbivore predators: Extraction of endogenous elicitor. *Journal of Chemical Ecology* 19:581–599.

Dirr, M. A. 1975. *Manual of woody landscape plants*. Champaign, IL: Stipes.

Dirzo, R., and J. L. Harper. 1982a. Experimental studies on slug-plant interactions: 3, Differences in the acceptability of individual plants of *Trifolium repens* to slugs and snails. *Journal of Ecology* 70:101–117.

————. 1982b. Experimental studies on slug-plant interactons: 4, The performance of cyanogenic and acyanogenic morphs of *Trifolium repens* in the field. *Journal of Ecology* 70:119–138.

Dittrich, H., T. Kutchan, and M. H. Zenk. 1992. The jasmonate precursor, 12-oxo-phytodienoic acid, induces phytoalexin synthesis in *Petroselinum hortense* cell cultures. *FEBS Letters* 309:33–36.

Dixon, A. F. G., and N. D. Barlow. 1979. Population regulation in the lime aphid. *Zoological Journal of the Linnean Society* 67:225–237.

Dixon, R. A., M. J. Harrison, and C. J. Lamb. 1994. Early events in the activation of plant defense responses. *Annual Review of Phytopathology* 32:479–501.

Doares, S., T. Syrovets, E. Weiler, and C. A. Ryan. 1995. Oligogalacturonides and chitosan activate plant defense genes through the octadecanoid pathway. *Proceedings of the National Academy of Sciences of the USA* 92:4095–4098.

Dobson, A., and M. Crawley. 1994. Pathogens and the structure of plant communities. *Trends in Ecology and Evolution* 9:393–398.

Dobson, H. E. M. 1994. Floral volatiles in insect biology. In *Insect-plant interactions*, ed. E. A. Bernays. Boca Raton, FL: CRC Press, 5:47–81.

Dodson, S. 1989. Predator-induced reaction norms. *Bioscience* 39:447–452.

Dolch, R., and T. Tscharntke. n.d. The talking tree debate revisited: Herbivory and induced resistance in *Alnus glutinosa*. Manuscript.

Donoghue, M. J. 1989. Phylogenies and the analysis of evolutionary sequences with examples from seed plants. *Evolution* 43:1137–1156.

Doughty, K. J., G. A. Kiddle, B. J. Pye, R. M. Wallsgrove, and J. A. Pickett. 1995. Selective induction of glucosinolates in oilseed rape leaves by methyl jasmonate. *Phytochemistry* 38:347–350.

Dreger-Jauffret, F., R. Bronner, and E. Westphal. 1990. *Caracterisation de la résistance aux Eriophyides (Acariens) chez la pomme de terre (var. Nicola)*. Paris: Association Nationale de Protection des Plantes.

Dreyer, D. L., B. C. Campbell, and K. C. Jones. 1984. Effect of bioregulator-treated sorghum on greenbug fecundity and feeding behavior: Implications for host-plant resistance. *Phytochemistry* 23:1593–1596.

Drukker, B., P. Scutareanu, and M. W. Sabelis. 1995. Do anthocorid predators respond to synomones from *Psylla*-infested pear trees under field conditions? *Entomologia Experimentalis et Applicata* 77:193–203.

Duffey, S. S., K. A. Bloem, and B. C. Campbell. 1986. Consequences of sequestration of plant natural products in plant-insect-parasitoid interactions. In *Interactions of plant resistance and parasitoids and predators of insects*, ed. D. J. Boethel and R. D. Eikenbarry. Chicester, England: Ellis Horwood, 31–60.

Duffey, S. S., and G. W. Felton. 1989. Plant enzymes in resistance to insects. In *Biocatalysis in agricultural biotechnology*, ed. J. R. Whitaker and P. E. Sonnet. Toronto: American Chemical Society, 289–313.

Duffey, S. S., and M. J. Stout. 1996. Antinutritive and toxic components of plant defense against insects. *Archives of Insect Biochemistry and Physiology* 32:3–37.

Dussourd, D. E., and R. F. Denno. 1991. Deactivation of plant defense: Correspondence between insect behavior and secretory canal architecture. *Ecology* 72:1383–1396.

———. 1994. Host range of generalist caterpillars: Trenching permits feeding on plants with secretory canals. *Ecology* 75:69–78.

Dussourd, D. E., and T. Eisner. 1987. Vein-cutting behavior: Insect counterploy to the latex defense of plants. *Science* 237:898–901.

Dyer, M. I., M. A. Acra, G. M. Wang, D.C. Coleman, D. W. Freckman, S. J. McNaughton, and B. R. Strain. 1991. Source-sink carbon relations in two *Panicum coloratum* ecotypes in response to herbivory. *Ecology* 72:1472–1483.

Ebel, J. 1986. Phytoalexin synthesis: The biochemical analysis of the induction process. *Annual Review of Phytopathology* 24:235–264.

Edelstein-Keshet, L., and M. Rausher. 1989. The effects of inducible plant defenses on herbivore populations: 1, Mobile herbivores in continuous time. *American Naturalist* 133:787–810.

Edmunds, G. F., and D. N. Alstad. 1978. Coevolution in insect herbivores and conifers. *Science* 199:941–945.

Edwards, P. J., and S. D. Wratten. 1982. Wound-induced changes in palatability in birch (*Betula pubescens* Ehrh. spp. pubescens). *American Naturalist* 120: 816–818.

―――. 1983. Wound induced defenses in plants and their consequences for patterns of insect grazing. *Oecologia* 59:88–93.

―――. 1985. Induced plant defenses against insect grazing: Fact or artefact? *Oikos* 44:70–74.

―――. 1987. Ecological significance of wound-induced changes in plant chemistry. In *Insects: Plants*, ed. V. Labeyrie, G. Fabres, and D. Lachaise. Dordrecht, The Netherlands: Dr W. Junk, 213–218.

Edwards, P. J., S. D. Wratten, and H. Cox. 1985. Wound-induced changes in the acceptability of tomato to larvae of *Spodoptera littoralis:* A laboratory bioassay. *Ecological Entomology* 10:155–158.

Edwards, P. J., S. D. Wratten, and R. M. Gibberd. 1991. The impact of inducible phytochemicals on food selection by insect herbivores and its consequences for the distribution of grazing damage. In *Phytochemical induction by herbivores*, ed. D. W. Tallamy and M. J. Raupp. New York: John Wiley, 205–221.

Edwards, P. J., S. D. Wratten, and S. Greenwood. 1986. Palatability of British trees to insects: Constitutive and induced defences. *Oecologia* 69:316–319.

Ehrlen, J., and O. Eriksson. 1993. Toxicity in fleshy fruits: A non-adaptive trait? *Oikos* 66:107–113.

Eisner, T. 1981. Leaf folding in a sensitive plant: A defensive thorn-exposure mechanism? *Proceedings of the National Academy of Sciences of the USA* 78:402–404.

Eisner, T., K. D. McCormick, M. Sakaino, M. Eisner, S. R. Smedley, D. J. Aneshansley, M. Deyrup, R. L. Myers, and J. Meinwald. 1990. Chemical defense of a rare mint plant. *Chemoecology* 1:30–37.

Elliger, C. A., D. F. Zinkel, B. G. Chan, and A. C. Waiss. 1976. Diterpene acids as larval growth inhibitors. *Experientia* 32:1364–1366.

Elton, C. S. 1924. Periodic fluctuations in the numbers of animals: Their causes and effects. *Journal of Experimental Biology* 2:119–163.

Endler, J. A. 1986. *Natural selection in the wild.* Princeton, NJ: Princeton University Press.

English, J., K. Bonner, and A. J. Haagen-Smit. 1939. Structure and synthesis of a plant wound hormone. *Science* 90:329.

English-Loeb, G. M. 1989. Nonlinear responses of spider mites to drought-stressed host plants. *Ecological Entomology* 14:45–55.

―――. 1990. Plant drought stress and outbreaks of spider mites: A field test. *Ecology* 71:1401–1411.

English-Loeb, G. M., and R. Karban. 1988. Negative interactions between Willamette mites and Pacific mites: Possible management strategies for grapes. *Entomologia Experimentalis et Applicata* 48:269–274.

―――. 1991. Consequences of mite feeding injury to beans on the fecundity and survivorship of the two-spotted spider mite (Acari: Tetranychidae). *Experimental and Applied Acarology* 11:125–136.

English-Loeb, G. M., R. Karban, and D. Hougen-Eitzman. 1993. Direct and indirect competition between spider mites feeding on grapes. *Ecological Applications* 3:699–707.

English-Loeb, G. M., R. Karban, and M. A. Walker. n.d. Genotypic variation in constitutive and induced resistance in grapes against spider mite herbivores. Manuscript.

Enyedi, A. J., and I. Raskin. 1993. Induction of UDP-glucose: Salicylic acid glucosyl-transferase activity in tobacco mosaic virus–inoculated tobacco (*Nicotiana tobacum*) leaves. *Plant Physiology* 101:1375–1380.

Enyedi, A. J., N. Yalpani, P. Silverman, and I. Raskin. 1992. Signal molecules in systemic plant resistance to pathogens and pests. *Cell* 70:879–886.

Ericsson, A., R. Gref, C. Hellqvist, and B. Lågstrom. 1988. Wound response of living bark of Scots pine seedlings and its influence on feeding by the weevil, *Hylobius abietis*. In *Mechanisms of woody plant defenses against insects*, ed. W. J. Mattson, J. Levieux, and C. Bernard-Dagan. New York: Springer-Verlag, 227–235.

Ernest, K. A. 1994. Resistance of creosote bush to mammalian herbivory: Temporal consistency and browsing-induced changes. *Ecology* 75:1684–1692.

Evans, J. 1984. *Silviculture of broadleaved woodland*. London: Her Majesty's Stationary Office.

Evans, J. R. 1989. Photosynthesis and nitrogen relationships in leaves of C_3 plants. *Oecologia* 78:9–19.

Evans, P. S. 1971. Root growth of *Lolium perenne* L.: 2, Effects of defoliation and shading. *New Zealand Journal of Agricultural Research* 14:552–562.

Faeth, S. H. n.d. Timing of oak defoliation: Effects on leaf miners and leaf chewers via phenological and chemical changes. Manuscript.

————. 1986. Indirect interactions between temporally separated herbivores mediated by the host plant. *Ecology* 67:479–494.

————. 1987. Community structure and folivorous insect outbreaks: The roles of vertical and horizontal interactions. In *Insect outbreaks*, ed. P. Barbosa and J. C. Schultz. San Diego: Academic Press, 135–171.

————. 1990. Structural damage to oak leaves alters natural enemy attack on a leaf miner. *Entomologia Experimentalis et Applicata* 57:57–63.

————. 1991. Variable induced responses: Direct and indirect effects on oak folivores. In *Phytochemical induction by herbivores*, ed. D. W. Tallamy and M. J. Raupp. New York: John Wiley, 293–323.

————. 1992a. Do defoliation and subsequent phytochemical responses reduce future herbivory on oak trees? *Journal of Chemical Ecology* 18:915–925.

————. 1992b. Interspecific and intraspecific interactions via plant responses to folivory: An experimental field test. *Ecology* 73:1802–1813.

Faeth, S. H., E. F. Connor, and D. Simberloff. 1981. Early leaf abscission: A neglected source of mortality for folivores. *American Naturalist* 117:409–415.

Fagerstrom, T., S. Larsson, and O. Tenow. 1987. On optimal defense in plants. *Functional Ecology* 1:73–81.

Falkenstein, E., B. Groth, A. Mithofer, and E. W. Weiler. 1991. Methyl jasmonate and α-linolenic acid are potent inducers of tendril coiling. *Planta* 185: 316–322.

Farmer, E. E. 1994. Fatty acid signalling in plants and their associated microorganisms. *Plant Molecular Biology* 26:1423–1437.

Farmer, E. E., D. Caldelari, G. Pearce, M. K. Walker-Simmons, and C. A. Ryan. 1994. Diethyldithiocarbamic acid inhibits the octadecanoid signaling pathway for the wound induction of proteinase inhibitors in tomato leaves. *Plant Physiology* 106:337–342.

Farmer, E. E., R. R. Johnson, and C. A. Ryan. 1992. Regulation of expression of proteinase inhibitor genes by methyl jasmonate and jasmonic acid. *Plant Physiology* 98:995–1002.

Farmer, E. E., and C. A. Ryan. 1990. Interplant communication: Airborne methyl jasmonate induces synthesis of proteinase inhibitors in plant leaves. *Proceedings of the National Academy of Sciences of the USA* 87:7713–7716.

———. 1992. Octadecanoid precursors of jasmonic acid activate the synthesis of wound-inducible proteinase inhibitors. *Plant Cell* 4:129–134.

Farrer, W. 1898. The making and improvement of wheats for Australian conditions. *Agricultural Gazette of New South Wales* 9:131–168.

Feeny, P. 1976. Plant apparency and chemical defense. *Recent Advances in Phytochemistry* 10:1–40.

———. 1977. Defensive ecology of the Cruciferae. *Annals of the Missouri Botanical Garden* 64:221–234.

Felton, G. W., J. L. Bi, C. B. Summers, A. J. Mueller, and S. S. Duffey. 1994. Potential role of lipoxygenases in defense against insect herbivory. *Journal of Chemical Ecology* 20:651–666.

Felton, G. W., K. Donato, R. J. Del Vecchio, and S. S. Duffey. 1989. Activation of plant foliar oxidases by insect feeding reduces nutritive quality of foliage for noctuid herbivores. *Journal of Chemical Ecology* 15:2667–2694.

Felton, G. W., C. B. Summers, and A. J. Mueller. 1994. Oxidative responses in soybean foliage herbivory by bean leaf beetle and three-cornered alfalfa hopper. *Journal of Chemical Ecology* 20:639–650.

Felton, G. W., J. Workman, and S. S. Duffey. 1992. Avoidance of antinutritive plant defense: Role of midgut pH in Colorado potato beetle. *Journal of Chemical Ecology* 18:571–583.

Fernandes, G. W. 1990. Hypersensitivity: A neglected plant resistance mechanism against insect herbivores. *Environmental Entomology* 19:1173–1182.

Fernandes, G. W., and T. G. Whitham. 1989. Selective fruit abscission by *Juniperus monosperma* as an induced defense against predators. *American Midland Naturalist* 121:389–392.

Ferree, D.C., and F. R. Hall. 1981. Influence of physical stress on photosynthesis and transpiration of apple leaves. *Journal of the American Society for Horticultural Science* 106:348–351.

Feth, F., R. Wagner, and K. G. Wagner. 1986. Regulation in tobacco callus of enzyme activities of the nicotine pathway: 1, The route ornithine to methylpyrroline. *Planta* 168:402–407.

Fineblum, W. L., and M. D. Rausher. 1995. Tradeoff between resistance and tolerance to herbivore damage in a morning glory. *Nature* 377:517–520.

Finerty, J. P. 1980. *The population ecology of cycles in small mammals.* New Haven: Yale University Press.

Fischer, D.C., M. Kogan, and P. Greany. 1990. Inducers of plant resistance to insects. In *Safer insecticides: Development and use,* ed. E. Hodgson and R. J. Kuhr. New York: Marcel Dekker, 257–280.

Fischer, D.C., M. Kogan, and J. Paxton. 1990. Effect of glyceollin, a soybean phytoalexin, on feeding by three phytophagous beetles (Coleoptera: Coccinellidae and Chrysomelidae): Dose versus response. *Environmental Entomology* 19:1278–1282.

Fischlin, A., and W. Baltensweiler. 1979. Systems analysis of the larch budmoth system: 1, The larch–larch budmoth relationship. *Mitteilungen der Schweizerischen Entomologischen Gesellschaft* 52:273–289.

Fisher, M. 1987. The effect of previously infested spruce needles on the growth of the green spruce aphid, *Elatobium abietinum,* and the effect of the aphid on the amino acid balance of the host plant. *Annals of Applied Biology* 111: 33–41.

Flaherty, D. L., and C. B. Huffaker. 1970. Biological control of Pacific mites and Willamettes mites in San Joaquin valley vineyards: 1, The role of *Metaseilus occidentalis. Hilgardia* 40:267–308.

Fletcher, J. T., and J. M. Rowe. 1975. Observations and experiments on the use of an avirulent mutant strain of tobacco mosaic virus as a means of controlling tomato mosaic. *Annals of Applied Biology* 81:171–179.

Flor, H. H. 1971. Current status of the gene-for-gene concept. *Annual Review of Phytopathology* 9:275–296.

Foggo, A., and M. R. Speight. 1993. Root damage and water stress: Treatments affecting the exploitation of the buds of common ash *Fraxinus excelsior* L., by larvae of the ash bud moth *Prays fraxinella* Bjerk. (Lep., Yponomeutidae). *Oecologia* 96:134–138.

Foggo, A., M. R. Speight, and J. Gregoire. 1994. Root disturbance of common ash, *Fraxinus excelsior* (Oleaceae), leads to reduced foliar toughness and increased feeding by a folivorous weevil, *Stereonychus fraxini* (Coleoptera, Curculionidae). *Ecological Entomology* 19:344–348.

Foott, W. H. 1962. Competition between two species of mites: 1, Experimental results. *Canadian Entomologist* 94:365–375.

———. 1963. Competition between two species of mites: 2, Factors influencing intensity. *Canadian Entomologist* 95:45–57.

Ford, R. G., and F. A. Pitelka. 1984. Resource limitation in populations of the California vole. *Ecology* 65:122–136.

Foster, M. A., J. C. Schultz, and M. D. Hunter. 1992. Modelling gypsy moth–virus–leaf chemistry interactions: Implications of plant quality for pest and pathogen dynamics. *Journal of Animal Ecology* 61:509–520.

Fowler, S. V., and J. H. Lawton. 1985. Rapidly induced defenses and talking trees: The devil's advocate position. *American Naturalist* 126:181–195.

Fowler, S. V., and M. MacGarvin. 1986. The effects of leaf damage on the performance of insect herbivores on birch, *Betula pubescens. Journal of Animal Ecology* 55:565–573.

Fox, J. F., and J. P. Bryant. 1984. Instability of the snowshow hare and woody plant interaction. *Oecologia* 63:128–135.

Fox, L. R. 1981. Defense and dynamics in plant-herbivore systems. *American Zoologist* 21:853–864.

Fraenkel, G. S. 1959. The raison d'etre of secondary plant substances. *Science* 129:1466–1470.

Franceschi, V. R., and H. D. Grimes. 1991. Low levels of atmospheric methyl jasmonate induce the accumulation of soybean vegatative storage proteins and anticyanins. *Proceedings of the National Academy of Sciences of the USA* 88:6745–6749.

Frank, S. A. 1993. A model of inducible defense. *Evolution* 47:325–327.

Freeland, W. J. 1974. Vole cycles: Another hypothesis. *American Naturalist* 108:238–245.

Frischknecht, P. M., M. Bättig, and T. W. Baumann. 1987. Effect of drought and wounding stress on indole alkaloid formation in *Catharanthus roseus. Phytochemistry* 26:707–710.

Fujiwara, M., H. Oku, and T. Shiraishi. 1987. Involvement of volatile substances in systemic resistance of barley against *Erysiphe graminis* f. sp. hordei induced by pruning of leaves. *Journal of Phytopathology* 120:81–84.

Fulton, R. W. 1986. Practices and precautions in the use of cross protection for plant virus disease control. *Annual Review of Phytopathology* 24:67–81.

Gabriel, D. W., and B. G. Rolfe. 1990. Working models of specific recognition in plant-microbe interactions. *Annual Review of Phytopathology* 28:365–391

Gaffney, T., L. Friedrich, B. Vernooij, D. Negrotto, G. Nye, S. Uknes, E. Ward, H. Kessmann, and J. Ryals. 1993. Requirement of salicylic acid for the induction of systemic acquired resistance. *Science* 261:754–756.

Gaines, M. S., N. C. Stenseth, M. L. Johnson, R. A. Ims, and S. Bondrup-Nielsen. 1991. A response to solving the enigma of population cycles with a multifactorial perspective. *Journal of Mammalogy* 72:627–631.

Gershenzon, J. 1994. The cost of plant chemical defense against herbivory: A biochemical perspective. In *Insect-plant interactions,* ed. E. A. Bernays. Boca Raton, FL: CRC Press, 5:105–173.

Gershenzon, J., and R. Croteau. 1991. Terpenoids. In *Herbivores: Their interactions with secondary plant metabolites,* 2d edition, ed. G. A. Rosenthal and M. R. Berenbaum. San Diego: Academic Press, 1:165–219.

Gershenzon, J., C. J. Murtagh, and R. Croteau. 1993. Absence of rapid terpene turnover in several diverse species of terpene-accumulating plants. *Oecologia* 96:583–592.

Giamonstaris, A., and R. Mithen. 1995. The effect of modifying the glucosinolate content of leaves of oilseed rape (*Brassica napus* ssp. *oleifera*) on its interaction with specialist and generalist pests. *Annals of Applied Biology* 126:347–363.

Giebel, J. 1982. Mechanisms of resistance to plant nematodes. *Annual Review of Phytopathology* 20:257–279.

Gijzen, M., E. Lewinsohn, and R. Croteau. 1991. Characterization of constitutive and wound-inducible monoterpene cyclases of grand fir (*Abies grandis*). Archives of Biochemistry and Biophysics 289:267–273.

Gilbert, L. E. 1975. Ecological consequences of a coevolved mutualism between butterflies and plants. In *Coevolution in animals and plants*, ed. L. E. Gilbert and P. H. Raven. Austin: University of Texas Press, 210–214.

Givnish, T. J. 1986. Economics of biotic interactions. In *On the economy of plant form and function*, ed. T. J. Givnish. Cambridge: Cambridge University Press.

Godiard, L., M. R. Grant, R. A. Dietrich, S. Kiedrowski, and J. L. Dangl. 1994. Perception and response in plant disease resistance. *Current Opinion in Genetics and Development* 4:662–671.

Goldstein, J. L., and T. Swain. 1963. Changes in tannins in ripening fruits. *Phytochemistry* 2:371–383.

Gonzalez, B., J. Boucaud, J. Salette, J. Langlois, and M. Duyme. 1989. Changes in stubble carbohydrate content during regrowth of defoliated perennial ryegrass (*Lolium perenne* L.) on two nitrogen levels. *Grass and Forage Science* 44:411–415.

Gonzalez-Coloma, A., C. S. Wisdom, and P. W. Rundel. 1990. Compound interactions effects of plant antioxidants in combination with carbaryl on performance of *Trichoplusia ni* (cabbage looper). *Journal of Chemical Ecology* 16: 887–899.

Gorlach, J., S. Volrath, G. Knaufbeiter, G, Hengy, U. Beckhove, K. Kogel, M. Oostendorp, T. Staub, E. Ward, H. Kessmann, and J. Ryals. 1996. Benzothiadiazole, a novel class of inducers of systemic acquired resistance, activates gene expression and disease resistance in wheat. *Plant Cell* 8:629–643.

Gould, F. 1988. Evolutionary biology and genetically engineered crops. *Bioscience* 38:26–32.

Gould, S. J., and R. C. Lewontin. 1979. The spandrels of San Marco and the Panglossian paradigm: A critique of the adaptationist programme. *Proceedings of the Royal Society of London B* 205:581–598.

Gould, S. J., and E. S. Vrba. 1982. Exaptation: A missing term in the science of form. *Paleobiology* 8:4–15.

Greany, P. D., R. E. McDonald, P. E. Shaw, W. J. Schroeder, D. F. Howard, T. T. Hatton, P. L. Davis, and G. K. Rasmussen. 1987. Use of gibberellic acid to reduce grapefruit susceptibility to attack by the Caribbean fruit fly *Anastrepha suspensa* (Diptera: Tephritidae). *Tropical Science* 27:261–270.

Greany, P. D., and J. P. Shapiro. 1993. Manipulating and enhancing citrus fruit fly resistance to the Caribbean fruit fly (Diptera: Tephritidae). *Florida Entomologist* 76:258–263.

Green, T. R., and C. A. Ryan. 1972. Wound-induced proteinase inhibitor in plant leaves: A possible defense mechanism against insects. *Science* 175:776–777.

Gref, R., and A. Ericsson. 1985. Wound-induced changes of resin acid concentrations in living bark of Scots pine seedlings. *Canadian Journal of Forest Research* 15:92–96.

Gregersen, P. L., and V. Smedegaard. 1989. Induction of resistance in barley against *Erysiphe graminis* f. sp. hordei after preinoculation with the saprophytic fungus, *Cladosporum macrocarpum*. *Journal of Phytopathology* 124:128–136.

Grime, J. P. 1979. *Plant strategies and vegetation processes.* Chichester, England: John Wiley.

Grime, J. P., J. C. Crick, and J. E. Rincon. 1986. The ecological significance of plasticity. In *Plasticity in plants,* ed. D. H. Jennings and A. J. Trewavas. Symposia of the Society for Experimental Biology, no. 40. Cambridge: Company of Biologists, 5–29.

Guedes, M. E. M., J. Kuc, R. Hammerschmidt, and R. Bostock. 1982. Accumulation of six sesquiterpenoid phytoalexins in tobacco leaves infiltrated with *Pseudomonas lachrymans. Phytochemistry* 12:2987–2988.

Gulmon, S. L., and H. A. Mooney. 1986. Costs of defense and their effects on plant productivity. In *On the economy of plant form and function,* ed. T. J. Givnish. Cambridge: Cambridge University Press, 681–698.

Gundlach, H., M. J. Muller, T. M. Kutchan, and M. H. Zenk. 1992. Jasmonic acid is a signal transducer in elicitor-induced plant cell cultures. *Proceedings of the National Academy of Sciences of the USA* 89:2389–2393.

Hadwiger, L. A., and J. M. Beckman. 1980. Chitosan as a component of pea–*Fusarium solani* interactions. *Plant Physiology* 66:205–211.

Haggstrom, H., and S. Larsson. 1995. Slow larval growth on a suboptimal willow results in high predation mortality in the leaf beetle *Galerucella lineola. Oecologia* 104:308–315.

Hahn, M. G., J. Cheong, R. Alba, and F. Cote. 1993. Oligosaccharide elicitors: Structures and signal transduction. In *Plant signals in interactions with other organisms,* ed. J. C. Schultz and I. Raskin. Rockville, MD: American Society of Plant Physiologists, 11:24–46.

Hain, R., H. Reif, E. Krause, R. Langebartels, H. Kindl, B. Vornam, W. Wiese, E. Schmelzer, P. H. Schreier, R. H. Stocker, and K. Stenzel. 1993. Disease resistance results from foreign phytoalexin expression in a novel plant. *Nature* 361:153–156.

Hairston, N. G., F. E. Smith, and L. B. Slobodkin. 1960. Community structure, population control, and competition. *American Naturalist* 94:421–425.

Hait, G. M., and A. K. Sinha. 1986. Protection of wheat seedlings for *Helminthosporium* infection by seed treatment with chemicals. *Journal of Phytopathology* 115:97–107.

———. 1987. Biochemical changes associated with induction of resistance in rice seedlings to *Helminthosporium oryzae* by seed treatment with chemicals. *Zeitschrift für Pflanzenkrankheiten und Pflanzenschutz* 94:360–368.

Hall, F. R. 1972. Influence of alar on populations of European red mite and apple aphid on apples. *Journal of Economic Entomology* 65:1751–1753.

Al-Hamdani, A. M., R. S. Lutchmeah, and R. C. Cooke. 1983. Biological control of *Pythium ultimum*–induced damping-off by treating cress seed with the mycoparasite *Pythium oligandrum. Plant Pathology* 32:449–454.

Hammerschmidt, R. 1993. The nature and generation of systemic signals induced by pathogens, arthropod herbivores, and wounds. In *Advances in plant pathology,* ed. J. H. Andrews and I. C. Tommerup. New York: Academic Press, 10:307–337.

Hanhimäki, S. 1989. Induced resistance in mountain birch: Defense against leaf-chewing insect guild and herbivore competition. *Oecologia* 81:242–248.

Hanounik, S. B., and W. W. Osborne. 1977. The relationship between the population density of *Meloidogyne incognita* and nicotine content of tobacco. *Nematologica* 23:147–152.

Hansson, L., and H. Henttonen. 1985. Gradients in density variations of small rodents: The importance of latitude and snow cover. *Oecologia* 67:394–402.

Harborne, J. B. 1988. *Introduction to ecological biochemistry.* 3d edition. London: Academic Press.

Hare, J. D. 1983. Manipulation of host suitability for herbivore pest management. In *Variable plants and herbivores in natural and managed systems,* ed. R. F. Denno and M. S. McClure. New York: Academic Press, 655–680.

Harris, P. 1960. Production of pine resin and its effects on survival of *Rhyacionia buoliana* (Schiff.) (Lepidoptera: Olethreutidae). *Canadian Journal of Zoology* 38:121–130.

Harrison, S. 1995. Lack of strong induced or maternal effects in tussock moths (*Orgyia vetusta*) on bush lupine (*Lupinus arboreus*). *Oecologia* 103:343–348.

Harrison, S., and R. Karban. 1986. Effects of an early-season folivorous moth on the success of a later-season species, mediated by a change in the quality of the shared host, *Lupinus arboreus* Sims. *Oecologia* 69:354–359.

Hart, S. V., M. Kogan, and J. D. Paxton. 1983. Effect of soybean phytoalexins on the herbivorous insects Mexican bean beetle and soybean looper. *Journal of Chemical Ecology* 9:657–672.

Hartley, S. E., and R. D. Firn. 1989. Phenolic biosynthesis, leaf damage, and insect herbivory in birch (*Betula pendula*). *Journal of Chemical Ecology* 15:275–283.

Hartley, S. E., and J. H. Lawton. 1987. Effects of different types of damage on the chemistry of birch foliage, and the responses of birch feeding insects. *Oecologia* 74:432–437.

———. 1991. Biochemical aspects and significance of the rapidly induced accumulation of phenolics in birch foliage. In *Phytochemical induction by herbivores,* ed. D. W. Tallamy and M. J. Raupp. New York: John Wiley, 105–132.

Hartmann, T. 1991. Alkaloids. In *Herbivores: Their interactions with secondary plant metabolites,* 2d edition, ed. G. A. Rosenthal and M. R. Berenbaum. San Diego: Academic Press, 1:79–121.

Harvell, C. D. 1986. The ecology and evolution of inducible defenses in a marine bryozoan: Cues, costs, and consequences. *American Naturalist* 128:810–823.

———. 1990a. The ecology and evolution of inducible defenses. *Quarterly Review of Biology* 65:323–340.

———. 1990b. The evolution of inducible defence. *Parasitology* 100:S53–S61.

———. 1992. Inducible defenses and allocation shifts in a marine bryozoan. *Ecology* 73:1567–1576.

Harvey, P. H., and M. D. Pagel. 1991. *The comparative method in evolutionary biology.* Oxford: Oxford University Press.

Haslam, E. 1986. Secondary metabolism: Fact and fiction. *Natural Product Reports* 3:217–249.

Hatanaka, A. 1993. The biogeneration of green odour by green leaves. *Phytochemistry* 34:1201–1218.

Hatcher, P. E., N. D. Paul, P. G. Ayres, and J. B. Whittaker. 1994. The effect of a foliar disease (rust) on the development of *Gastrophysa viridula* (Coleoptera: Chrysomelidae). *Ecological Entomology* 19:349–360.

Haukioja, E. 1980. On the role of plant defences in the fluctuation of herbivore populations. *Oikos* 35:202–213.

———. 1982. *Inducible defences of white birch to a geometrid defoliator, Epirrita autumnata.* Proceedings of the 5th International Symposium on Insect-Plant Relationships, Wageningen, Pudoc, Wageningen, 199–203.

———. 1990a. Induction of defenses in trees. *Annual Review of Entomology* 36:25–42.

———. 1990b. Positive and negative feedbacks in insect-plant interactions. In *Population dynamics of forest insects,* ed. A. D. Watt. Andover, Hampshire: Intercept, 113–122.

———. 1990c. Toxic and nutritive substances as plant defence mechanisms against invertebrate herbivores. In *Pests, pathogens, and plant communities,* ed. J. J. Burdon and S. R. Leather. Oxford: Blackwell Scientific, 219–231.

Haukioja, E., and T. Hakala. 1975. Herbivore cycles and periodic outbreaks: Formulation of a general hypothesis. *Reports from the Kevo Subarctic Research Station* 12:1–9.

Haukioja, E., and S. Hanhimäki. 1985. Rapid wound-induced resistance in white birch foliage to the geometrid *Epirrita autumnata:* A comparison of trees and moths within and outside the outbreak range of the moth. *Oecologia* 65:223–228.

Haukioja, E., S. Hanhimäki, and G. H. Walter. 1994. Can we learn about herbivory on eucalypts from research on birches, or how general are general plant-herbivore theories? *Australian Journal of Ecology* 19:1–9.

Haukioja, E., and T. Honkanen. 1994. Why tree responses to herbivory are so variable? Proceedings of the International Union of Forestry Research Organizations meeting, Maui, Hawaii, February 2–6.

Haukioja, E., K. Kapiainen, and P. Niemelä. 1983. Plant availability hypothesis and other explanations of herbivore cycles: Complementary or exclusive alternatives? *Oikos* 40:419–432.

Haukioja, E., and S. Neuvonen. 1985. Induced long-term resistance of birch foliage against defoliators: Defensive or incidental? *Ecology* 66:1303–1308.

———. 1987. Insect population dynamics and induction of plant resistance: The testing of hypotheses. In *Insect outbreaks,* ed. P. Barbosa and J. C. Schultz. San Diego: Academic Press, 411–432.

Haukioja, E., and P. Niemelä. 1977. Retarded growth of a geometrid larva after mechanical damage to leaves of its host tree. *Annales Zoololici Fennici* 14:48–52.

———. 1979. Birch leaves as a resource for herbivores: Seasonal occurrence of increased resistance in foliage after mechanical damage of adjacent leaves. *Oecologia* 39:151–159.

Haukioja, E., K. Ruohomäki, J. Senn, J. Suomela, and M. Walls. 1990. Consequences of herbivory in the mountain birch (*Betula pubescens* ssp tortuosa): Importance of the functional organization of the tree. *Oecologia* 82:238–247.

Haukioja, E., J. Suomela, and S. Neuvonen. 1985. Long-term inducible resistance in birch foliage: Triggering cues and efficacy on a defoliator. *Oecologia* 65:363–369.

Havel, J. E. 1987. Predator-induced defenses: A review. In *Predation: Direct and indirect impacts on aquatic communities,* ed. W. C. Kerfoot and A. Sih. Hanover, NH: University Press of New England, 263–278.

Heath, M. C. 1995. Thoughts on the role and evolution of induced resistance in natural ecosystems, and its relationship to other types of plant defenses against disease. In *Induced resistance to disease in plants,* ed. R. Hammerschmidt and J. Kuc. Dordrecht, The Netherlands: Kluwer, 141–151.

Heath-Pagliuso, S., S. A. Matlin, N. Fang, R. H. Thompson, and L. Rappaport. 1992. Stimulation of furanocoumarin accumulation in celery and celeriac tissues by *Fusaruim oxysporum. Phytochemistry* 31:2683–2688.

Hedin, P. A. 1990. Bioregulator-induced changes in allelochemicals and their effects on plant resistance to pests. *Critical Reviews in Plant Sciences* 9:371–379.

Hedin, P. A., J. N. Jenkins, A. C. Thompson, J. C. McCarty, D. H. Smith, W. L. Parrott, and R. L. Shepherd. 1988. Effects of bioregulators on flavonoids, insect resistance, and yield of seed cotton. *Journal of Agricultural and Food Chemistry* 36:1055–1061.

Heftmann, E., and S. Schwimmer. 1972. Degradation of tomatine to 3b-hydroxy-5d-pregn-16-en-one by ripe tomatoes. *Phytochemistry* 11:2783–2787.

Heinrichs, E. A., ed. 1988. *Plant stress–insect interactions.* New York: John Wiley.

Helland, A. 1921. *Topografisk-statistisk beskrivelse over Sondre Bergenhus amt/efter offentlig foranstaltning udgivet ved Amund Helland.* Oslo: Kristiania.

Henderson, C. F., and J. K. Holloway. 1942. Influence of leaf age and feeding injury on the citrus red mite. *Journal of Economic Entomology* 35:683–686.

Hendrix, S. D., and E. J. Trapp. 1981. Plant-herbivore interactions: Insect-induced changes in host plant sex expression and fecundity. *Oecologia* 49:119–122.

Hendry, G. 1986. Why do plants have cytochrome P-450? Detoxification versus defense. *New Phytologist* 102:239–247.

Herms, D. A., and W. J. Mattson. 1992. The dilemma of plants: To grow or defend. *Quarterly Review of Biology* 67:283–335.

Heslop-Harrison, J. W. 1924. Sex in the Saliaceae and its modification by eriophiid mites and other influences. *British Journal of Experimental Biology* 1:445–472.

Hessen, D. S., and E. van Donk. 1993. Morphological changes in *Scenedesmus* induced by substances released by *Daphnia. Archiv für Hydrobiologie* 127:129–140.

Hibi, N., S. Higashiguchi, T. Hashimoto, and Y. Yamada. 1994. Gene expression in tobacco low-nicotine mutants. *Plant Cell* 6:723–735.

Hightshoe, G. L. 1988. *Native trees, shrubs, and vines for urban and rural America.* New York: Van Nostrand Reinhold.

Hildebrand, D. F., G. C. Brown, D. M. Jackson, and T. R. Hamilton-Kemp. 1993. Effects of some leaf-emitted volatile compounds on aphid population increase. *Journal of Chemical Ecology* 19:1875–1887.

Hildebrand, D. F., J. G. Rodriguez, G. C. Brown, K. J. Luu, and C. S. Volden. 1986. Peroxidative response of leaves in two soybean genotypes injured by two-spotted spider mites (Acari: Tetranychidae). *Journal of Economic Entomology* 79:1459–1465.

Hildebrand, D. F., J. G. Rodriguez, G. C. Brown, and C. S. Volden. 1986. Two-spotted spider mite (Acari: Tetranychidae) infestations on soybeans: Effect on composition and growth of susceptible and resistant cultivars. *Journal of Economic Entomology* 79:915–921.

Hilder, V. A., A. M. R. Gatehouse, S. E. Sheerman, R. F. Barker, and D. Boulter. 1987. A novel mechanism of insect resistance engineered into tobacco. *Nature* 330:160–163.

Hildmann, T., M. Ebneth, H. Peña-Cortés, J. J. Sánchez-Serrano, L. Willmitzer, and S. Prat. 1992. General roles of abscisic and jasmonic acids in gene activation as a result of mechanical wounding. *Plant Cell* 4:1157–1170.

Hill, R. E., and N. D. Hastie. 1987. Accelerated evolution in the reactive centre regions of serine protease inhibitors. *Nature* 326:96–99.

Holling, C. S. 1959. The components of predation as revealed by a study of small mammal predation of the European pine sawfly. *Canadian Entomologist* 91:293–320.

Honkanen, T., E. Haukioja, and J. Suomela. 1994. Effects of simulated defoliation and debudding on needle and shoot growth in Scots pine (*Pinus sylvestris*): Implications of plant source/sink relationships for plant-herbivore studies. *Functional Ecology* 8:631–639.

Horsfall, J., and E. B. Cowling, eds. 1980. *Plant disease.* New York: Academic Press.

Hougen-Eitzman, D., and R. Karban. 1995. Mechanisms of interspecific competition that result in successful control of Pacific mites following inoculations of Willamette mites on grapevines. *Oecologia* 103:157–161.

Hunter, M. D. 1987. Opposing effects of spring defoliation on late season oak caterpillars. *Ecological Entomology* 12:373–382.

Hurlbert, S. 1984. Pseudoreplication and the design of ecological field experiments. *Ecological Monographs* 54:187–211.

Inouye, D. W. 1982. The consequences of herbivory: A mixed blessing for *Jurinea mollis* (Asteraceae). *Oikos* 39:269–272.

Jacobson, S. E., and N. E. Olszewski. 1996. Gibberellins regulate the abundance of RNAs with sequence similarity to proteinase inhibitors, dioxygenases, and dehydrogenases. *Planta* 198:78–86.

Jakobek, J. L., and P. B. Lindgren. 1993. Generalized induction of defense responses in bean is not correlated with the induction of the hypersensitive response. *Plant Cell* 5:49–56.

Janzen, D. H. 1974. Tropical blackwater rivers, animals, and mast fruiting by the Dipterocarpaceae. *Biotropica* 6:69–103.

———. 1979. New horizons in the biology of plant defenses. In *Herbivores: Their interaction with secondary plant metabolites,* ed. G. A. Rosenthal and D. H. Janzen. New York: Academic Press, 331–350.

Jeker, T. B. 1983. Effect of past defoliation and foliage age on larval development of *Melasoma aenea* (Coleoptera, Chrysomelidae). *Mitteilungen der Schweizerischen Entomologischen Gesellschaft* 56:237–244.

Jenkins, J. N., P. A. Hedin, W. L. Parott, J. C. McCarty, and W. H. White. 1983. Cotton allelochemics and growth of tobacco budworm larvae. *Crop Science* 23:1195–1198.

Johannsen, W. 1911. The genotype conception of heredity. *American Naturalist* 45:129–159.

Johnson, N. D., and B. L. Bentley. 1988. Effects of dietary protein and lupine alkaloids on growth and survivorship of *Spodoptera eridania*. *Journal of Chemical Ecology* 14:1391–1403.

Johnson, N. D., and S. A. Brain. 1985. The response of leaf resin to artificial herbivory in *Eriodictyon californicum*. *Biochemical Systematics and Ecology* 13:5–9.

Johnson, N. D., B. Liu, and B. L. Bentley. 1987. The effects of nitrogen fixation, soil nitrate, and defoliation on the growth, alkaloids, and nitrogen levels of *Lupinus succulentus* (Fabaceae). *Oecologia* 74:425–431.

Johnson, N. D., L. Rigney, and B. L. Bentley. 1988. Short-term changes in alkaloid levels following leaf damage in lupines with and without symbiotic nitrogen fixation. *Journal of Chemical Ecology* 15:2425–2434.

———. 1989. Short-term induction of alkaloid production in lupines: Differences between N_2-fixing and nitrogen-limited plants. *Journal of Chemical Ecology* 15:2425–2434.

Johnson, R., J. Narvaez, G. An, and C. Ryan. 1989. Expression of proteinase inhibitors I and II in transgenic tobacco plants: Effects on natural defense against *Manduca sexta* larvae. *Proceedings of the National Academy of Sciences of the USA* 86:9871–9875.

Jones, C. G., and R. D. Firn. 1991. On the evolution of plant secondary chemical diversity. *Philosophical Transactions of the Royal Society of London B* 333:273–280.

Jones, C. G., R. F. Hopper, J. S. Coleman, and V. A. Krischik. 1993. Control of systemically induced herbivore resistance by plant vascular architecture. *Oecologia* 93:452–456.

Jones, D. A. 1972. Cyanogenic glycosides and their function. In *Phytochemical ecology,* ed. J. B. Harborne. London: Academic Press, 103–124.

Jones, D. A., and A. D. Ramnani. 1985. Altruism and movement of plants. *Evolutionary Theory* 7:143–148.

de Jong, G., and A. J. van Noordwijk. 1992. Acquisition and allocation of resources: Genetic (co)variances, selection, and life histories. *American Naturalist* 139:749–770.

Jongsma, M. A., P. L. Bakker, J. Peters, D. Bosch, and W. J. Stiekema. 1995. Adaptation of *Spodoptera exigua* larvae to plant proteinase inhibitors by induction of gut proteinase activity insensitive to inhibition. *Proceedings of the National Academy of Sciences of the USA* 92:8041–8045.

Jongsma, M. A., P. L. Bakker, W. J. Stiekema, and D. Bosch. 1995. Phage display of a double-headed inhibitor: Analysis of the binding domains of proteinase inhibitor, 2. *Molecular Breeding* 1:181–191.

Jongsma, M. A., P. L. Bakker, B. Visser, and W. J. Stiekema. 1994. Trypsin inhibitor activity in mature tobacco and tomato plants is mainly induced locally in response to insect attack, wounding, and virus infection. *Planta* 195:29–35.

Kahn, D. M., and H. V. Cornell. 1983. Early leaf abscission and folivores: Comments and considerations. *American Naturalist* 122:428–432.

Kakes, P. 1989. An analysis of the costs and benefits of the cyanogenic system in *Trifolium repens* L. *Theoretical and Applied Genetics* 77:111–118.

Karban, R. 1983. Induced responses of cherry trees to periodical cicada oviposition. *Oecologia* 59:226–231.

———. 1985. Resistance against spider mites in cotton induced by mechanical abrasion. *Entomologia Experimentalis et Applicata* 37:137–141.

———. 1986a. Induced resistance against spider mites in cotton: Field verification. *Entomologia Experimentalis et Applicata* 42:239–242.

———. 1986b. Interspecific competition between folivorous insects on *Erigeron glaucus*. *Ecology* 67:1063–1072.

———. 1987. Environmental conditions affecting the strength of induced resistance against mites in cotton. *Oecologia* 73:414–419.

———. 1988. Resistance to beet armyworms (*Spodoptera exigua*) induced by exposure to spider mites in cotton. *American Midland Naturalist* 119:77–82.

———. 1989. Fine-scale adaptation of herbivorous thrips to individual host plants. *Nature* 340:60–61.

———. 1990. Herbivore outbreaks on only young trees: Testing hypotheses about aging and induced resistance. *Oikos* 59:27–32.

———. 1991a. Induced resistance of *Gossypium australe* against its most abundant folivore, *Bucculatrix gossypii*. *Australian Journal of Ecology* 16:501–506.

———. 1991b. Inducible resistance in agricultural systems. In *Phytochemical induction by herbivores*, ed. D. W. Tallamy and M. J. Raupp. New York: John Wiley, 403–419.

———. 1993a. Costs and benefits of induced resistance and plant density for a native shrub, *Gossypium thurberi*. *Ecology* 74:9–19.

———. 1993b. Induced resistance and plant density of a native shrub, *Gossypium thurberi*, affect its herbivores. *Ecology* 74:1–8.

Karban, R., R. Adamchak, and W. C. Schnathorst. 1987. Induced resistance and interspecific competition between spider mites and a vascular wilt fungus. *Science* 235:678–680.

Karban, R., and F. R. Adler. 1996. Induced resistance to herbivores and the information content of early season attack. *Oecologia* 107:379–385.

Karban, R., A. A. Agrawal, and M. Mangel. 1997. The benefits of induced defenses against herbivores. *Ecology*. In press.

Karban, R., and J. R. Carey. 1984. Induced resistance of cotton seedlings to mites. *Science* 225:53–54.

Karban, R., and G. M. English-Loeb. 1990. A vaccination of Willamette spider mites (Acari: Tetranychidae) to prevent large populations of Pacific spider mites on grapevines. *Journal of Economic Entomology* 83:2252–2257.

Karban, R., G. M. English-Loeb, and D. Hougen-Eitzman. 1997. Mite vaccinations for sustainable management of spider mites in vineyards. *Ecological Applications*. In press.

Karban, R., D. Hougen-Eitzman, and G. English-Loeb. 1994. Predator-mediated apparent competition between two herbivores that feed on grapevines. *Oecologia* 97:508–511.

Karban, R., and J. H. Myers. 1989. Induced plant responses to herbivory. *Annual Review of Ecology and Systematics* 20:331–348.

Karban, R., and C. Niiho. 1995. Induced resistance and susceptibility to herbivory: Plant memory and altered plant development. *Ecology* 76:1220–1225.

Ke, D., and M. E. Saltveit. 1989. Wound-induced ethylene production, phenolic metabolism, and susceptibility to russet spotting in iceberg lettuce. *Physiologia Plantarum* 76:412–418.

Keating, S. T., M. D. Hunter, and J. C. Schultz. 1990. Leaf phenolic inhibition of gypsy moth nuclear polyhedrosis virus: Role of polyhedral inclusion body aggregation. *Journal of Chemical Ecology* 16:1445–1457.

Keith, L. B. 1963. *Wildlife's ten year cycle*. Madison: University of Wisconsin Press.

Kendall, D. M., and L. B. Bjostad. 1990. Phytohormone ecology: Herbivory by *Thrips tabaci* induces greater ethylene production in intact onions than mechanical damage alone. *Journal of Chemical Ecology* 16:981–991.

Kendra, D. F., D. Christian, and L. A. Hadwiger. 1989. Chitosan oligomers from *Fusarium solani*/pea interactions, chitosan/β-glucanase digestion of sporelings and from fungal wall chitin actively inhibit fungal growth and enhance disease resistance. *Physiological and Molecular Plant Physiology* 35:215–230.

Kernan, A., and R. W. Thornburg. 1989. Auxin levels regulate the expression of a wound-inducible proteinase inhibitor II-chloramphenicol acetyl transferase gene fusion in vitro and in vivo. *Plant Physiology* 91:73–78.

Kessmann, H., T. Staub, C. Hofmann, T. Maetzke, J. Herzog, E. Ward, S. Uknes, and J. Ryals. 1994. Induction of systemic acquired disease resistance in plants by chemicals. *Annual Review of Phytopathology* 32:439–459.

Khan, M. B., and J. B. Harborne. 1990. Induced alkaloid defence in *Atropa acuminate* in response to mechanical and herbivore leaf damage. *Chemoecology* 1:77–80.

Kielkiewicz, M. 1988. Susceptibility of previously damaged strawberry plants to mite attack. *Entomologia Experimentalis et Applicata* 47:201–203.

Kitcher, P. 1985. *Vaulting Ambition: Sociobiology and the Quest for Human Nature*. Cambridge: MIT Press.

Kittock, D. L., H. F. Arle, T. J. Henneberry, L. A. Bariola, and V. T. Walhood. 1980. Timing late-season fruiting termination of cotton with potassium 3,4-dichloroisothiazole-5-carboxylate. *Crop Science* 20:330–333.

Klein, J. 1982. *Immunology: The science of self-nonself discrimination*. New York: John Wiley.

Kloepper, J. W., S. Tuzin, and J. A. Kuc. 1992. Proposed definitions related to induced disease resistance. *Biocontrol Science and Technology* 2:349–351.

Koda, Y., Y. Kikuta, T. Kithara, T. Nishi, and K. Mori. 1992. Comparisons of various biological activities of sterioisomers of methyl jasmonate. *Phytochemistry* 31:1111–1114.

Kogan, M., and D.C. Fischer. 1991. Inducible defenses in soybean against herbivorous insects. In *Phytochemical induction by herbivores*, ed. D. W. Tallamy and M. J. Raupp. New York: John Wiley, 347–378.

Kogan, M., and J. Paxton. 1983. Natural inducers of plant resistance to insects. In *Plant resistance to insects*, ed. P. A. Hedin. Washington, DC: American Chemical Society, 153–171.

Kolaczyk, A., and K. Wiackowski. 1997. Induced defence in the ciliate *Euplotes octocarinatus* is reduced when alternative prey is available to the predator. *Acta Protozoologica*. In press.

Kolodny-Hirsch, D. M., and F. P. Harrison. 1986. Yield loss relationships of tobacco and tomato hornworms (Lepidoptera: Sphingidae) at several growth stages of Maryland tobacco. *Journal of Economic Entomology* 79:731–735.

Kolodny-Hirsch, D. M., J. A. Saunders, and F. P. Harrison. 1986. Effects of simulated tobacco hornworm (Lepidoptera: Sphingidae) defoliation in growth dynamics and physiology of tobacco as evidence of plant tolerance to leaf consumption. *Environmental Entomology* 15:1137–1144.

Kommedahl, T., and C. E. Windels. 1981. Introduction of microbial antagonists to specific courts of infection: Seeds, seedlings, and wounds. In *Biological control in crop production*, ed. G. C. Papavizas. Totowa, NJ: Allanheld, Osmun, 227–248.

Kouki, J. 1991a. Interaction between a specialist herbivore, *Galerucella nymphaeae*, and its host plant, *Nuphar lutea*. Ph.D. dissertation. University of Helsinki, Finland.

———. 1991b. Tracking spatially variable resources: An experimental study on the oviposition of the water-lily beetle. *Oikos* 61:243–249.

Krause, S. C., and D. F. Raffa. 1995. Defoliation intensity and larval age interact to affect sawfly performance on previously injured *Pinus resinosa*. *Oecologia* 102:24–30.

Krebs, C. J., S. Boutin, R. Boonstra, A. R. E. Sinclair, J. N. M. Smith, M. R. T. Dale, K. Martin, and R. Turkington. 1995. Impact of food and predation on the snowshoe hare cycle. *Science* 269:1112–1115.

Krebs, C. J., and J. H. Myers. 1974. Populations cycles in small mammals. *Advances in Ecological Research* 8:267–399.

Krischik, V. A., P. Barbosa, and C. Reichelderfer. 1988. Three trophic level interactions: Allelochemicals, *Manduca sexta* (L.) and *Bacillus thuringiensis* var. kurstaki Berliner. *Environmental Entomology* 17:476–482.

Krischik, V. A., and R. F. Denno. 1983. Individual, population, and geographic patterns in plant defense. In *Variable plants and herbivores in natural and managed systems*, ed. R. F. Denno and M. S. McClure. New York: Academic Press, 463–512.

Kuc, J. 1987. Plant immunization and its applicability for disease control. In *Innovative approaches to plant disease control*, ed. I. Chet. New York: John Wiley, 255–274.

———. 1995. Induced systemic resistance: An overview. In *Induced resistance to disease in plants,* ed. R. Hammerschmidt and J. Kuc. Dordrecht, The Netherlands: Kluwer, 169–175.

Kuc, J., and F. L. Caruso. 1977. Activated coordinated chemical defense against disease in plants. In *Host plant resistance to pests,* ed. P. A. Hedin. Washington, DC: American Chemical Society, 78–89.

Kuenen, D. J. 1948. The fruit tree red spider (*Metatetranychus ulmi* Koch, Tetranychidae, Acari) and its relation to its host plant. *Tijdschrift voor Entomologie* 91:83–102.

Kuenen, D. J., and A. Post. 1958. Influence of treatments on predators and other limiting factors of *Metatetranychus ulmi* (Koch). In *Proceedings of the Tenth International Congress of Entomology,* ed. E. C. Becker. Montreal, 4:611–615.

Kuhlmann, H.-W., and K. Heckmann. 1985. Interspecific morphogens regulating prey-predator relationships in protozoa. *Science* 227:1347–1349.

———. 1994. Predation risk of typical ovoid and winged morphs of *Euplotes* (Protozoa, Ciliophora). *Hydrobiologia* 284:219–227.

Kuhn, D. N., J. Chappell, A. Boudet, and K. Hahlbrock. 1984. Induction of phenylalanine ammonia lyase and 4-coumarate: CoA ligase mRNAs in cultured plant cells by UV light or fungal elicitor. *Proceedings of the National Academy of Sciences of the USA* 81:1102–1106.

Kusch, J. 1993. Behavioral and morphological changes in ciliates induced by the predator *Amoeba proteus. Oecologia* 96:354–359.

———. 1995. Adaptation of inducible defense in *Euplotes daidaleos* (Ciliophora) to predation risks by various predators. *Microbial Ecology* 30:79–88.

Kusch, J., and H.-W. Kuhlmann. 1994. Cost of *Stenostomum*-induced morphological defence in the ciliate *Euplotes octocarinatus. Archiv für Hydrobiologie* 130:257–267.

Lack, D. 1954a. Cyclic mortality. *Journal of Wildlife Management* 18:25–37.

———. 1954b. *The Natural Regulation of Animal Numbers.* Oxford: Oxford University Press.

Lampert, W. 1989. The adaptive significance of diel vertical migration of zooplankton. *Functional Ecology* 3:21–27.

Lampert, W., K. O. Rothhaupt, and E. von Elert. 1994. Chemical induction of colony formation in a green alga (*Scenedesmus acutus*) by grazers (*Daphnia*). *Limnology and Oceanography* 39:1543–1550.

Landsberg, J. 1990. Dieback of rural eucalypts: Responses of foliar dietary quality and herbivory to defoliation. *Australian Journal of Ecology* 15:89–96.

Lanza, J. 1988. Ant preference for *Passiflora* nectar mimics that contain amino acids. *Biotropica* 20:341–344.

Lapinjoki, S. P., H. A. Elo, and H. T. Taipale. 1991. Development and structure of resin glands on tissues of *Betula pendula* Roth. during growth. *New Phytologist* 117:219–223.

Larkin, J. C., D. G. Oppenheimer, A. M. Lloyd, E. T. Paparozzi, and M. D. Marks. 1994. Roles of GLABROUS and TRANSPARENT TESTA genes in *Arabidopsis* trichome development. *Plant Cell* 6:1065–1976.

Lawton, K., S. L. Potter, S. Uknes, and J. Ryals. 1994. Acquired resistance signal transduction in *Arabidopsis* is ethylene independent. *Plant Cell* 6:581–588.

Lawton, K., B. Vernooij, L. Friedrich, T. Gaffney, D. Alexander, D. Negrotto, J. P. Metraux, H. Kessmann, M. G. Rella, S. Uknes, E. Ward, and J. Ryals. 1993. Signal transduction in systemic acquired resistance. In *Plant signals in interactions with other organisms*, ed. J. C. Schultz and I. Raskin. Rockville, MD: American Society of Plant Physiologists, 11:126–133.

Leather, S. R. 1993. Early season defoliation of bird cherry influences autumn colonization by the bird cherry aphid, *Rhopalosiphum padi*. *Oikos* 66:43–47.

Leather, S. R., A. D. Watt, and G. I. Forrest. 1987. Insect-induced chemical changes in young lodgepole pine (*Pinus contorta*): The effect of previous defoliation on oviposition, growth, and survival of the pine beauty moth, *Panolis flammea*. *Ecological Entomology* 12:275–281.

Lee, H., J. Leon, and I. Raskin. 1995. Biosynthesis and metabolism of salicylic acid. *Proceedings of the National Academy of Sciences of the USA* 92:4076–4079.

Lerdau, M., M. Litvak, and R. Monson. 1994. Plant chemical defense: Monoterpenes and the growth-differentiation balance hypothesis. *Trends in Ecology and Evolution* 9:58–61.

Levins, R. 1968. *Evolution in changing environments*. Princeton, NJ: Princeton University Press.

Lewinsohn, E., M. Gijzen, and R. Croteau. 1991. Defense mechanisms of conifers: Differences in constitutive and wound-induced monoterpene biosynthesis among species. *Plant Physiology* 96:44–49.

Lewis, A. C. 1984. Plant quality and grasshopper feeding: Effects of sunflower condition on preference and performance in *Melanoplus differentialis*. *Ecology* 65:836–843.

Lidicker, W. Z. 1973. Regulation of numbers in an island population of the California vole: A problem in community dynamics. *Ecological Monographs* 43: 271–302.

Lienk, S. E., and P. J. Chapman. 1951. Influence of the presence or absence of the European red mite on two-spotted spider mite abundance. *Journal of Economic Entomology* 44:623.

Lin, H., M. Kogan, and D. Fischer. 1990. Induced resistance in soybean to the Mexican bean beetle (Coleoptera: Coccinellidae): Comparisons of inducing factors. *Environmental Entomology* 19:1852–1857.

Lindow, S. E. 1985. Integrated control and the role of antibiosis in biological control of fire blight and frost injury. In *Biological control on the phylloplane*, ed. C. E. Windels and S. E. Lindow. St. Paul: American Phytopathology Society, 83–115.

Lindroth, R. L. 1991. Differential toxicity of plant allelochemicals to insects: Roles of enzymatic detoxification systems. In *Insect-plant interactions*, ed. E. A. Bernays. Baton Rouge, Louisiana: CRC Press, 3:1–33.

Lively, C. M. 1986a. Canalization versus developmental conversion in a spatially variable environment. *American Naturalist* 128:561–572.

———. 1986b. Competition, comparative life histories, and maintenance of shell dimorphism in a barnacle. *Ecology* 67:858–864.

Lloyd, D. G. 1984. Variation strategies of plants in heterogeneous environments. *Biological Journal of the Linnean Society* 21:357–385.

Long, E. 1990. Plant "vaccine" within five years. *Grower* 114 (24): 9.

Loomis, W. E. 1953. Growth and differentiation: An introduction and summary. In *Growth and differentiation in plants*. Ames: Iowa State College Press, 1–17.

Loose, C. J., and P. Dawidowicz. 1994. Trade-offs in diel vertical migration by zooplankton: The costs of predator avoidance. *Ecology* 75:2255–2263.

Loper, G. M. 1968. Effect of aphid infestation on the coumestrol content of alfalfa varieties differing in aphid resistance. *Crop Science* 8:104–106.

Lorio, P. L. J. 1988. Growth differentiation-balance relationships in pines affect their resistance to bark beetles (Coleoptera: Scolytidae). In *Mechanisms of woody plant defenses against insects*, ed. W. J. Mattson, J. Levieux, and C. Bernard-Dagan. New York: Springer-Verlag, 73–92.

Louda, S. M., K. H. Keeler, and R. D. Holt. 1990. Herbivore influences on plant performance and competitive interactions. In *Perspectives in plant competition*, ed. J. B. Grace and D. Tilman. New York: Academic Press, 413–444.

Loughrin, J. H., D. A. Potter, and T. R. Hamilton-Kemp. 1995. Volatile compounds induced by herbivory act as aggregation kairomones for the Japanese beetle (*Popillia japonica* Newman). *Journal of Chemical Ecology* 21:1457–1467.

Lowenberg, G. J. 1997. Effects of floral herbivory, limited pollination, and intrinsic plant characteristics on phenotypic gender in *Sanicula arctopoides*. *Oecologia*. In press.

Lowman, M. D. 1982. Effects of different rates and methods of leaf area removal on rain forest seedlings of coachwood (*Ceratopetalum apetalum*). *Australian Journal of Botany* 30:477–483.

Lundberg, S., J. Jaremo, and P. Nilsson. 1994. Herbivory, inducible defence, and population oscillations: A preliminary theoretical analysis. *Oikos* 71:537–540.

McCloud, E. S., and I. T. Baldwin. n.d. Oral secretions trigger the JA cascade. Manuscript.

McCloud, E. S., D. W. Tallamy, and F. T. Halaweish. 1995. Squash beetle trenching behavior: Avoidance of cucurbitacin induction or mucilaginous plant sap? *Ecological Entomology* 20:51–59.

McDonald, R. E., P. E. Shaw, P. D. Greany, T. T. Hatton, and C. W. Wilson. 1987. Effect of gibberellic acid on certain physical and chemical properties of grapefruit. *Tropical Science* 27:17–22.

McGurl, B., G. Pearce, M. Orozca-Cardenas, and C. A. Ryan. 1992. Structure, expression, and antisense inhibition of the systemin precursor gene. *Science* 255:1570–1573.

McIntyre, J. L., J. A. Dodds, and J. D. Hare. 1981. Effects of localized infections on *Nicotiana tobacum* by tobacco mosaic virus on systemic resistance against diverse pathogens and an insect. *Phytopathology* 71:297–301.

McKenzie, J. A., J. M. Dearn, and M. J. Whitten. 1982. Genetic basis of resistance to diazinon in Victorian populations of the Australian sheep blowfly, *Lucilia cuprina*. *Australian Journal of Biological Science* 33:85–95.

McKey, D. 1974. Adaptive patterns in alkaloid physiology. *American Naturalist* 108:305–320.

———. 1979. The distribution of secondary compounds within plants. In *Herbivores: Their interaction with secondary plant metabolites*, ed. G. A. Rosenthal and D. H. Janzen. New York: Academic Press, 56–133.

McMurtry, J. A. 1970. Some factors of foliage condition limiting population growth of *Oligonychus punicae* (Acarina: Tetranychidae). *Annals of the Entomological Society of America* 63:406–412.

McNaughton, S. J. 1985. Interactive regulation of grass yield and chemical properties by defoliation, a salivary chemical, and inorganic nutrition. *Oecologia* 65:478–486.

McNaughton, S. J., and J. L. Tarrants. 1983. Grass leaf silicification: Natural selection for an inducible defense against herbivores. *Proceedings of the National Academy of Sciences of the USA* 80:790–791.

Maddox, G. D., and R. B. Root. 1987. Resistance to 16 diverse species of herbivorous insects within a population of goldenrod, *Solidago altissima:* Genetic variation and heritability. *Oecologia* 72:8–14.

Malamy, J., J. P. Carr, D. F. Klesig, and I. Raskin. 1990. Salicylic acid: A likely endogenous signal in the resistance response of tobacco to viral infection. *Science* 250:1002–1004.

Malcolm, S. B., and M. P. Zalucki. 1996. Milkweed latex and cardenolide induction may resolve the lethal plant defence paradox. *Entomologia Experimentalis et Applicata* 80:193–196.

Marpeau, A., J. Walter, J. Launay, J. Charon, and P. Baradat. 1989. Effects of wounds on the terpene content of twigs of maritime pine (*Pinus pinaster* Ait.): 2, Changes in the volatile terpene hydrocarbon composition. *Trees* 4:220–226.

Marquis, R. J. 1992. A bite is a bite is a bite? Constraints on response to folivory in *Piper arieianum* (Piperaceae). *Ecology* 73:143–152.

Martin, M. M., and J. S. Martin. 1984. Surfactants and their role in preventing the precipitation of proteins by tannins in insect guts. *Oecologia* 61:342–345.

Marutani, M., and R. Muniappan. 1991. Interactions between *Chromolaena odorata* (Asteraceae) and *Pareuchaetes pseudoinsulata* (Lepidoptera: Arctiidae). *Annals of Applied Biology* 119:227–237.

Maschinski, J., and T. G. Whitham. 1989. The continuum of plant responses to herbivory: The influence of plant association, nutrient availability, and timing. *American Naturalist* 134:1–19.

Mason, R. R. 1987. Non-outbreak species of forest Lepidoptera. In *Insect outbreaks*, ed. P. Barbosa and J. C. Schultz. San Diego: Academic Press, 31–57.

Masson, C., and H. Mustaparta. 1990. Chemical information processing in the olfactory systems of insects. *Physiological Reviews* 70:199–245.

Matson, P. A., and F. P. Hain. 1987. Host conifer defense strategies: A hypothesis. In *The role of the host in the population dynamics of insects, Proceedings, IUFRO Conference*, ed. L. Safranyik. Banff, Alberta: Forestry Service and USDA Forest Service, 33–42.

Mattiacci, L., M. Dicke, and M. A. Posthumus. 1995. β-Glucosidase: An elicitor of herbivore-induced plant odor that attracts host-searching parasitic wasps. *Proceedings of the National Academy of Sciences of the USA* 92:2036–2040.

Mattson, W. J., R. K. Lawrence, R. A. Haack, D. A. Herms, and P. Charles. 1988. Defensive strategies of woody plants against different insect-feeding guilds in relation to plant ecological strategies and intimacy of association with insects. In *Mechanisms of woody plant defenses against insects*, ed. W. J. Mattson, J. Levieux, and C. Bernard-Dagan. New York: Springer-Verlag, 3–38.

Mattson, W. J., N. Lorimer, and R. A. Leary. 1982. Role of plant variability (trait vector dynamics and diversity) in plant/herbivore interactions. In *Resistance to diseases and pests in forest trees*, ed. H. M. Heybroek, B. R. Stephan, and K. von Weissenberg. Wageningen, The Netherlands: Centre for Agricultural Publishing, 295–303.

Mattson, W. J., and S. R. Palmer. 1988. Changes in levels of foliar minerals and phenolics in trembling aspen *Populus tremuloides*, in response to artifical defoliation. In *Mechanisms of woody plant defenses against insects*, ed. W. J. Mattson, J. Levieux, and C. Bernard-Dagan. New York: Springer-Verlag, 157–169.

Mattson, W. J., G. A. Simmons, and J. A. Witter. 1988. The spruce budworm in eastern North America. In *Dynamics of forest insect populations*, ed. A. A. Berryman. New York: Plenum, 309–330.

Mauch, F., L. A. Hadwiger, and T. Boller. 1984. Ethylene: Symptom, not signal for the induction of chitinase and β-1,3-glucanase in pea pods by pathogens and elicitors. *Plant Physiology* 76:607–611.

Mauricio, R., M. D. Bowers, and F. A. Bazzaz. 1993. Pattern of leaf damage affects fitness of the annual plant *Raphanus sativus* (Brassicaceae). *Ecology* 74:2066–2071.

May, R. M. 1973. *Stability and complexity in model ecosystems*. Princeton, NJ: Princeton University Press.

———. 1981. Models for single populations. In *Theoretical ecology*, 2d edition. Sunderland, MA: Sinauer, 5–29.

Maynard Smith, J., R. Burian, S. Kauffman, P. Alberch, J. Campbell, B. Goodwin, R. Lande, D. Raup, and L. Wolpert. 1985. Developmental constraints and evolution. *Quarterly Review of Biology* 60:265–287.

Messina, F. J., T. A. Jones, and D.C. Nelson. 1993. Performance of Russian wheat aphid (Homoptera: Aphididae) on perennial range grasses: Effects of previous defoliation. *Environmental Entomology* 22:1349–1354.

Metcalf, R. L., and R. L. Lampman. 1989. The chemical ecology of Diabroticites and Cucurbitaceae. *Experientia* 45:240–247.

Metcalf, R. L., and E. R. Metcalf. 1991. *Plant kairomones in insect biology and control*. New York: Chapman and Hall.

Métraux, J. P., H. Signer, J. Ryals, E. Ward, M. Wyss-Benz, J. Gaudin, K. Raschdorf, E. Schmid, W. Blum, and B. Inverardi. 1990. Increase in salicylic acid at the onset of systemic acquired resistance in cucumber. *Science* 250:1004–1006.

Mihaliak, C. A., J. Gershenzon, and R. Croteau. 1991. Lack of rapid monoterpene turnover in rooted plants: Implications for theories of plant chemical defense. *Oecologia* 87:373–376.

Miles, D. B., and A. E. Dunham. 1993. Historical perspectives in ecology and evolutionary biology: The use of phylogenetic comparative analyses. *Annual Review of Ecology and Systematics* 24:587–619.

Milewski, A. V., T. P. Young, and D. Madden. 1991. Thorns as induced defenses: Experimental evidence. *Oecologia* 86:70–75.

Mizukami, H., Y. Tabira, and B. E. Ellis. 1993. Methyl jasmonate–induced rosmarinic acid biosynthesis in *Lithospermum erythrorhizon* cell suspension cultures. *Plant Cell Reports* 12 (12): 706–709.

Mizusaki, S., Y. Tanabe, M. Noguchi, and E. Tamaki. 1973. Activities of ornithine decarboxylase, putrescine N-methyltransferase and N-methylputrescine oxidase in tobacco roots in relation to nicotine biosynthesis. *Plant and Cell Physiology* 14:103–110.

Mole, S. 1994. Trade-offs and constraints in plant-herbivore defense theory: A life-history perspective. *Oikos* 71:3–12.

Mole, S., and P. G. Waterman. 1985. Stimulatory effects of tannins and cholic acid on tryptic hydrolysis of proteins: Ecological implications. *Journal of Chemical Ecology* 11:1323–1332.

Mook, J. H., and J. van der Toorn. 1985. Delayed response of common reed *Phragmites australis* to herbivory as a cause of cyclic fluctuations in the density of the moth *Archanara geminipuncta. Oikos* 44:142–148.

Mooney, H. A., and S. L. Gulmon. 1982. Constraints on leaf structure and function in reference to herbivory. *Bioscience* 32:198–206.

Moore, L. W., and G. Warren. 1979. *Agrobacterium radiobacter* strain 84 and biological control of crown gall. *Annual Review of Phytopathology* 17:163–179.

Moran, N. A., and W. D. Hamilton. 1980. Low nutritive quality as defense against herbivores. *Journal of Theoretical Biology* 86:247–254.

Morse, S., S. D. Wratten, P. J. Edwards, and H. M. Niemeyer. 1991. The effect of maize leaf damage on the survival and growth rate of *Rhopalosiphum padi. Annals of Applied Biology* 119:251–256.

Mueller, M. J., and W. Brodschelm. 1994. Quantification of jasmonic acid by capillary gas chromatography-negative chemical ionization mass spectrometry. *Analytical Biochemistry* 218:425–435.

Muller, K. O. 1959. Hypersensitivity. In *Plant pathology: An advanced treatise,* ed. J. G. Horsfall and A. E. Dimond. New York: Academic Press, 1:469–519.

Murdoch, W. W. 1966. Community structure, population control, and competition: A critique. *American Naturalist* 100:219–226.

Mutikainen, P., and M. Walls. 1995. Growth, reproduction, and defence in nettles: Responses to herbivory modified by competition and fertilization. *Oecologia* 104:487–495.

Myers, J. H. 1981. Interactions between western tent caterpillars and wild rose: A test of some general plant herbivore hypotheses. *American Naturalist* 50:11–25.

———. 1988. The induced defense hypothesis: Does it apply to the population dynamics of insects? In *Chemical mediation of coevolution,* ed. K. C. Spencer. San Diego: Academic Press, 345–365.

———. 1990. Population cycles of western tent caterpillars: Experimental introductions and synchrony of fluctuations. *Ecology* 71:986–995.

Myers, J. H., and D. Bazely. 1991. Thorns, spines, prickles, and hairs: Are they stimulated by herbivory and do they deter herbivores? In *Phytochemical induction by herbivores,* ed. D. W. Tallamy and M. J. Raupp. New York: John Wiley, 325–344.

Myers, J. H., and K. S. Williams. 1984. Does tent caterpillar attack reduce the food quality of red alder foliage? *Oecologia* 62:74–79.

———. 1987. Lack of short or long term inducible defenses in the red alder–western tent caterpillar system. *Oikos* 48:73–78.

Naaranlahti, T., S. Auriola, and S. Lapinjoki. 1991. Growth-related dimerization of vindoline and cantharanthine in *Cantharanthus roseus* and effect of wounding on the process. *Phytochemistry* 30:1451–1453.

Narvaez-Vasquez, J., M. Orozco-Cardenas, and C. A. Ryan. 1994. A sulfhydryl reagent modulates systemic signaling for wound-induced and systemin-induced proteinase inhibitor synthesis. *Plant Physiology* 105:725–730.

National Academy of Sciences. 1986. *Pesticide resistance: Strategies and tactics for management.* Washington, DC: National Academy Press.

Nef, L. 1988. Interactions between the leaf miner, *Phyllocnistis suffusella,* and poplars. In *Mechanisms of woody plant defenses against insects,* ed. W. J. Mattson, J. Levieux, and C. Bernard-Dagan. New York: Springer-Verlag, 239–251.

Nelson, R. S., P. Powell Abel, and R. N. Beachy. 1987. Lesions and virus accumulation in inoculated transgenic tobacco plants expressing the coat protein of tobacco mosaic virus. *Virology* 158:126–132.

Neuvonen, S., S. Hanhimaki, J. Suomela, and E. Haukioja. 1988. Early season damage to birch foliage affects the performance of a late season herbivore. *Journal of Applied Entomology* 105:182–189.

Neuvonen, S., and E. Haukioja. 1984. Low nutritive quality as defence against herbivores: Induced responses in birch. *Oecologia* 63:71–74.

———. 1985. How to study induced plant resistance? *Oecologia* 66:456–457.

———. 1991. The effects of inducible resistance in host foliage on birch-feeding herbivores. In *Phytochemical induction by herbivores,* ed. D. W. Tallamy and M. J. Raupp. New York: John Wiley, 277–291.

Neuvonen, S., E. Haukioja, and A. Molarius. 1987. Delayed inducible resistance against a leaf-chewing insect in four deciduous tree species. *Oecologia* 74: 363–369.

Nitao, J. K., and A. R. Zangerl. 1987. Floral development and chemical defense allocation in wild parsnip (*Pastinaca sativa*). *Ecology* 68:521–529.

Nobel, P. S. 1978. Surface temperatures of cacti: Influences of environmental and morphological factors. *Ecology* 59:986–996.

Nojiri, H., H. Yamane, H. Seto, I. Yamaguchi, N. Murofushi, T. Yoshihara, and H. Shibaoka. 1992. Qualitative and quantitative analysis of endogenous jasmonic acid in bulbing and non-bulbing onion plants. *Plant Cell Physiology* 33:1225–1231.

Nooden, L. D., and A. C. Leopold. 1988. *Senescence and aging in plants.* New York: Academic Press.

Norris, D. M., and M. Kogan. 1980. Biochemical and morphological bases of resistance. In *Breeding plants resistant to insects,* ed. F. G. Maxwell and P. R. Jennings. New York: John Wiley, 23–61.

Oaks, A. 1992. A re-evaluation of nitrogen assimilation in roots. *Bioscience* 42:103–111.

Oaks, A., and B. Hirel. 1985. Nitrogen metabolism in roots. *Annual Review of Plant Physiology* 36:345–365.

Ohgushi, T. 1992. Resource limitation on insect herbivore populations. In M. D. Hunter, T. Ohgushi, and P. W. Price, *Effects of resource distribution on animal-plant interactions.* San Diego: Academic Press, 199–241.

Ohnmeiss, T. E., and I. T. Baldwin. 1994. The allometry of nitrogen allocation to growth and an inducible defense under nitrogen-limited growth. *Ecology* 75:995–1002.

Ohnmeiss, T. E., E. S. McCloud, G. Y. Lynds, and I. T. Baldwin. n.d. Within-plant relationships among wounding, jasmonic acid, and nicotine: Implications for defense in *Nicotiana sylvestris*. Manuscript.

Olson, M. M., and C. R. Roseland. 1991. Induction of the coumarins scopoletin and ayapin in sunflower by insect-feeding stress and effects of coumarins on the feeding of sunflower beetle (Coleoptera: Chrysomelidae). *Environmental Entomology* 20:1166–1172.

O'Neill, M. J., S. A. Adesanya, M. F. Roberts, and I. R. Pantry. 1986. Inducible isoflavanoids from the lima bean, *Phaseolus lunatus*. *Phytochemistry* 25:1315–1322.

Orozco-Cardenas, M., B. McGurl, and C. A. Ryan. 1993. Expression of an antisense prosystemin gene in tomato plants reduces resistance toward *Manduca sexta* larvae. *Proceedings of the National Academy of Sciences of the USA* 90:8273–8276.

Ourry, A., J. Boucaud, and J. Salette. 1988. Nitrogen mobilization from stubble and roots during re-growth of defoliated perennial ryegrass. *Journal of Experimental Botany* 39:803–809.

Ourry, A., B. Gonzalez, and J. Boucaud. 1989. Osmoregulation and role of nitrate during regrowth after cutting of ryegrass (*Lolium perenne*). *Physiologia Plantarum* 76:177–182.

Overhulser, D., R. I. Gara, and R. Johnsey. 1972. Emergence of *Pissodes strobi* (Coleoptera: Curculionidae) from previously attacked Sitka spruce. *Annals of the Entomological Society of America* 65:1423–1424.

Padilla, D. K., and S. C. Adolph. 1996. Plastic inducible morphologies are not always adaptive: The importance of time delays in a stochastic environment. *Evolutionary Ecology* 10:105–117.

Paige, K. N., and T. G. Whitham. 1987. Overcompensation in response to mammalian herbivory: The advantage of being eaten. *American Naturalist* 129:407–416.

Painter, R. H. 1958. Resistance of plants to insects. *Annual Review of Entomology* 3:267–290.

Palaniswamy, P., and R. J. Lamb. 1993. Wound induced antixenotic resistance to flea beetles in crucifers. *Canadian Entomologist* 125:903–912.

Papavizas, G. C. 1985. *Trichomderma* and *Glioclaium*: Biology, ecology, and potential for biological control. *Annual Review of Phytopathology* 23:23–54.

Parker, M. A. 1992. Constraints on the evolution of resistance to pests and pathogens. In *Pest and pathogens: Plant responses to foliar attack*, ed. P. G. Ayres. Oxford: Bios, 181–197.

Parr, J. C., and R. Thurston. 1972. Toxicity of nicotine in synthetic diets to larvae of the tobacco hornworm. *Annals of the Entomological Society of America* 65:1185–1188.

Paul, V. J., and K. L. van Alstyne. 1992. Activation of chemical defenses in the tropical green algae *Halimeda* spp. *Journal of Experimental Marine Biology and Ecology* 160:191–203.

Pearce, G., C. A. Ryan, and D. Liljegren. 1988. Proteinase inhibitors I and II in fruit of wild tomato species: Transient components of a mechanism for defense and seed dispersal. *Planta* 175:527–531.

Pearce, G., D. Strydom, S. Johnson, and C. A. Ryan. 1991. A polypeptide from tomato leaves induces wound-inducible proteinase inhibitor proteins. *Science* 253:895–898.

Pellmyr, O., W. Tang, I. Groth, G. Bergstrom, and L. B. Thien. 1991. Cycad cones and angiosperm floral volatiles: Inferences for the evolution of insect pollination. *Biochemical Systematics and Ecology* 19:623–627.

Pellmyr, O., and L. B. Thien. 1986. Insect reproduction and floral fragrances: Keys to the evolution of the angiosperms? *Taxon* 35:76–85.

Peña-Cortés, H., T. Albrecht, S. Prat, E. W. Weiler, and L. Willmitzer. 1993. Aspirin prevents wound-induced gene expression in tomato by blocking jasmonic acid biosynthesis. *Planta* 19:123–128.

Peña-Cortés, H., J. J. Sánchez-Serrano, R. Mertens, L. Willmitzer, and S. Prat. 1989. Abscisic acid is involved in the wound-induced expression of the proteinase inhibitor II gene in potato and tomato. *Proceedings of the National Academy of Sciences of the USA* 86:9851–9855.

Peña-Cortés, H., L. Willmitzer, and J. J. Sánchez-Serrano. 1991. Abscisic acid mediates wound induction but not developmental-specific expression of the proteinase inhibitor II gene family. *Plant Cell* 3:963–972.

Pickard, B. G. 1973. Action potentials in higher plants. *Botanical Review* 39: 172–201.

Poethig, R. S. 1990. Phase change and the regulation of shoot morphogenesis in plants. *Science* 250:923–930.

Popay, A. J., and D. D. Rowan. 1994. Endophytic fungi as mediators of plant-insect interactions. In *Insect-plant interactions,* ed. E. A. Bernays. Boca Raton, FL: CRC Press, 5:83–103.

Porat, R., A. Borochov, and A. H. Halevy. 1993. Enhancement of petunia and dendrobium flower senescence by jasmonic acid methyl ester is via the promotion of ethylene production. *Plant Growth Regulation* 13:297–301.

Potter, D. A., and C. T. Redmond. 1989. Early spring defoliation, secondary leaf flush, and leafminer outbreaks on American holly. *Oecologia* 81:192–197.

Powell Abel, P., R. S. Nelson, B. De, N. Hoffman, S. G. Rogers, R. T. Fraley, and R. N. Beachy. 1986. Delay of disease development in transgenic plants that express the tobacco mosaic virus coat protein gene. *Science* 232:738–743.

Preszler, R. W., and P. W. Price. 1993. The influence of *Salix* leaf abscission on leaf-miner survival and life history. *Ecological Entomology* 18:150–154.

Price, J. F. 1981. Response of two-spotted spider mites, leaf miners, and their parasitoids and other arthropods to newly developed pesticides in chrysanthemums. *Proceedings of the Florida State Horticultural Society* 94:80–83.

Price, P. W. 1980. *Evolutionary biology of parasites.* Princeton, NJ: Princeton University Press.

———. 1984. *Insect ecology.* 2d edition. New York: Wiley.

———. 1986. Ecological aspects of host plant resistance and biological control: Interactions among three trophic levels. In *Interactions of plant resistance and parasitoids and predators of insects,* ed. D. J. Boethel and R. D. Eikenbarry. Chichester, England: Ellis Harwood, 11–30.

Price, P. W., C. E. Bouton, P. Gross, B. A. McPheron, J. N. Thompson, and A. E. Weis. 1980. Interactions among three trophic levels: Influence of plants on interactions between insect herbivores and natural enemies. *Annual Review of Ecology and Systematics* 11:41–65.

Prins, A. H., B. Verboom, and J. Verboom. 1987. On the relationship between *Ethmia bipunctella* and its host plant *Cynoglossum officinale* L. *Mededelingen van de Faculteit Landbouwwetenschappen Rijksuniversiteit Gent* 52:1335–1341.

Prins, A. H., H. J. Verkaar, and M. van den Herik. 1989. Responses of *Cynoglossum officinale* L. and *Senecio jacobaea* L. to various degrees of defoliation. *New Phytologist* 111:725–731.

Provenza, F. D. 1995. Postingestive feedback as an elementary determinant of food preference and intake in ruminants. *Journal of Range Management* 48:2–17.

Pullin, A. S. 1987. Changes in leaf quality following clipping and regrowth of *Urtica dioica,* and consequences for a specialist insect herbivore, *Aglais urticae. Oikos* 49:39–45.

Pullin, A. S., and J. E. Gilbert. 1989. The stinging nettle, *Urtica dioica,* increases trichome density after herbivore and mechanical damage. *Oikos* 54:275–280.

Qin, B., X. Zhang, G. Wu, and P. Tien. 1992. Plant resistance to fungal diseases induced by the infection of cucumber mosaic virus attenuated by satellite RNA. *Annals of Applied Biology* 120:361–366.

Raffa, K. F. 1991. Induced defensive reactions in conifer–bark beetle systems. In *Phytochemical induction by herbivores,* ed. D. W. Tallamy and M. J. Raupp. New York: John Wiley, 245–276.

Raffa, K. F., and A. A. Berryman. 1982a. Accumulation of monoterpenes and associated volatiles following inoculation of grand fir with a fungus transmitted by the fir engraver, *Scolytus ventralis* (Coleoptera: Scolytidae). *Canadian Entomologist* 114:797–810.

———. 1982b. Physiological differences between lodgepole pines resistant and susceptible to the mountain pine beetle and associated microogranisms. *Environmental Entomology* 11:486–492.

———. 1987. Interacting selective pressures in conifer–bark beetle systems: A basis for reciprocal adaptations. *American Naturalist* 129:234–262.

Raskin, I. 1992. Role of salicylic acid in plants. *Annual Review of Plant Physiology and Plant Molecular Biology* 43:439–463.

Rasmussen, J. B., R. Hammerschmidt, and M. N. Zook. 1991. Systemic induction of salicylic acid accumulation in cucumber after inoculation with *Pseudomonas syringae* pv. *syringae. Plant Physiology* 97:1342–1347.

Rast, A. T. B. 1972. MII-16, an artificial symptomless mutant of tobacco mosaic virus for seedling inoculation of tomato crops. *Netherlands Journal of Plant Pathology* 78:110–112.

Raupp, M. J. 1985. Effects of leaf toughness on the mandibular wear of the leaf beatle, *Plagiodera versicolora*. *Ecological Entomology* 10:73–79.

Raupp, M. J., and R. F. Denno. 1984. The suitability of damaged willow leaves as food for the leaf beetle, *Plagiodera versicolora*. *Ecological Entomology* 9:443–448.

Raupp, M. J., and C. S. Sadof. 1991. Responses of leaf beetles to injury-related changes in their salicaceous hosts. In *Phytochemical induction by herbivores*, ed. D. W. Tallamy and M. J. Raupp. New York: John Wiley, 183–204.

Rausher, M. D. 1978. Search image for leaf shape in a butterfly. *Science* 200:1071–1073.

———. 1992. Natural selection and the evolution of plant-insect interactions. In *Insect chemical ecology*, ed. B. D. Roitberg and M. B. Isman. New York: Chapman and Hall, 20–88.

Rausher, M. D., K. Iwao, E. L. Simms, N. Ohsaki, and D. Hill. 1993. Induced resistance in *Ipomoea purpurea*. *Ecology* 74:20–29.

Ray, J. 1901. Les maladies cryptogamiques des végétaux. *Revue Générale de Botanique* 13:145–151.

Reeve, H. K., and P. W. Sherman. 1993. Adaptation and the goals of evolutionary research. *Quarterly Review of Biology* 68:1–32.

Reinbothe, S., B. Mollenhauer, and C. Reinbothe. 1994. JIPs and RIPs: The regulation of plant gene expression by jasmonates in response to environmental cues and pathogens. *Plant Cell* 6:1197–1209.

Reymond, P., S. Grunberger, K. Paul, M. Muller, and E. E. Farmer. 1995. Oligogalacturonide defense signals in plants: Large fragments interact with the plasma membrane in vitro. *Proceedings of the National Academy of Sciences of the USA* 92:4145–4149.

Reynolds, G. W., and M. C. Smith. 1985. Effects of leaf position, leaf wounding, and plant age of two soybean genotypes on soybean looper (Lepidoptera: Noctuidae) growth. *Environmental Entomology* 14:475–478.

Rhoades, D. F. 1979. Evolution of plant chemical defense against herbivores. In *Herbivores: Their interaction with secondary plant metabolites*, ed. G. A. Rosenthal and D. H. Janzen. New York: Academic Press, 3–54.

———. 1983. Responses of alder and willow to attack by tent caterpillars and webworms: Evidence for pheromonal sensitivity of willows. In *Plant resistance to insects*, Symposium Series 208, ed. P. A. Hedin. Washington, DC: American Chemical Society, 55–68.

———. 1985a. Offensive-defensive interactions between herbivores and plants: Their relevance in herbivore population dynamics and ecological theory. *American Naturalist* 125:205–238.

———. 1985b. Pheromonal communication between plants. In *Chemically mediated interactions between plants and other organisms*, ed. G. A. Cooper-Driver, T. Swain, and E. E. Conn. New York: Plenum, 195–218.

Rhoades, D. F., and R. G. Cates. 1976. Toward a general theory of plant antiherbivore chemistry. *Recent Advances in Phytochemistry* 10:168–213.

Rhodes, M. J. C. 1994. Physiological roles for secondary metabolites in plants: Some progress, many outstanding problems. *Plant Molecular Biology* 24:1–20.

Ricard, J. L. 1981. Commercialization of a *Trichoderma* based mycofungicide: Some problems and solutions. *Biocontrol News and Information* 2:95–98.

Ridley, M. 1983. *The explanation of organic diversity.* Oxford: Oxford University Press.

Riessen, H. P. 1992. Cost-benefit model for the induction of an antipredator defense. *American Naturalist* 140:349–362.

Rigby, N. M., A. J. MacDougall, P. W. Needs, and R. R. Selvendran. 1994. Phloem translocation of a reduced oligogalacturonide in *Ricinus communis* L. *Planta* 193:536–541.

Risch, S. J. 1985. Effects of induced chemical changes on interpretation of feeding preference tests. *Entomologia Experimentalis et Applicata* 39:81–84.

Rishbeth, J. 1975. Stump inoculation: A biological control of *Fomes annosus:* 3, Inoculation with *Peniophora gigantea. Annals of Applied Biology* 52:63–77.

———. 1988. Biological control of air-borne pathogens. *Philosophical Transactions of the Royal Society of London B* 318:265–281.

Roberts, K. 1992. Potential awareness of plants. *Nature* 360:14–15.

Rockwood, L. L. 1974. Seasonal changes in the susceptibility of *Crescentia alata* leaves to the flea beetle, *Oedionychus* sp. *Ecology* 55:142–148.

Rohfritsch, O. 1981. A defense mechanism of *Picea excelsa* L. against the gall former *Chermes abietis* L. (Homoptera, Adelgidae). *Zeitschrift für Angewandte Entomologie* 92:18–26.

Roland, J., and J. H. Myers. 1987. Improved insect performance from host-plant defoliation: Winter moth on oak and apple. *Ecological Entomology* 12:409–414.

Root, R. B. 1973. Organization of a plant-arthropod association in simple and diverse habitats: The fauna of collards (*Brassica oleracea*). *Ecological Monographs* 43:95–124.

Rosenthal, G. A., M. A. Berge, A. J. Ozinskas, and C. G. Hughes. 1988. Ability of L-canavanine to support nitrogen metabolism in the jack bean, *Canavalia ensiformis* (L.) DC. *Journal of Agricultural and Food Chemistry* 36:1159–1163.

Rosenthal, J. P., and P. M. Kotanen. 1994. Terrestrial plant tolerance to herbivory. *Trends in Ecology and Evolution* 9:145–148.

Rossiter, M., J. C. Schultz, and I. T. Baldwin. 1988. Relationships among defoliation, red oak phenolics, and gypsy moth growth and reproduction. *Ecology* 69:267–277.

Roth, L. M., and T. Eisner. 1962. Chemical defenses of arthropods. *Annual Review of Entomology* 7:107–136.

Rottger, V. U., and F. Klinghauf. 1976. Änderung im stoffwechsel von zuckerrübenblättern durch befall mit *Pegomya betae* Curt. (Muscidae: Anthomyidae). *Zeitschrift für Angewandte Entomologie* 82:220–227.

Ruess, R. W. 1988. The interaction of defoliation and nutrient uptake in *Sporobolus kertrophyllus,* a short-grass species from the Serengeti plains. *Oecologia* 77:550–556.

Ruohomäki, K., S. Hanhimäki, E. Haukioja, L. Iso-Iivari, S. Neuvonen, P. Niemelä, and J. Suomela. 1992. Variability in the efficacy of delayed inducible resistance in mountain birch. *Entomologia Experimentalis et Applicata* 62:107–115.

Ryals, J., E. Ward, P. Ahl-Goy, and J. P. Métraux. 1992. Systemic acquired resistance: An inducible defence mechanism in plants. In *Inducible plant proteins,* ed. J. L. Wray. Cambridge: Cambridge University Press, 205–229.

Ryan, C. A. 1983. Insect-induced chemical signals regulating natural plant protection responses. In *Variable plants and herbivores in natural and managed systems,* ed. R. F. Denno and M. S. McClure. New York: Academic Press, 43–60.

———. 1987. Oligosaccharide signalling in plants. *Annual Review of Cell Biology* 3:295–317.

———. 1989. Proteinase inhibitor gene families: Strategies for transformation to improve plant defenses against herbivores. *Bioessays* 10:20–24.

———. 1990. Protease inhibitors in plants: Genes for improving defenses against insects and pathogens. *Annual Review of Phytopathology* 28:425–429.

———. 1992. The search for the proteinase inhibitor–inducing factor, PIIF. *Plant Molecular Biology* 19:123–133.

Saha, D.C., J. M. Johnson-Cicalese, P. M. Halisky, M. I. van Heemstra, and C. R. Funk. 1987. Occurrence and significance of endophyte fungi in the fine fescue. *Plant Disease* 71:1021–1024.

Sahashi, N., H. Tsuji, and J. Shishiyama. 1989. Barley plants grown under germ-free conditions have increased susceptibility to two powdery mildew fungi. *Physiological and Molecular Plant Pathology* 34:163–170.

Sánchez-Serrano, J., M. Keil, A. O'Connor, J. Schell, and L. Willmitzer. 1987. Wound-induced expression of a potato proteinase inhibitor II gene in transgenic tobacco plants. *EMBO Journal* 6:303–306.

Sano, H., and Y. Ohashi. 1995. Involvement of small GTP-binding proteins in defense signal-transduction pathways of higher plants. *Proceedings of the National Academy of Sciences of the USA* 92:4138–4144.

Saunders, J. A. 1979. Investigations of vacuoles isolated from tobacco. *Plant Physiology* 64:74–78.

Saunders, J. W., and L. P. Bush. 1979. Nicotine biosynthetic enzyme activities in *Nicotiana tabacum* genotypes with different alkaloids levels. *Plant Physiology* 64:236–240.

Schaller, A., D. R. Bergey, and C. A. Ryan. 1995. Induction of wound response genes in tomato leaves by Bestatin, an inhibitor of aminopeptidases. *Plant Cell* 7:1893–1898.

Schlicting, C. D. 1986. Evolution of phenotypic plasticity in plants. *Annual Review of Ecology and Systematics* 17:667–693.

Schmalhausen, I. I. 1949. *Factors of evolution.* Philadelphia: Blakiston Press.

Schoonhoven, L. M., and J. Meerman. 1978. Metabolic cost of changes in diet and neutralization of allelochemics. *Entomologia Experimentalis et Applicata* 24:489–493.

Schrag, S. J., and V. Perrot. 1996. Reducing antibiotic resistance. *Nature* 381: 120–121.

Schultz, D. E., and D.C. Allen. 1977. Characteristics of sites with high black cherry mortality due to bark beetles following defoliation by *Hydria punivorata. Environmental Entomology* 6:77–81.

Schultz, J. C. 1983. Habitat selection and foraging tactics of caterpillars in heterogeneous environments. In *Variable plants and herbivores in natural and managed systems*, ed. R. F. Denno and M. S. McClure. New York: Academic Press, 61–90.

———. 1988. Plant responses induced by herbivores. *Trends in Ecology and Evolution* 3:45–49.

Schultz, J. C., and I. T. Baldwin. 1982. Oak leaf quality declines in response to defoliation by gypsy moth larvae. *Science* 217:149–151.

Schultz, J. C., and S. T. Keating. 1991. Host-plant-mediated interactions between the gypsy moth and a baculovirus. In *Microbial mediation of plant-herbivore interactions*, ed. P. Barbosa, V. A. Krischik, and C. G. Jones. New York: John Wiley, 489–506.

Scriber, J. M. 1984. Host plant suitability. In *Chemical ecology of insects*, ed. W. J. Bell and R. T. Carde. New York: Chapman and Hall, 159–204.

Scutareanu, P., B. Drukker, J. Bruin, M. A. Posthumus, and M. W. Sabelis. 1996. Leaf volatiles and polyphenols in pear trees infested by *Psylla pyricola*: Evidence for simultaneously induced responses. *Chemoecology* 7:34–38.

Seigler, D. S. 1977. Primary roles for secondary compounds. *Biochemical Systematics and Ecology* 5:195–199.

Seigler, D. S., and P. W. Price. 1976. Secondary compounds in plants: Primary functions. *American Naturalist* 110:101–105.

Seldal, T. 1994. Proteinase inhibitors in plants and fluctuating populations of herbivores. Ph.D. dissertation, University of Bergen, Norway.

Seldal, T., K.-J. Andersen, and G. Hogstedt. 1994. Grazing-induced proteinase inhibitors: A possible cause for lemming population cycles. *Oikos* 70:3–11.

Seldal, T., E. Dybwad, K.-J. Andersen, and G. Hogstedt. 1994. Wound-induced proteinase inhibitors in grey alder (*Alnus incana*): A defense mechanism against attacking insects. *Oikos* 71:239–245.

Sembdner, G., and B. Parthier. 1993. The biochemistry and the physiological and molecular actions of jasmonates. *Annual Review of Plant Physiology and Plant Molecular Biology* 44:569–589.

Senn, J., and E. Haukioja. 1994. Reactions of mountain birch to bud removal: Effects of severity and timing, and implications for herbivores. *Functional Ecology* 8:494–501.

Shain, L., and W. E. Hillis. 1972. Ethylene production in *Pinus radiata* in response to *Sirex-Amylostereum* attack. *Phytopathology* 62:1407–1409.

Shanks, C. H. Jr., and R. P. Doss. 1989. Population fluctuations of two-spotted spider mite (Acari: Tetranychidae) on strawberry. *Environmental Entomology* 18:641–645.

Shapiro, A. M., and J. E. DeVay. 1987. Hypersensitivity reaction of *Brassica nigra* L. (Cruciferae) kills eggs of *Pieris* butterflies (Lepidoptera: Pieridae). *Oecologia* 71:631–632.

Sherman, P. W. 1988. The levels of analysis. *Animal Behavior* 36:616–619.

Siedow, J. N. 1991. Plant lipoxygenase: Structure and function. *Annual Review of Plant Physiology and Plant Molecular Biology* 42:145–188.

Siegel, M. R., M. L. Dalhlman, and L. P. Bush. 1989. The role of endophytic fungi in grasses: New approaches to biological control of pests. In *Integrated pest management for turfgrass and ornamentals*, ed. A. R. Leslie and R. L. Metcalf. Washington, DC: Environmental Protection Agency, 169–186.

Sijmons, P. C., H. J. Atkinson, and U. Wyss. 1994. Parasitic strategies of root nematodes and associated host cell responses. *Annual Review of Phytopathology* 32:235–259.

Silkstone, B. E. 1987. The consequence of leaf damage for subsequent insect grazing on birch (*Betula* spp.): A field experiment. *Oecologia* 74:149–152.

Sillen-Tullberg, B. 1988. Evolution of gregariousness in aposematic butterfly larvae: A phylogenetic analysis. *Evolution* 42:293–305.

———. 1993. The effect of biased inclusion of taxa on the correlation between discrete characters in phylogenetic trees. *Evolution* 47:1182–1191.

Simberloff, D., and P. Stiling. 1987. Larval dispersion and survivorship in a leafmining moth. *Ecology* 68:1647–1657.

Simms, E. L. 1992. Costs of plant resistance to herbivores. In R. S. Fritz and E. L. Simms, *Plant Resistance to Herbivores and Pathogens: Ecology, Evolution, and Genetics*. Chicago: University of Chicago Press, 392–425.

Simms, E. L., and M. D. Rausher. 1987. Costs and benefits of plant resistance to herbivory. *American Naturalist* 130:570–581.

———. 1989. The evolution of resistance to herbivory in *Ipomoea purpurea*: 2, Natural selection by insects and the cost of resistence. *Evolution* 43:573–585.

Sinha, A. K. 1984. A new concept in plant disease control. *Science and Culture* 50:181–186.

Sinha, A. K., and G. N. Hait. 1982. Host sensitization as a factor in induction of resistance in rice against *Drechslera* by seed treatment with phytoalexin inducers. *Transactions of the British Mycological Society* 79:213–219.

Slansky, F., and P. Feeny. 1977. Stabilization of nitrogen accumulation by larvae of the cabbage butterfly on wild and cultivated food plants. *Ecological Monographs* 47:209–228.

Slansky, F., and J. G. Rodrigues. 1987. *Nutritional ecology of insects, mites, spiders, and related invertebrates*. New York: John Wiley.

Slansky, F., and G. S. Wheeler. 1992. Caterpillars' compensatory feeding response to diluted nutrients leads to toxic allelochemical dose. *Entomologia Experimentalis et Applicata* 65:171–186.

Sluss, R. R. 1967. Population dynamics of the walnut aphid, *Chromaphis juglandicola* (Kalt.) in northern California. *Ecology* 48:41–58.

Smedegaard-Petersen, V., and O. Stolen. 1981. Effects of energy-requiring defense reactions on yield and grain quality in a powdery mildew–resistant barley cultivar. *Phytopathology* 71:396–399.

Smith, L. L., J. Lanza, and G. C. Smith. 1990. Amino acid concentrations in extrafloral nectar of *Impatiens sultani* increase after simulated herbivory. *Ecology* 71:107–115.

Smith, N. G. 1982. Population irruptions and periodic migrations in the dayflying moth *Urania fulgens*. In *The ecology of a tropical forest*, ed. E. G. Leigh, A. S. Rand, and D. M. Windsor. Washington, DC: Smithsonian Institution Press, 331–344.

―――. 1983. Host plant toxicity and migration in the day-flying moth *Urania*. *Florida Entomologist* 66:76–85.

Song, W. C., C. D. Funk, and A. R. Brash. 1993. Molecular cloning of an allene ocide synthase: A cytochrome P450 specialized for the metabolism of fatty acid hydroperoxides. *Proceedings of the National Academy of Sciences of the USA* 90:8519–8523.

Spitze, K. 1992. Predator-mediated plasticity of prey life history and morphology: *Chaoborus americanus* predation on *Daphnia pulex*. *American Naturalist* 139: 229–247.

Sprugel, D. G., T. M. Hinckley, and W. Schaap. 1991. The theory and practice of branch autonomy. *Annual Review of Ecology and Systematics* 22:309–334.

Staswick, P. E. 1992. Jasmonate, genes, and fragrant signals. *Plant Physiology* 99:804–807.

―――. 1994. Storage proteins of vegetative plant tissues. *Annual Review of Plant Physiology and Plant Molecular Biology* 45:303–322.

Staswick, P. E., W. P. Su, and S. H. Howell. 1992. Methyl jasmonate inhibition of root growth and induction of a leaf protein are decreased in an *Arabidopsis thaliana* mutant. *Proceedings of the National Academy of Sciences of the USA* 89:6837–6840.

Steinberg, S., M. Dicke, and L. E. Vet. 1993. Relative importance of infochemicals from first and second trophic level in long-range host location by the larval parasitoid *Cotesia glomerata*. *Journal of Chemical Ecology* 19:47–59.

Stemberger, R. S. 1988. Reproductive costs and hydrodynamic benefits of chemically induced defenses in *Keratella testudo*. *Limnology and Oceanography* 33:593–606.

Stenseth, N. C. 1995. Showshoe hare populations: Squeezed from below and above. *Science* 269:1061–1062.

Stenseth, N. C., and R. A. Ims. 1993. The biology of lemmings: A conclusion with a look to future challenges. In *The biology of lemmings*, ed. N. C. Stenseth and R. A. Ims. London: Academic Press, 521–531.

Stephenson, A. G. 1982. The role of the extrafloral nectaries of *Catalpa speciosa* in limiting herbivory and increasing fruit production. *Ecology* 63:663–669.

Stephenson, A. G., and R. I. Bertin. 1983. Male competition, female choice, and sexual selection in plants. In *Pollination biology*, ed. L. Real. Orlando, FL: Academic Press, 109–149.

Stiling, P., and D. Simberloff. 1989. Leaf abscission: Induced defense against pests or response to damage? *Oikos* 55:43–49.

Stock, W. D., D. Le Roux, and F. Van der Heyden. 1993. Regrowth and tannin production in woody and succulent karoo shrubs in response to simulated browsing. *Oecologia* 96:562–568.

Stockhoff, B. A. 1993a. Diet heterogeneity: Implications for growth of a generalist herbivore, the gypsy moth. *Ecology* 74:1939–1949.

―――. 1993b. Protein intake by gypsy moth larvae on homogeneous and heterogeneous diets. *Physiological Entomology* 18:409–419.

―――. 1994. Maximization of daily canopy photosynthesis: Effects of herbivory on optimal nitrogen distribution. *Journal of Theoretical Biology* 169:209–220.

Stout, M. J., and S. S. Duffey. 1996. Characterization of induced resistance in tomato plants. *Entomologia Experimentalis et Applicata* 79:273–283.

Stout, M. J., J. Workman, and S. S. Duffey. 1994. Differential induction of tomato foliar proteins by arthropod herbivores. *Journal of Chemical Ecology* 20:2575–2594.

———. 1996. Identity, spatial distribution, and variability of induced chemical responses in tomato plants. *Entomologia Experimentalis et Applicata* 79:255–271.

Strauss, S. Y. 1991. Direct, indirect, and cumulative effects of three native herbivores on a shared host plant. *Ecology* 72:543–558.

Strong, D. R., J. H. Lawton, and R. Southwood. 1984. *Insects on plants: Community patterns and mechanisms.* Cambridge: Harvard University Press.

Strong, F. E., and E. Kruitwagon. 1967. Traumatic acid: An accelerator of abcision in cotton explants. *Nature* 215:1380–1381.

Sultan, S. E. 1987. Evolutionary implications of phenotypic plasticity in plants. *Evolutionary Biology* 21:127–178.

———. 1992. Phenotypic plasticity and the neo-Darwinian legacy. *Evolutionary Trends in Plants* 6:61–71.

Suomela, J., and M. P. Ayres. 1994. Within-tree and among-tree variation in leaf characteristics of mountain birch and its implications for herbivory. *Oikos* 70:212–222.

Ta, T. C., F. D. H. Macdowall, and M. A. Faris. 1990. Utilization of carbon and nitrogen reserves of alfalfa roots in supporting N_2-fixation and shoot regrowth. *Plant and Soil* 127:231–236.

Tabashnik, B. E., N. L. Cushing, N. Finson, and M. W. Johnson. 1990. Field development of resistance to *Bacillus thuringiensis* in diamondback moth (Lepidoptera: Plutellidae). *Journal of Economic Entomology* 83:1671–1676.

Takabayashi, J., S. Takahashi, M. Dicke, and M. A. Posthumus. 1995. Developmental stage of the herbivore *Pseudaletia seperata* affects the production of herbivore-induced synomone by corn plants. *Journal of Chemical Ecology* 21:273–287.

Tallamy, D. W. 1985. Squash beetle feeding behavior: An adaptation against induced cucurbit defenses. *Ecology* 66:1574–1579.

Tallamy, D. W., and V. A. Krischik. 1989. Variation and function of cucurbitacins in *Cucurbita:* An examination of current hypotheses. *American Naturalist* 133:766–786.

Tallamy, D. W., and E. S. McCloud. 1991. Squash beetles, cucumber beetles, and inducible cucurbit responses. In *Phytochemical induction by herbivores,* ed. D. W. Tallamy and M. J. Raupp. New York: John Wiley, 155–181.

Tamarin, R. H. 1978. A defense of single-factor models of population regulation. In *Populations of small mammals under natural conditions,* ed. D. P. Snyder. Linesville, PA: Pymatuning Laboratory of Ecology, University of Pittsburgh, 5:159–162.

Tenhaken, R., A. Levine, L. F. Brisson, R. A. Dixon, and C. Lamb. 1995. Function of the oxidative burst in hypersensitive disease resistance. *Proceedings of the National Academy of Sciences of the USA* 92:4158–4163.

Thaler, J., and R. Karban. 1997. A phylogenetic reconstruction of constitutive and induced resistance in *Gossypium. American Naturalist.* In press.

Thompson, J. N. 1988. Evolutionary ecology of the relationship between oviposition preference and performance of offspring in phytophagous insects. *Entomologia Experimentalis et Applicata* 47:3–14.

———. 1994. *The coevolutionary process.* Chicago: University of Chicago Press.

Thornburg, R. W., and X. Li. 1991. Wounding *Nicotiana tabacum* leaves causes a decline in endogenous indole-3-acetic acid. *Plant Physiology* 96:802–805.

Thornburg, R. W., S. Park, and X. Li. 1993. Hormonal regulation of wound-inducible proteinase inhibitor II genes. In *Control of plant gene expression,* ed. D. P. Verma. Boca Raton, FL: CRC Press, 91–101.

Threlfall, D. R., and I. A. Whitehead. 1988. Co-ordinated inhibition of squalene synthetase and induction of enzymes of sesquiterpenoid phytoalexin biosynthesis in cultures of *Nicotiana tabacum. Phytochemistry* 27:2567–2580.

Tien, P., and G. Wu. 1991. Satellite RNA for the biocontrol of plant disease. *Advances in Virus Research* 39:321–339.

Tien, P., X. Zhang, B. Qiu, B. Qin, and G. Wu. 1987. Satellite RNA for the control of plant diseases caused by cucumber mosaic virus. *Annals of Applied Biology* 111:143–152.

Till, I. 1987. Variability of expression of cyanogenesis in white clover (*Trifolium repens* L.). *Heredity* 59:265–271.

Till-Bottraud, I., and P. Gouyon. 1992. Intra- versus interplant Batesian mimicry? A model on cyanogenesis and herbivory in clonal plants. *American Naturalist* 139:509–520.

Tollrian, R. 1995. Predator-induced morphological defenses: Costs, life history shifts, and maternal effects in *Daphnia pulex. Ecology* 76:1691–1705.

Tomczyk, A. 1989. *Physiological and biochemical responses of different host plants to infestation by spider mites (Acarina: Tetranychidae).* Warsaw: Warsaw Agricultural University Press.

Tomlin, E. S., and M. K. Sears. 1992. Indirect competition between the Colorado potato beetle (Coleoptera: Chrysomelidae) and the potato leafhopper (Homoptera: Cicadellidae) on potato: Laboratory study. *Environmental Entomology* 21:787–792.

Trumble, J. T., D. M. Kolodny-Hirsch, and I. P. Ting. 1993. Plant compensation for arthropod herbivory. *Annual Review of Entomology* 38:93–119.

Tuomi, J., T. Fagerstrom, and P. Niemelä. 1991. Carbon allocation, phenotypic plasticity, and induced defenses. In *Phytochemical induction by herbivores,* ed. D. W. Tallamy and M. J. Raupp. New York: John Wiley, 85–104.

Tuomi, J., P. Niemelä, F. S. Chapin, J. P. Bryant, and S. Sirén. 1988. Defensive responses of trees in relation to their carbon/nutrient balance. In *Mechanisms of woody plant defenses against insects,* ed. W. J. Mattson, J. Levieux, and C. Bernard-Dagan. New York: Springer-Verlag, 57–72.

Tuomi, J., P. Niemelä, E. Haukioja, S. Sirén, and S. Neuvonen. 1984. Nutrient stress: An explanation for plant anti-herbivore responses to defoliation. *Oecologia* 61:208–210.

Tuomi, J., P. Niemelä, M. Rousi, S. Sirén, and T. Vuorisalo. 1988. Induced accumulation of foliage phenols in mountain birch: Branch response to defoliation. *American Naturalist* 132:602–608.

Tuomi, J., P. Niemelä, and S. Sirén. 1990. The Panglossian paradigm and delayed inducible accumulation of foliar phenolics in mountain birch. *Oikos* 59:399–410.

Turchin, P. 1990. Rarity of density dependence or population regulation with lags? *Nature* 344:660–663.

Turlings, T. C. J., P. J. McCall, H. T. Alborn, and J. H. Tumlinson. 1993. An elicitor in caterpillar oral secretions that induces corn seedlings to emit chemical signals attractive to parastic wasps. *Journal of Chemical Ecology* 19:411–425.

Turlings, T. C. J., and J. H. Tumlinson. 1991. Do parasitoids use herbivore-induced plant chemical defenses to locate hosts? *Florida Entomologist* 74:42–50.

Turlings, T. C. J., J. H. Tumlinson, R. R. Heath, A. T. Proveaux, and R. E. Doolittle. 1991. Isolation and identification of allelochemicals that attract the larval parasitoid, *Cotesia marginiventris* (Cresson), to the microhabitat of one of its hosts. *Journal of Chemical Ecology* 17:2235–2251.

Turlings, T. C. J., J. H. Tumlinson, and W. J. Lewis. 1990. Exploitation of herbivore-induced plant odors by host-seeking parasitic wasps. *Science* 250:1251–1253.

Ueda, J., T. Mizumoto, and J. Kato. 1991. Quantitative changes of abscisic acid and methyl jasmonate correlated with vernal leaf abscission of *Ficus superba* var. japonica. *Biochemie und Physiologie der Pflanzen* 187:203–210.

Uritani, I., T. Saito, H. Honda, and W. K. Kim. 1975. Induction of furanoterpenoids in sweet potato roots by the larval components of the sweet potato weevils. *Agricultural and Biological Chemistry* 37:1857–1862.

Valentine, H. T., W. E. Wallner, and P. M. Wargo. 1983. Nutritional changes in host foliage during and after defoliation, and their relation to the weight of gypsy moth pupae. *Oecologia* 57:298–302.

van Alstyne, K. L. 1988. Herbivore grazing increases polyphenolic defenses in the intertidal brown alga *Fucus distichus*. *Ecology* 69:655–663.

van Dam, N. M., E. van der Meijden, and R. Verpoorte. 1993. Induced responses in three alkaloid-containing plant species. *Oecologia* 95:425–430.

van Dam, N. M., and K. Vrieling. 1994. Genetic variation in constitutive and inducible pyrrolizidine alkaloid levels in *Cynoglossum officinale* L. *Oecologia* 99:374–378.

van den Bosch, R. 1978. *The pesticide conspiracy.* Garden City, NY: Doubleday.

van der Meijden, E., M. van Bemmelen, R. Kooi, and B. J. Post. 1984. Nutritional quality and chemical defence in the ragwort–cinnabar moth interaction. *Journal of Animal Ecology* 53:443–453.

van der Meijden, E., M. Wijn, and H. J. Verkaar. 1988. Defense and regrowth: Alternative plant strategies in the struggle against herbivores. *Oikos* 51: 355–363.

van de Vrie, M., A. Tomczyk, J. F. Price, and D. Kropczynska. 1988. Interactions between the two-spotted spider mite (*Tetranychus urticae* Koch) and leaf miner (*Liriomyza trifolii* Burgess) on chrysanthemum leaves. *Mededelingen van de Faculteit Landbouwwetenschappen Rijksuniversiteit Gent* 53 (2b): 811–819.

van Emden, H. F. 1969. Plant resistance to aphids induced by chemicals. *Journal of the Science of Food and Agriculture* 20:385–387.

Van Etten, C. H., and H. L. Tookey. 1979. Chemistry and biological effects of glucosinolates. In *Herbivores: Their interaction with secondary plant metabolites*, ed. G. A. Rosenthal and D. H. Janzen. New York: Academic Press, 471–500.

Van Etten, H. D., D. F. Matthews, and P. S. Matthews. 1989. Phytoalexin detoxification: Importance for pathogenicity and practical implications. *Annual Review of Phytopathogy* 27:143–164.

van Noordwijk, A. J. 1989. Reaction norms in genetical ecology. *Bioscience* 39: 453–458.

van Someren, V. G. L. 1937. Chemical changes in the food-plant: A cause of failure in rearing larvae. *Proceedings of the Royal Entomological Society of London A* 12:10.

Vaughn, S. F., and H. W. Gardner. 1993. Lipoxygenase-derived aldehydes inhibit fungi pathogenic on soybean. *Journal of Chemical Ecology* 19:2337–2345.

Verma, D. P. S., and K. Nadler. 1984. Legume-rhizobium symbiosis: Host's point of view. In *Genes involved in microbe-plant interactions*, ed. D. P. S. Verma and T. Holm. Vienna: Springer-Verlag, 57–93.

Vet, L. E. M., and M. Dicke. 1992. Ecology of infochemical use by natural enemies in a tritrophic context. *Annual Review of Entomology* 37:141–172.

Vezina, A., and R. M. Peterman. 1985. Tests of the role of a nuclear polyhedrosis virus in the population dynamics of its host, douglas-fir tussock moth, *Orgyia pseudotsugata* (Lepidoptera: Lymantriidea). *Oecologia* 67:260–266.

Via, S., R. Gomulkiewicz, G. De Jong, S. M. Scheiner, C. D. Schlicting, and P. H. Van Tienderen. 1995. Adaptive phenotypic plasticity: Consensus and controversy. *Trends in Ecology and Evolution* 10:212–217.

Vicari, M., and D. R. Bazely. 1993. Do grasses fight back? The case for antiherbivore defences. *Trends in Ecology and Evolution* 8:137–141.

Wagner, M. R. 1988. Induced defenses in ponderosa pine against defoliating insects. In *Mechanisms of woody plant defenses against insects*, ed. W. J. Mattson, J. Levieux, and C. Bernard-Dagan. New York: Springer-Verlag, 141–155.

Wagner, M. R., and P. D. Evans. 1985. Defoliation increases nutritional quality and allelochemics of pine seedlings. *Oecologia* 67:235–237.

Wagner, R., F. Feth, and K. G. Wagner. 1986. The regulation of enzyme activities of the nicotine pathway in tobacco. *Physiologia Plantarum* 68:667.

Wallace, M. M. H. 1970. The biology of the jarrah leaf miner, *Perthida glyphopa* Common (Lepidoptera: Incurvariidae). *Australian Journal of Zoology* 18:91–104.

Wallner, W. E., and G. S. Walton. 1979. Host defoliation: A possible determinant of gypsy moth population quality. *Annals of the Entomological Society of America* 72:62–67.

Walter, J., J. Charon, A. Marpeau, and J. Launay. 1989. Effects of wounding on the terpene content of twigs of maritime pine (*Pinus pinaster* Ait.). *Trees* 3:210–219.

Wareing, P. F., M. M. Khalifa, and K. J. Treharne. 1968. Rate limiting processes in photosynthesis at saturating light intensities. *Nature* 220:453–457.

Waterman, P. G., and S. Mole. 1989. Extrinsic factors influencing production of secondary metabolites in plants. In *Insect-plant interactions,* ed. E. A. Bernays. Boca Raton, FL: CRC Press, 1:107–134.

Watson, M. A., and B. B. Casper. 1984. Morphogenetic constraints on patterns of carbon distribution in plants. *Annual Review of Ecology and Systematics* 15:233–258.

Webb, J. W., and V. C. Moran. 1978. The influence of the host plant on the population dynamics of *Acizzia russellae* (Homoptera: Psyllidae). *Ecological Entomology* 3:313–321.

Weiler, E. W., T. Albrecht, B. Groth, Z. Q. Xia, M. Luxem, H. Liss, L. Andert, and P. Spengler. 1993. Evidence for the involvement of jasmonates and their octadecanoid precursors in the tendril coiling response of *Bryonia dioica. Phytochemistry* 32:591–600.

Weiler, E. W., T. M. Kutcham, T. Gorba, W. Brodschelm, U. Niesel, and F. Buflitz. 1994. The *Pseudomonas* phytotoxin coronatine mimics octadecanoid signaling molecules of higher plants. *FEBS Letters* 345:9–13.

Welter, S. C. 1989. Arthropod impact on plant gas exchange. In *Insect-plant interactions,* ed. E. A. Bernays. Boca Raton, FL: CRC Press, 1:135–150.

Welter, S. C., P. S. McNally, and D. S. Farnham. 1989. Willamette mite (Acari: Tetranychidae) impact on grape productivity and quality: A reappraisal. *Environmental Entomology* 18:408–411.

Werner, R. A. 1979. Influence of host foliage on development, survival, fecundity, and oviposition of the spear-marked black moth, *Rheumaptera hastata* (Lepidoptera: Geometridae). *Canadian Entomologist* 111:317–322.

Werner, R. A., and B. Illman. 1994. Response of Lutz, Sitka, and white spruce to attack by *Dendroctonus rufipennis* (Coleoptera: Scolytidae) and blue stain fungi. *Environmental Entomology* 23:472–478.

West, C. 1985. Factors underlying the late seasonal appearance of the lepidopterous leaf-mining guild on oak. *Ecological Entomology* 10:111–120.

Westigard, P. H., P. B. Lombard, R. B. Allen, and J. G. Strang. 1980. Pear psylla: Population suppression through host plant modification using daminozide. *Environmental Entomology* 9:275–277.

Westphal, E., F. Dreger, and R. Bronner. 1990. The gall mite *Aceria cladophthirus:* 1, Life-cycle, survival outside the gall, and symptoms' expression on susceptible or resistant *Solanum dulcamara* plants. *Experimental and Applied Acarology* 9:183–200.

Wheeler, G. S., and F. Slansky. 1991. Effect of constitutive and herbivore-induced extractables from susceptible and resistant soybean foliage on nonpest noctuid caterpillars. *Journal of Economic Entomology* 84:1068–1079.

Whipps, J. M. 1996. Interactions between fungi and plant pathogens in soil and the rhizosphere. In *Multitrophic interactions in terrestrial systems,* ed. A. C. Gange and V. K. Brown. Oxford: Blackwell Scientific.

White, J. 1981. Flagging: Hosts defences versus oviposition strategies in periodical cicadas (*Magicicada* spp., Cicadidae, Homoptera). *Canadian Entomologist* 113:727–738.

White, R. F., and J. F. Antoniw. 1989. The use of plant viruses as inoculants.

In *Microbial inoculation of crop plants*, ed. R. Campbell and R. M. Macdonald. Oxford: Oxford University Press, 79–87.

Whitham, T. G. 1983. Host manipulation of parasites: Within-plant variation as a defense against rapidly evolving pests. In *Variable plants and herbivores in natural and managed systems*, ed. R. F. Denno and M. S. McClure. New York: Academic Press, 15–41.

Whitham, T. G., J. Maschinski, K. C. Larson, and K. N. Paige. 1991. Plant responses to herbivory: The continuum from negative to positive and underlying physiological mechanisms. In *Plant-animal interactions: Evolutionary ecology in tropical and temperate regions*, ed. P. W. Price, T. W. Lewinsohn, G. W. Fernandes, and W. W. Benson. New York: John Wiley, 227–256.

Whitham, T. G., and S. Mopper. 1985. Chronic herbivory: Impacts on architecture and sex expression of pinyon pine. *Science* 228:1089–1091.

Whitham, T. G., A. G. Williams, and A. M. Robinson. 1984. The variation principle: Individual plants as temporal and spatial mosaics of resistance to rapidly evolving pests. In *A new ecology: Novel approaches to interactive systems*, ed. P. W. Price, C. N. Slobodchikoff, and W. S. Gaud. New York: John Wiley, 15–51.

Wiens, J. A. 1989. Spatial scaling in ecology. *Functional Ecology* 3:385–397.

Wilcox, A., and M. J. Crawley. 1988. The effects of host plant defoliation and fertilizer application on larval growth and oviposition behaviour in cinnabar moth. *Oecologia* 76:283–287.

Wildon, D.C., J. F. Thain, P. E. H. Minchin, I. R. Gubb, A. J. Reilly, Y. D. Skipper, H. M. Doherty, P. J. O'Donnell, and D. J. Bowles. 1992. Electrical signalling and systemic proteinase inhibitor induction in the wounded plant. *Nature* 360:62–65.

Williams, A. G., and T. G. Whitham. 1986. Premature leaf abscission: An induced plant defense against gall aphids. *Ecology* 67:1619–1627.

Williams, K. S., and J. H. Myers. 1984. Previous herbivore attack of red alder may improve food quality for fall webworm larvae. *Oecologia* 63:166–170.

Williams, M. M. II, N. Jordan, and C. Yerkes. 1995. The fitness cost of triazine resistance in jimsonweed (*Datura stramonium* L.). *American Midland Naturalist* 133:131–137.

Williams, S., L. Friedrich, S. Dincher, N. Carozzi, H. Kessmann, E. Ward, and J. Ryals. 1992. Chemical regulation of *Bacillus thuringiensis* d-endotoxin expression in transgenic plants. *Bio-Technology* 10:540–543.

Willson, M. F., and N. Burley. 1983. *Mate choice in plants: Tactics, mechanisms, and consequences*. Princeton, NJ: Princeton University Press.

Wilson, T. M. A. 1993. Strategies to protect crop plants against viruses: Pathogen-derived resistance blossoms. *Proceedings of the National Academy of Sciences of the USA* 90:3134–3141.

Wink, M. 1983. Wounding-induced increase of quinolizidine alkaloid accumulation in lupin leaves. *Zeitschrift für Naturforschung C* 38:905–909.

———. 1984. Chemical defense of *Leguminosae*: Are quinolizidine alkaloids part of the antimicrobial defense system of lupins. *Zeitschrift für Naturforschung C* 39:548–552.

————. 1987. Chemical ecology of quinolizidine alkaloids. In *Allelochemicals: Role in agriculture and forestry*, ed. G. R. Waller. Washington, DC: American Chemical Society, 326–333.

Wolfson, J. L. 1991. The effects of induced plant proteinase inhibitors on herbivorous insects. In *Phytochemical induction by herbivores*, ed. D. W. Tallamy and M. J. Raupp. New York: John Wiley, 223–243.

Wolfson, J. L., and L. L. Murdock. 1987. Suppression of larval Colorado potato beetle growth and development by digestive proteinase inhibitors. *Entomologia Experimentalis et Applicata* 44:235–240.

————. 1990. Growth of *Manduca sexta* on wounded tomato plants: Role of induced proteinase inhibitors. *Entomologia Experimentalis et Applicata* 54:257–264.

Wool, D., and D. F. Hales. 1996. Previous infestation affects recolonization of cotton by *Aphis gossypii* Glover: Induced resistance or plant damage? *Phytoparasitica* 24:39–48.

Workman, J., G. W. Felton, D. Sternberg, and S. S. Duffey. n.d. Inactivation of plant protease inhibitors by wound-activated phytochemicals. Manuscript.

Wratten, S. D., P. J. Edwards, and I. Dunn. 1984. Wound-induced changes in the palatability of *Betula pubescens* and *B. pendula*. *Oecologia* 61:372–375.

Wrensch, D. L., and S. S. Y. Young. 1975. Effects of quality of resource and fertilization status on some fitness traits in the two-spotted spider mite, *Tetranychus urticae* Koch. *Oecologia* 18:259–267.

Xu, Y., P. L. Chang, D. Llu, M. L. Naraslmhan, K. G. Raghothama, P. M. Hasegawa, and R. A. Bressan. 1994. Plant defense genes are synergistically induced by ethylene and methyl jasmonate. *Plant Cell* 6:1077–1085.

Yao, K., V. DeLuca, and N. Brisson. 1995. Creation of a metabolic sink for tryptophan alters the phenylpropanoid pathway and the susceptibility of potato to *Phytophthora infestans*. *Plant Cell* 7:1787–1799.

Yoshikawa, M., N. Yamaoka, and Y. Takeuchi. 1993. Elicitors: Their significance and primary modes of action in the induction of plant defense reactions. *Plant Cell Physiology* 34:1163–1173.

Yoshinori, W., S. Watanabe, and S. Kuroda. 1967. Changes in photosynthetic activities and chlorophyll contents of growing tobacco leaves. *Botanical Magazine Tokyo* 80:123–129.

Zangerl, A. R. 1990. Furanocoumarin induction in wild parsnip: Evidence of an induced defense against herbivores. *Ecology* 71:1926–1932.

Zangerl, A. R., and M. R. Berenbaum. 1990. Furanocoumarin induction in wild parsnip: Genetics and population variation. *Ecology* 71:1933–1940.

————. 1995. Spatial, temporal, and environmental limits on xanthotoxin induction in wild parsnip foliage. *Chemoecology* 5–6:37–42.

Zangerl, A. R., and C. E. Rutledge. 1996. The probability of attack and patterns of constitutive and induced defense: A test of optimal defense theory. *American Naturalist* 147:599–608.

Zeringue, H. J. 1987. Changes in cotton leaf chemistry induced by volatile elicitors. *Phytochemistry* 26:1357–1360.

Zhang, Z.-P., G. Y. Lynds, and I. T. Baldwin. n.d. Transport of [2-^{14}C]-JA from leaves to roots mimics wound-induced changes in endogenous JA pools in *Nicotiana sylvestris*. Manuscript.

Zou, J., and R. G. Cates. 1994. Role of douglas fir (*Pseudotsuga menziesii*) carbohydrates in resistance to budworm (*Choristoneura occidentalis*). *Journal of Chemical Ecology* 20:395–405.

Index